D1212110

Plantation Workers

Plantation Workers
Resistance and Accommodation

EDITED BY

**Brij V. Lal, Doug Munro, and
Edward D. Beechert**

1993

University of Hawaii Press
HONOLULU

93 94 95 96 97 98 5 4 3 2 1

Library of Congress Cataloging-in-Publication Data
Plantation workers : resistance and accommodation edited by Brij
V. Lal, Doug Munro, and Edward D. Beechert.
p.   cm.
Includes bibliographical references and index.
ISBN 0-8248-1496-7
1. Plantation workers—Pacific Area—Case studies.   2. Plantation
workers—Latin America—Case studies.   3. Plantations—Pacific Area—
History—Case studies.   4. Plantations—Latin America—History—
Case studies.   I. Lal, Brij V.   II. Munro, Doug.   III. Beechert,
Edward D.
HD8039.P496P166   1993   93-10913
305.5'63—dc20   CIP

Designed by Kenneth Miyamoto

Contents

Preface

The genesis of this collection of essays lay in discussions in the mid-1980s between Brij Lal and Edward Beechert, then both at the University of Hawaii at Manoa. Since then, Lal has joined the Australian National University and Beechert has retired. They were later joined by Doug Munro, then at Bond University and now at the University of the South Pacific. As historians of labor, Lal and Beechert both felt the need for a comparative volume that focused on the actual working experiences of plantation laborers. In Pacific Islands historiography, much had been written on the question of labor recruitment in the wider region, especially the extent to which recruiting and migration were coerced or voluntary arrangements. But rather less had been written about the workers' experiences on the plantations and the ways in which they coped with their new and unfamiliar situation.

This book attempts to redress this imbalance. We offer it as a contribution both to Pacific Islands historiography and to the comparative study of plantation labor. In particular, we have sought to unify the discrete literatures on plantation laborers in various parts of the world.

For reasons inherent in putting together any collection of essays, this volume has long been in the making. We thank the dispersed group of contributors for their patience and understanding in the face of unavoidable delays. We are especially grateful to Pamela Kelley, our sponsoring editor, for her support and encouragement all along. She showed us what editorial professionalism is all about.

Thanks are also due to Marie Baker, Barbara Triffett, Celeste Warren, and Urmila Prakash for typing parts of the manuscript.

The comparative nature of this volume is reflected by the fact that no single contributor is personally known to all the others. Indeed, two of the editors (Munro and Beechert) have never been in the same room together. For the record, Lal recruited Bennett, Firth, Moore, and Munro; Beechert recruited Gonzales; and Munro enlisted Joseph, McCreery, and Wells.

1

Patterns of Resistance
and Accommodation

Doug Munro

As the world capitalist system developed during the nineteenth century non-slave labour became a commodity that circulated around the globe and contributed to capital accumulation in metropolitan centres.[1]

The plantation was introduced into certain countries of the Third World by the metropolitan nations of the North Atlantic for the benefit of those nations.[2]

I do not know if coffee and sugar are essential to the happiness of Europe, but I well know that these two products have accounted for the unhappiness of two great regions of the world: America has been depopulated so as to have land on which to plant them; Africa has been depopulated so as to have the people to cultivate them.[3]

The three quotations link the plantation mode of production with the needs of the industrial world for agricultural products, raw materials, and mass beverages and sweeteners. Plantations may be roughly defined as large-scale agricultural units in tropical and semitropical areas, typically engaged in monoculture or dual cropping, export oriented, under foreign ownership and management, and dependent on a sizable, servile, and low-paid labor force.[4] The literature on the diagnostic features of plantation production and plantation systems is extensive, and different writers have formulated their more precise definitions with different criteria in mind. C. C. Goldthorpe, for example, regards plantation agriculture in organizational terms as "the production of commodity crops in large-scale agribusiness organizations termed plantations which have a strict bureaucratic internal structure."[5] Edgar T. Thompson, by contrast, stresses more the plantation as an all-pervasive social system.

The plantation [writes Thompson] is an institution in just as real a sense as the Catholic Church. It arises to deal with certain seem-

ingly eternal problems of an ordered society. . . . It is made up of
people but, like the church, it is an impersonal and implacable
automatism having a set of norms which control all the people
who constitute it, planters and labourer alike, who "belong" to
the estate, as though the estate were something existing apart
from its people: the plantation demands and dictates.[6]

In addition to this "internal dimension," the West Indian scholar
George Beckford points to an

external dimension [which] derives from two characteristics of
plantation production—its export orientation and foreign ownership. . . . [P]lantations were, and are, the product of metropolitan capital and enterprise. This and its orientation to overseas
markets ties it to the wider world economic community in very
precise ways. Plantations are only one part of a much wider world
economic system consisting of a set of relations which meet at a
metropolitan and industrial centre far removed from the plantations.[7]

Beckford was mainly referring to Latin America and the Caribbean,
Africa, and Asia, but his words apply as much to plantation systems
in the Pacific Islands. There, too, extensive areas of land had to be
put aside by one means or another for plantations, the same problems of capital formation and markets for the crop applied, and
plantations in the Pacific were no less dependent than elsewhere
on procuring a plentiful supply of low-paid labor. The present volume is mainly concerned with the Pacific but contains contributions on other areas to provide a comparative dimension that will
offset the insularity of Pacific historiography.

Extensive plantation systems in the Pacific only became widespread in the 1860s, when the American Civil War disrupted supplies of raw materials to the textile factories of Britain and Europe.
The result was the rapid spread of cotton plantations in places as
far afield as Samoa, Fiji, Tahiti, northern New South Wales, and
Queensland. Ultimately, cotton was a transitional crop that boomed
for about a decade, depending on locale, and then collapsed. Sugar
and copra, or a combination of both, then emerged as staples, again
depending on locale. It is a commonplace that the critical limiting
factor of plantation agriculture was not so much the availability of
land or even capital, but labor. It was usually difficult to obtain,
often expensive, not always dependable, sometimes intractable,
and constantly in need of replenishment. Planters were initially
confronted by the near-universal unwillingness of Pacific Islanders

to work on expatriate plantations near their own villages. So in order to close the gap between labor needs and supply, planters were forced to look further afield. The Melanesian Islands became the most significant labor reserve within the region and served the needs of British planters in Fiji and Queensland, German planters in Samoa, and predominantly French planters in the New Hebrides (Vanuatu) and New Caledonia. When the needs of planters (and miners) went beyond the resources of the Melanesian labor reserve, Asian laborers were introduced to make good the shortfall. The exception to this pattern of recruiting was Hawaii, whose sugar industry depended primarily on Hawaiian and part-Hawaiian laborers until as late as 1875. The dramatic expansion of the sugar industry over the next half century was met by importing successive waves of Asian workers from China, Japan, Korea, and the Philippines, who comprised the overwhelming proportion of the industry's labor force. Very few Pacific Islanders from other parts of the region were employed. A further contrast is evident in that several thousand Europeans were introduced as plantation workers (see Beechert in this volume, and note 39 of this chapter).

In terms of production, plantation systems in the Pacific were unimportant by comparison with other regions; so too was the size of the plantation work force within the region.[8] While some 1,120,000 persons from the Indian subcontinent crossed the "black waters" as migrant laborers, only 61,000, or approximately 5 percent of the total, traveled to Fiji.[9] Queensland received the largest numbers of plantation workers from the Pacific Islands, about 62,000, but this figure pales by comparison with the 100,000 local Indians whom coffee planters in Guatemala regularly mobilized for each harvest during the 1890s.[10] Nevertheless, the scale of the plantation work force within the Pacific, although the precise number can never be accurately determined, was still appreciable. The most recent headcount reveals that by the outbreak of World War II, some 1.5 million Pacific Islanders and a further 500,000 Asians and their families were recruited, mostly for plantation work.[11] The upshot of labor migration into and within the Pacific Islands has been a legacy of multiracial populations in places such as Hawaii, New Caledonia, and Fiji.

The defining feature of nineteenth-century plantation labor in the Pacific was the indenture system. With the abolition of slavery in the British Empire in 1834 and its eventual suppression in other parts of the world, the employment of servile labor fell under new contractual arrangements, and the most common was indenture— or the binding of one person to work for another for a given period of time, which ranged from two years (in the British Solomon Islands Protectorate) to five years (Indians in Fiji) in the Pacific, in return for wages and other specified conditions of labor. Indentured service usually derived its legal authority from the various Masters and Servants Acts, such as those of Queensland and Hawaii, which ostensibly set out mutual rights and obligations but, in reality, provided criminal punishment for breaches of contract by workers—hence the indenture system sometimes being called the penal contract system. The purpose of indentured service, then, was twofold: in providing for a fixed term of service, it stabilized the work force by preventing a high turnover, while the penal sanctions placed in employers' hands a blunt instrument of discipline. These restrictive and discriminatory workings of contracts of indenture provide the context within which the resistance and accommodation of plantation workers was played out.

In recent years, the Pacific Islands labor trade has been the focus of considerable scholarly attention; it has become one of the most extensively researched topics in Pacific history. Indeed, the labor trade was a major building block in the edifice of the new Pacific historiography, which gave greater recognition to the roles of Pacific Islanders in making their own history and shaping the outcome of events. Developed in the 1950s at the Australian National University by Professor J. W. Davidson and his colleagues, and thus often termed the "Canberra school," it sought to diminish the role of the imperial factor as an organizing theme and to concentrate instead on writing "island-centered" histories of the region that would give greater agency to Pacific Islanders in the shaping of events and their outcomes.[12] A prominent contribution to the new historiography was Dorothy Shineberg's classic study of the sandalwood trade in Melanesia, which firmly stressed that Melanesians had acted from their own motives rather than being simply reactive to Europeans.[13] Shineberg also linked the employment of Melanesians on sandalwood shore stations with the large-scale recruiting for Queensland and Fiji that commenced in the early 1860s. As the supplies of sandalwood were depleted, the Europeans involved often turned to plantation agriculture. Needing a labor force for their fledgling enterprises, the planters knew where to find workers already accustomed to notions of contract labor and labor migration.

Labor trade studies featured prominently in the shift from an imperial to an island-centered Pacific history. Deryck Scarr and Peter Corris, in correcting the older view that recruiting for plantations was characterized by kidnapping and fraud, showed that the business was less European dominated than had previously been allowed. Not only that, the Islanders had their own reasons for recruiting, and indeed, the labor trade would not have been possible without a substantial degree of voluntary participation. Scarr and Corris also dealt in greater depth than their predecessors on the actual plantation experience and left open the possibility that Pacific Islanders exercised a considerable measure of control over their working lives.[14]

Later research followed the revisionary efforts of Scarr and Corris by focusing on the experience of plantation workers in specific places. One development was the revisionists themselves being revised in various studies on the Queensland sugar industry by academic and nonacademic writers. Although this corpus is too varied in content and approach to be considered a coherent counterrevisionist school of thought, it has a common thread in that the Melanesians are once again cast in the role of an exploited and oppressed work force.[15] Other writers came to, or had already arrived at, the same broad conclusions about the working lives of nineteenth-century plantation workers in places as far apart as Fiji, Samoa, German New Guinea, and Hawaii.[16] Similar revision and debate occurred in Latin American studies following the survey in 1979 on rural labor by Arnold J. Bauer. Arguing that rural labor systems in that part of the world were less restrictive and coercive than had hitherto been allowed, Bauer may be regarded as the equivalent of a Scarr or a Corris.[17] The ebb and flow of debate continued with Clive Moore's finely grained study of Melanesian (especially Malaitan) workers in Mackay, the major sugar district in Queensland, which swung the pendulum back in support of the original "revisionists."[18]

The sharp focus of Moore's study served to highlight the tendency toward increasing specialization already evident in scholarly work on the labor trade studies and in Pacific history generally.[19] The research that was done typically concentrated on specific areas and sometimes even specific island groups in even more specific localities. Nevertheless, the scope of enquiry had broadened from the administrative organizational and legal requirements bearing on the labor trade[20] and the structure of the sugar industry through to questions of retention and adaptation of culture in the alien environment of a plantation,[21] and the extent to which the private lives of individuals could be separated from their working lives.[22] As early as 1980, Colin Newbury identified the two broad thrusts of labor trade studies:

The work of Deryck Scarr, Peter Corris, and others emphasizes
the active participation of Melanesian and Micronesian societies
in early labor migration. There is less on kidnapping and more on
collaboration by indigenous middlemen, less on legislation which
was difficult to enforce, and more on the preconditions of mobility—scarce resources, changes in agricultural technology, local
warfare and the impact of missionaries, recruiters, and naval
patrols. Secondly, a quite different emphasis is given to the history of socio-economic change in Melanesia in the works of Stewart Firth, K. L. Gillion, and Pierre Gascher who stress the role
of the colonial state as recruiter and developer through land
alienation, taxation, and subsidies to private employers.[23]

A subsequent development has been the detailed quantitative
research by Ralph Shlomowitz on the economic and demographic
aspects of the labor trade.[24] In all this, however, very little attention was paid to how workers resisted the plantation regime, and
what forms this took. Although the question of resistance received
scattered mention, no systematic attempt had been made to address this quite central question, either in specific cases or across
geographic boundaries. The past studies have enriched our understanding of specific plantation experiences, but the time has come
to go a step further and to attempt comparative work.

This book seeks to fill a gap in the literature by providing analytical
content to issues relating to resistance/accommodation, whereas
previous scholars of publications on plantation life have been
largely descriptive in their treatment of these matters. In the
present case, the essays are sufficiently varied in range and scope
to enable comparative evaluation. The difficulty of comparison
should not be underestimated, especially when cross-cultural analysis and an interdisciplinary approach are involved. Such difficulties are compounded when the attempt is made to put a human face
on the question of labor rather than treating labor as an abstraction, or by applying a broad brush and placing the emphasis on
migration trends and labor systems at the expense of the human
dimension. The contents of other collections of essays on labor
migration serve to underline these observations. Take, for example,
the very worthwhile anthology on international labor migration
edited by Shula Marks and Peter Richardson, which originated in a
seminar series on labor migration in the British Empire held in London at the Institute of Commonwealth Studies between 1978 and
1982. It makes important statements about the political economy of
labor migration in its wider aspects rather than dealing with the

conditions of labor at the work place. This macro-approach is particularly evident in the three comparative contributions to the volume, and it was perhaps only feasible to make comparisons across geographic areas by concentrating on scope rather than detail, and by drawing largely on published instead of documentary sources.[25]

The present anthology departs in several respects from its predecessors. The other collections are more global in their coverage, touching only peripherally on the Pacific. The emphasis of this book, by contrast, is on the Pacific with contributions from other areas, namely Guatemala (by David McCreery), Yucatán (by Allen Wells and Gilbert M. Joseph), and northern Peru (by Michael Gonzales). Rather than being generalized in its approach, this book pursues the particular theme and in so doing addresses an important question that has received insufficient attention in the literature. The chapters in this collection show certain common themes as well as the inevitable particularities of time and place, which will enable a better understanding of the responses of laborers to the plantation regime.

The approaches taken in much of the previously published work by Pacific historians on resistance and accommodation are open to criticism. Ronald Takaki, in his book on plantation life and labor in Hawaii, has a chapter entitled "Contested Terrain: Patterns of Resistance," in which he provides an undifferentiated listing of the forms that resistance might take. This array of devices includes desertion, assault and murder, shirking, malingering, feigning incomprehension of orders, and destruction of crops and employers' property. In another chapter ("A New World of Labor: From Siren to Siren") Takaki stresses the strong hand of authority, which included the use of police and the legal system as well as more direct methods of discipline. He also discusses the use of indirect methods of worker control, such as the extension of credit and the bonus system and, more contentiously, the planters' preference for a multiethnic work force to hinder the development of conspiracy and worker solidarity. Indeed, it is far from clear whether he regards labor management on plantations as coercive or paternalistic. Moreover, by treating resistance and accommodation as distinct and separable issues, instead of as opposite sides of the same coin, Takaki puts forward seemingly contradictory interpretations. To read the chapter on planter control is to realize that power lay firmly in the employers' hands; yet the chapter on resistance imparts the impression that far from being cowed and overworked, the laborers evaded plantation discipline and fought the system in

grand style. What Takaki's shopping list approach cannot convey is
a clear picture of the circumstances under which worker resistance
could be mounted and, conversely, employer pressure sustained.
To regard absconding in the same breath as exodus from the plantations once the worker's contract had expired, and then to lump
these together with temporary escapism through the use of alcohol
and opium on the one hand and, on the other hand, the more permanent variety afforded by suicide, is to conflate fundamentally
different reactions to the plantation regime.[26]

Another published writer on the subject is Kay Saunders, who
identifies three major categories of resistance in colonial Queensland: passive or surreptitious resistance, such as malingering, shirking, and feigning ignorance; outer-directed resistance or overt
industrial actions, whether it be disobedience, destruction of property, or physical retaliation; and inner-directed aggression, such as
suicide, maiming themselves, and attacks on other workers. These
modes of resistance are then subdivided into individual and collective manifestations, which is a problematic distinction at best.[27]
Saunders avoids one of Takaki's conceptual pitfalls; rather, she
identifies some of the circumstances when resistance was most
likely and what form it might take. Thus, laborers who were illegally recruited were more likely to be intractable, as was the case
on Fiji plantations,[28] and those who came from societies that valued
physical prowess and other tough, male virtues were more prone to
assaulting masters and overseers.[29] Culturally conditioned modes of
resistance are evident also among Indian laborers on Fiji plantations, for example, where suicide was rare indeed among Moslem
laborers, whose religion proscribed death by one's own hand, and
more prevalent among Hindu workers, whose religion lacked a
clearly defined ethical position regarding self-induced death. Furthermore, it was the minority group of Madrasis from southern
India who "were especially prone to depression and suicide when
placed among northern Indians with their different customs and
language," who somewhat despised the Madrasis.[30]

Saunders, like Takaki, takes a timeless ramble, showing little
appreciation of chronology. Nor does she sufficiently take into
account alterations over time to the structural conditions of plantation life and labor, the changing nature of the labor force, and the
effects of these developments on worker resistance and accommodation. There were fundamental structural differences between
the 1870s, when labor regulations were rudimentary and the plantation mode of production in full swing, and the 1890s, when small-
farm operators were integral to the sugar industry and when the

labor trade in Melanesians "was governed by 7 Acts of Parliament, 18 schedules, 54 regulations and 38 instructions."[31] More to the point, the laborers by this time were no longer largely raw recruits but included a high proportion of time-expired workers, who were often very adroit in handling employers and in their knowledge of the system. Free to pick and choose employers and assertive in pursuing their interests, they also commanded far higher wages than their first-indenture counterparts and opted for short, often seasonal, contracts. Furthermore, they were preferred by the small-farm operators, a sentiment that was reciprocated, since the time-expired laborers as a rule considered conditions on the small holdings as superior to those on plantations.[32]

Saunders also falls into the same trap as Takaki in treating resistance as something quite separable from accommodation. Scholars of labor systems in Latin American and the Caribbean have long been aware of the pitfalls of such a dichotomized approach.[33] Her remark about the "consummate ease . . . [with which Melanesian workers] hindered the untroubled and efficient operation of the plantation" sits uncomfortably with her other writings, which unremittingly press the case that they endured harsh and oppressive conditions and had no option but to acquiesce to the uncompromising and all-pervasive system of plantation discipline with its battery of legal and extralegal supports.[34] Problems of this sort will occur if a dichotomous model is presented with the only alternatives being resistance or accommodation. If the story is restricted to separate dialogues and only told from one perspective or the other at any one time, then the interactions and relationships will not be fully drawn.

An even more restrictive approach is adopted by the most recent contributor to the field, who regards all types of action and inaction as resistance to the almost complete exclusion of accommodation. Writing about Indian women laborers in Fiji, Shaista Shameem takes the view that "most women resisted oppression and exploitation every step of the way," especially "the physical violation by all categories of males on the plantations" whether these be managers, overseers, or other Indian laborers.[35] This uncomplicated, almost triumphalist, approach results in her taking rather extreme positions. The high rate of absenteeism among the Indian women, for example, is perceived as an indication of their "struggle." While there might have been some women who expressed their unwillingness to cooperate in this way, many others suffered from debilitating ailments, such as anemia, that were directly attributable to conditions of life and labor on the plantations, including the prevalence of overtasking.[36] In finding heroic deeds that seldom occurred and then winning on paper the battles that were lost long ago in reality—if they ever happened at all—Shameem brings to mind the words of David McCreery (in this volume): "Oppressed peoples have no obligation to act in ways academics find dramatic or exciting, but rather to survive and endure and to ensure the survival of their families and communities in the face of what threaten to be literally overwhelming pressures."

Her conclusions are in direct opposition to those of an earlier writer, Brij V. Lal. Whereas Shameem considers that resistance was a strategy for survival for Indian women, Lal (referring to both male and female laborers) argues that their strategy for survival was accommodation. But the differences go deeper, because Lal is the first writer on the Pacific labor trade to recognize the interrelatedness between accommodation and resistance and how the two are inseparable, being a continuum of one activity. In particular, he demonstrates that each is the product of specific historical circumstances:

> The primary reason for the paucity of active protest . . . was the authoritarian nature of the plantation system in Fiji. The frequency and effectiveness with which the employers were able to prosecute their laborers, year after year, starkly underlined the workers' vulnerability and reinforced the futility of overt action. Nonresistance, thus, became a strategy for survival. But the indentured laborers' difficulties were also compounded by serious problems of organization within the nascent Indian community itself. Their diverse social and cultural background, their differing aspirations and motivations for migrating to Fiji, their varying individual experiences on the plantations, and the absence of institutional structures within the indentured Indian community, that could have become avenues for mobilization, also helped to reduce the potential for collective action.[37]

Lal's work has proved to be seminal in Pacific historiography; the dialogue and connections between resistance and accommodation are points that repeatedly emerge in the contributions to the present volume. From this comparative sampling, which includes case studies bearing on areas outside the Pacific, the diagnostic features of resistance and accommodation, and their interrelationships, are more readily identified.

Central to the whole question of workers' responses, and with it their ability to resist, was the authoritarian character of the plantation as an institution, which depended in large part for its success on coercive ability. Whatever else it may be, a plantation is structured around the exercise of power in quite explicit ways; and this is symbolized by the planter's house usually being set on elevated ground overlooking the laborers' quarters on the plain below; by the planter's house being off limits to workers, so that social distance was maintained; and by overseers often being on horseback, not simply for mobility and safety but to reinforce some of the basic facts pertaining to the social relations of production. The whip can also be seen as an extension of a mounted overseer's own person, reaching out well beyond arm's length—almost a form of remote control. Referring to the cacao planters in German Samoa, Stewart Firth has noted that flogging "was of symbolic importance to them, the minimum demand of a group . . . renowned for the contempt they had of the[ir Chinese] labourers,"[38] who were required, moreover, to wear identification badges. The racial dimension suggested by Firth's passage was also quite explicit in that managers and overseers were generally white skinned, while the laborers were overwhelmingly colored. As Ian Campbell has pointed out, "Race stood in a fair way to becoming a definition of social class"[39] (see also Beechert, McCreery, and Gonzales in this volume).

The authoritarian nature of the plantation was reinforced by the production routines and the need to maintain output and productivity; these imposed unfamiliar and onerous work routines. So even if plantation life was an improvement to "a labourer who had known hunger, had slept in the open or in a mud hovel, and encountered little but abuse from his betters," as Gillion writes of Indian laborers in Fiji,[40] plantation life and work was still an unattractive proposition. As well as material conditions being typically harsh and impersonal, the work itself involved drudgery and tedium. It is not without reason that the plantation mode of production has variously been termed the "factory in the field" or the *plantation industrielle*—especially sugar plantations, where the sequence of events from first planting through to weeding, cutting, loading, and transporting the cane to a nearby mill for immediate crushing and refining is similar to production-line manufacture in an industrial factory. The transition from the familiar "task-oriented" labor patterns of traditional village life to the regimented and monotonous "time-oriented" routine of plantation work is likewise akin to the Industrial Revolution, which imposed "a regularity, routine and monotony quite unlike pre-industrial patterns" that were organized around seasons, the needs of livestock, and the whims of inclination. What Eric Hobsbawm describes as "the tyranny of the clock" with respect to mechanized factory labor becomes, in the words of a Filipino plantation laborer in Hawaii, working "from siren to siren."[41]

Enforcing these unaccustomed routines and maintaining the pace of work "from siren to siren" required a highly regulated authority structure. The essentially adversarial relationship between planter and worker, with the overseer at the interface, was heightened because the laborer had no stake in the enterprise and therefore no incentive to exertion except to keep out of trouble. Coercion was also necessary to keep laborers on the plantation for the full term of their contracts rather than to seek alternative sources of wage employment. There could be no question that plantations would be other than coercive and that physical control would always be the last resort.

The ways that this worked in practice varied, but the common features are too pervasive to be accidental. Kusha Haraksingh has powerfully illustrated the mechanics of oppression in Trinidad, and he uses the phrases "interlocking incarceration" to encapsulate the "view of the labour force being held 'captive' at several different levels." In addition to outright physical coercion, there was the use of indebtedness, the equally obnoxious use of drivers or foremen from among the ranks of the workers ("the very agency which ought to have been an avenue for registering complaints and protest was subverted to other ends"), the low standards of amenities, nutrition, and life expectancy, "which together should have induced an inclination to escape, [but] served only to create an oppressive sense of inertia."[42]

The mechanics of oppression are particularly evident in the case study on Samoa, where a strict discipline was maintained on German plantations by means of a reign of near-terror (see Munro and Firth in this volume). Physical coercion, the crudest form of discipline and punishment, was generally the most common on plantations, wherever the locale. It ranged from flogging and assault and battery to milder expressions, such as a cuff across the ear while being tipped out of bed in the morning. Variations included periods of confinement in lockups and the use of foot-irons on potential absconders. In some cases, plantations had their own police force (see Gonzales in this volume). Humiliating treatment was a frequent reinforcement to physical punishment, such as practiced by the major plantation enterprise in Samoa (the *Deutsche Handels- und Plantagen-Gesellschaft* [DHPG]), which sometimes exhibited offenders with a signboard giving details of their misdemeanors, not to mention the use of a device that prevented miscreant laborers from lying down or reaching their faces, in order that they might be the object of ridicule.[43] Of course there were the awkward ones—those intransigent individuals who refused to bow to authority. They can be found in any coercive arrangement, including prisons and concentration camps as well as plantations and haciendas. Sometimes inspirational in their actions, they received punishments that more often served to deter others so inclined. Galant van der Caab, a slave in the Cape of Good Hope Colony, is a classic example of an intransigent whose spirit was never broken, but he did not survive long into adulthood; he was hanged in 1825.[44]

It is difficult to generalize about the use of physical force, or the threat of it, as a means of backing up orders and demands, but it had a ubiquitous though varied presence. Or as Eric Wolf has said,

> The powerless are easy victims. . . . The poor peasant or the landless laborer who depends on a landlord for the largest part of his livelihood, or the totality of it, has no tactical power: he is completely within the power domain of his employer, without sufficient reserves of his own to serve him as resources in the power struggle. Poor peasants, and landless labourers, therefore, are unlikely to pursue the course of rebellion, *unless* they are able to challenge the power which constrains them.[45]

Nevertheless, to regard the plantation exclusively as "an instrument of force"[46] is to miss the point that there were limits to which planters were prepared or able to impose their will: some of the flavor and complexity of the matter is starkly revealed by the following two passages relating to the British Solomon Islands Protectorate in the early decades of the present century:

> Europeans treated their labourers and servants with a mixture of racist contempt and paternalism, depending on their temperaments, backgrounds and circumstances. Brutal treatment of islanders by planters was common and is clearly remembered by old ex-plantation workers today. However, this usually stopped short of killing or maiming: an assault which resulted in an islander being seriously injured or killed was likely to come to the notice of the District Officer or the Inspector of Labour and to bring a punishment (usually a fine, but possibly expulsion from the Protectorate). More usually, ill-treatment would be of the kind reportedly entered in the diary of a San Cristobal planter: "Belted that black swine Tohoni today. Bread a failure—when will that bloody animal learn to cook!"[47]

But it was not a one-way traffic in violence. The notion of "contested terrain" is also worth bearing in mind. As Judith Bennett points out:

Planters regularly gave laborers a blow over the ear or a kick in the backside to make them do what they were told. Violence was so much a part of the expected disciplinary methods of the planters that a white man had to prove himself competent in it to be a success. It was common practice to ask prospective overseers and managers if they could fight. The "new chum" overseer was always "tried out" by plantation laborers. Men would openly defy the overseer and expect him to hit them or lose face. Sometimes they would try to get him down to the labor lines at night when he would be an easy target. Everyone involved in plantations, including the government, knew that fisticuffs, or a brawl, were a standard and acceptable way of enforcing discipline—provided the laborer was in the wrong, knew it, and was not caused serious physical harm. If, however, the laborers felt they had been wronged they would frequently gang up on an overseer and beat him.[48]

These two descriptions point to a sometimes overlooked factor: that the internal politics of each plantation is defined by the interaction between the influence of wider considerations, such as the constraining hand of government authority and the labor regulations, and also by the circumstances specific to each plantation. This implies the notion of a watchful, neverending contest—a them/us situation—which is moderated by both sides recognizing the limits of each other's power and tolerance, but with each ever attempting to gain ground on the other in accordance with tacitly understood rules of the game. This latter idea of a "moral economy" has been neatly expressed by Peter Kolchin in his study of resistance to bondage in Imperial Russia and the United States South when he speaks of

an ongoing process of give-and-take between masters and servants. Neither slavery nor serfdom was rigidly defined by the laws, rules, attitudes and actions of the dominant class; the responses of the bound population served to influence the nature of—and set limits to—its own oppression. In both countries the antagonistic classes were involved in a process of mutual groping toward a tolerable *modus vivendi*, a process in which the slaves' and serfs' ideas about their proper relationship with their owners were of the utmost importance. . . . [Such] confrontations reflected the bondsmen's firmly developed—if never precisely articulated—notions of what constituted acceptable treatment within the generally unacceptable systems of forced labor in which they were held. . . . [T]he very existence of such standards among them implied a tacit if unrecognised acceptance of their own unfreedom, of the idea that there was such a thing as an acceptable relationship within slavery and serfdom.[49]

There were limits to the power that planters could wield over their work force, and compromises were made between planters and workers in the interests of harmony and productivity. While these compromises obviously varied over time and place, they were nevertheless "grossly unfair and one-sided . . . with the master holding a big edge" in the power relationship.[50]

Controlling and disciplining the plantation labor force involved more than the actual or threatened use of physical and emotional brutality. The practice of the company store charging laborers for goods at inflated prices or advancing credit against wages, sometimes resulting in debts exceeding wages and consequently an extension of the period of service, was a common enough occurrence and a very useful, if unprincipled, device. According to the company's own records, the DHPG in Samoa managed to cut total labor expenses by 11 percent through the sale of store items to its workers, to give one example, while sugar planters in Queensland affected considerable savings through a combination of deferred wages and truck.[51] In Hawaii, planters were required to withhold 15 percent of a laborer's wages until the termination *and cancellation* of the contract. It was a type of enforced saving, but it also served to discourage desertion. But when the institution of indenture in Hawaii was legally terminated in 1900, many planters refused to cancel the contracts, in the hope of keeping laborers bound to the plantation.[52] Like the use of physical coercion, however, the employers' resort to economic pressure and sanctions varied over time and place. Michael Gonzales (in this volume) argues that debt peonage was counterproductive and rarely practiced in Peru, and here he is addressing a debate as significant in Latin American historiography as is the "kidnapping myth" in Pacific labor trade studies, because debt servitude was certainly a feature of some other parts of Latin America, such as Guatemala and southern Mexico during the same period, and was in some circumstances a means of economic security for workers rather than being an instrument of coercion and control.[53]

Crucial to the enforcement of discipline and control were the provisions of the law, which weighted power in the planters' favor. The restrictive and discriminatory aspects of the contracts of indenture were often tightened by a capricious application of the law as an instrument of domination. In Hawaii, for example, the protective provisions of the original Masters and Servants Act of 1850 were progressively eroded by strengthening the penal clauses, resulting in the institution of indenture shifting from contract labor law to a system of servitude and peonage.[54] Lal's chapter in this

volume clearly demonstrates how effective the planters in Fiji were
in their use of one-sided Labour Ordinances. Whereas employers
secured convictions for 82 percent of the charges they laid, those
same courts upheld only 35 percent of laborers' complaints against
employers for breaches of the labor regulations (see also Moore in
this volume). Not only did employers enjoy a higher success but
they laid far more complaints before the courts—almost 10,000
between 1890 and 1897, for example, as against 311 by laborers
during the same period, which indicates the latter's sense of futility
in attempting to go through judicial channels.[55] An even less satisfactory situation obtained on German plantations in Samoa,
because there were no labor laws until 1897. In this almost unregulated setting of company rule, the laborers had no avenues for
redress or complaint. Instead, the German consul functioned as the
official arm of the DHPG, endorsing a harsh system of discipline
and overwhelmingly concerned to keep down the firm's labor costs
at the expense of the laborers themselves.[56]

Lal also demonstrates the extent to which the use of wider state
power—especially when there is an affinity of interests between
planters and the colonial state—could be deployed to stifle worker
resistance: the presence of a dominant plantation company; a succession of colonial governors who hesitated to oppose the company
even had they wanted to; a less than impartial judiciary; a general
tenor of contempt toward Indian plantation workers in white society; and the action of hounding out of the colony those Inspectors
of Labor who sought to protect what few rights the laborers had,
weighted power heavily on the employers' side. Had the sugar
industry in Fiji been structured around a large number of small
plantations, the regulations could have been enforced by denying
errant planters further supplies of labor. The authorities exercised
this option in the Assam tea gardens of northeast India, where plantations with a death rate among first-indenture laborers of over
seventy per thousand per year were refused new recruits.[57] But if
such action had been taken in Fiji, it would have financially ruined
the colony.

The extent of state power on the wider rural society from which
the plantation labor force was drawn is even more starkly revealed
by David McCreery's chapter on Guatemala. The independence Liberals, or neo-Liberals, who took over the government in 1871, were
faced with the question of dubious legitimacy. Their own violent
coming to power impressed upon them the limits of state control, so
they set about installing or augmenting the apparatus of repression
and control necessary to keep them there—roads, a telegraph system, army, and militia. Lacking legitimacy, and anxious for eco-

nomic development, the state "fell back on coercion, on political hegemony, in its most bald and undisguised form," and McCreery speaks of the "devastating violence of the state" in maintaining order and control of the countryside and in mobilizing the indigenous population for Guatemala's expanding coffee industry through forced labor drafts as well as conscription into the army.[58] In the face of the state's awesome and capricious power, large-scale uprisings virtually ceased. These same considerations also resulted in plantation workers, and the Indian communities from where they were drawn, shifting their resistance strategies "away from the often open confrontation characteristic of the century before 1870 to more subtle forms of avoidance and mitigation. . . . Guatemala's Indians adopted modes of resistance—from deception to evasion and escape to the forcing up of debts and the resort to violence on the rare occasions—after 1871 best calculated to extract from a system into which they had been drawn voluntarily the best result at the least risk, and they waited for a better day." It may not have been very brave but it was sensible not to openly antagonize, for this would achieve nothing except to bring down retribution.

Impediments to worker resistance and incentives for planters to exert strong-handed authority were intensified from time to time by depressed market prices for tropical products. Copra dropped from £22 in the late 1870s to just over £14 in 1886, and market prices rose only because of the world shortage created in the early 1900s by the Philippine crop being afflicted by a blight.[59] Most of the chapters in this volume deal with sugar, which suffered an even more precipitous fall in price: in 1883–1884 the world price fell by one-third, from 19 shillings per hundredweight to 13 shilling 3 pennies per hundredweight, from which the crop did not fully recover for another thirty years. What made the slump even more severe was that the 1883 price was a depressed one.[60] The exception involved Hawaii, which was shielded from depressed prices by the Reciprocity Agreement of 1877; it gave Hawaiian sugar tariff-free access to the United States market, but this advantage was offset to some extent by mainland pressure to substitute for Asian laborers a more expensive labor force of European extraction (see Beechert in this volume).

Faced with severe profitability restraints, it was as well that planters could take advantage of the applied research of that previous decade to increase per unit yields; and it is probably true to say that the agricultural scientist in the field and the industrial chemist in

the mill were as much the saviors of the sugar industry as were the
workers who provided the muscle power in an age before mechanization.[61] The distinction has been drawn between the "industrial
revolution" based on coal, iron, and the use of machines and the
"scientific revolution" based on steel, electricity, oil, and chemi-
cals, which was "far more prodigious in its results."[62] Sugar production was an obvious beneficiary of the scientific revolution in
the sense that per acre yields expanded dramatically through the
application of scientific method, although this may also have contributed to overproduction and therefore to the depressed market
prices. Nevertheless, the capital outlays that underwrote the scientific revolution in the sugar industry were only possible because of
the increasingly corporate character of plantation agriculture.
Plantation companies, and sometimes also mining companies, were
the only viable means by which colonial governments could guarantee their finances in an age when colonies were expected to pay
their own way. Thus there was an essential congruence of interests
between local administrations and the large plantation companies,
such as the Colonial Sugar Refinery Company in Fiji, the DHPG in
Samoa, Levers in the Solomon Islands, and the "Big Five" in
Hawaii; and the logic of the situation led everywhere to official
support for the large firm as the mechanism of economic development.[63]

The point of this digression is to say that profitability constraints
resulting from depressed market prices led to increasingly harsh
material conditions for plantation workers as employers sought to
affect economies by such means as overtasking laborers, scrimping
on rations, housing, and health services, deducting from wages,
and extending contracts on flimsy pretexts, and backing up their
demands with a heavy hand.[64] Most of the case studies in this volume are concerned with these difficult decades of low market
prices. At the same time, the increasing dependence of colonial
administrations on plantation enterprise contributed to a disinclination in official circles to ameliorate the conditions of labor,
since this would place further constraints on profitability and
therefore on colonial revenues. The authorities were generally prepared to countenance the employers' resort to a variety of legal and
extralegal devices to get more work out of indentured plantation
workers and to minimize the ongoing costs of labor in the name of
"progress." In Fiji, for example, the daily task was unilaterally
redefined by employers, in other words increased. This accounts
for the effectiveness with which management was able to use legal
mechanisms to defeat workers' demands or attempts at redress. For

the state to have done otherwise would have been to court its own financial disaster. "Were the affairs of the Colonial Sugar Company to become crooked," remarked a colonial official in Fiji, "the Colony would utterly collapse."[65] There could also be overtly political motives behind plantation development, as in Samoa, where the DHPG's estates provided Germany's sole claim to annexing those islands to the Reich.[66]

Only in Queensland were working conditions unrelated to the fall in sugar prices. Instead, the late 1880s and 1890s witnessed greatly improved plantation conditions. There is no paradox: the heavy recruiting in the 1880s depleted the Pacific reservoir of potential recruits, making labor scarcer, more expensive to obtain, and so more valuable; to procure and hold labor, employers had to, and often did, offer more attractive working conditions. Improved material conditions thus occurred through competition for labor, and this process occurred despite the fall in sugar prices. Add to this the presence of the large pool of assertive time-expired Melanesians in Queensland (between 42 percent and 67 percent during the 1890s), and an unprecedented degree of legislative protection, and it is clear that they were operating from a position of relative strength vis-à-vis their employers.[67]

When circumstances did not favor the application of coercive measures of a direct sort, planters resorted to more sophisticated methods of social control. The shift from a repressive to a more paternalist regime was usually brought about by protective labor regulations or else by the need to make conditions more attractive in order to maintain one's labor force. A clear demonstration of the latter approach is evident in the anxiety of the Planters' Association in Fiji in 1912—when it became clear that further supplies of laborers from India were at risk—to have the labor laws amended in ways it had opposed only a few years before.[68] Michael Gonzales' chapter in this volume on the sugar plantations of the Aspíllagas family in northern Peru provides a comparable example. The Aspíllagas needed more laborers to meet increased production and discovered that they could only keep their labor force intact if in a community setting, even though this involved greater expenditure in housing, medical care, and recreational facilities. Another aspect of replacing the stick with a carrot was the introduction of a system of rewards and bonuses, which further raised the costs of labor (although not necessarily unit labor costs, if more labor-effort was elicited by the rewards and bonuses).[69] The family's approach provides an apt enough illustration of the adaptability of enlightened management, but their motives were entirely pragmatic. Although the Aspíllagas "adopted a conciliatory attitude toward workers' grievances, reasoning that it was better to grant concessions than to risk a major confrontation that could poison labor relations for decades," they nevertheless controlled workers through manipulation, and the threat of repression and violence was always there as a last resort. The element of expediency underlying the Aspíllagas' motives is also evident in the contrast between their harsh treatment of imported Chinese laborers, who were bound by their contracts of indenture, and the more benign conditions enjoyed by local Indian laborers, who had a greater degree of freedom in the choice of employers and indeed whether or not to work on plantations at all.[70]

Another method of social control was to maintain an ethnically mixed labor force in the hope that intergroup rivalries would inhibit the development of worker solidarity and the possibility of conspiracy. Even when a racial mix of laborers was forced on employers, as at the phosphate mines of Ocean Island and Nauru, the effect was the same.[71] Furthermore, planters sometimes encouraged, or at least condoned, the development of worker institutions that would create the appropriate social divisions. The organized recreational clubs, or *Cabildos de Nacion,* which were formed by slaves in Cuba, were initially tolerated by both planters and the authorities because they reduced the tensions of plantation life while at the same time inhibiting the integration of African slaves with the rest of Cuban society. Increasingly, however, the *Cabildos* fell from favor, largely because they were seen as a focus for the preservation of African culture, which would, in turn, serve to discourage the Hispanicization of freed slaves.[72] The heterogeneous composition of a given labor force need not be a deliberate policy of employers, even though it is likely to achieve the same effect of inhibiting the development of a sense of community within the work force. But divide-and-rule tactics of this sort were not always possible even when contrived at: "Faced with the political realities of the world supply of cheap labor," writes Edward Beechert about Hawaii, "the planters were never able to achieve a multiethnic work force of sufficient magnitude which would enable them to substitute workers as a means of controlling wage demands. They took what labor was available."[73] What is more apparent is that the power structure of the plantation, with its battery of legal and extralegal devices in the interests of worker control, was generally effective in suppressing large-scale worker resistance. There was nothing in the Pacific comparable to the great slave rebellions in Jamaica and Brazil, and it is notable that the big plantation strikes in the Pacific Islands occurred after the institution of indenture had been abolished, and in the case of Queensland only after the Pacific Islanders had been replaced by a white labor force.[74]

This suggests an inverse relationship between the suppression of workers and their resistance, and in many cases such a formula holds true. But even in extremely coercive labor systems, resistance in one form or another was sometimes the order of the day. That an excessively vigorous disciplinary regime could be counterproductive, both in the short- and the long-term, is illustrated by Robert Ross in his study of slavery in the Colony of the Cape of Good Hope. Slaves often reacted to the pain and indignity of being flogged by absconding soon after. There was also a discernable cause-and-effect process whereby escalating punishment tended only to fuel the slaves' resolve to resistance, until in the end the "Cape rulers were forced to create a whole series of institutions to ensure that they maintained their grip on the Colony."[75] Wells and Joseph (in this volume) address similar issues with regard to the social relations of production on henequen plantations in Yucatán. Both planters and government authorities made constant resort to such repressive mechanisms as bounty hunters, secret police, national guard troops, and local government officials to maintain control; the planters, moreover, instituted a thoroughgoing debt peonage to bind workers to plantations; and still a large level of resistance prevailed, not only among plantation laborers but by peasant villagers resisting the encroachment of plantations onto their lands and being drafted into the national guard. In contrast to neighboring Guatemala (see McCreery in this volume), there was a breakdown in social order as the state not only lost its grip on the henequen zone and its periphery but could no longer isolate and suppress violent disputes, even on the haciendas. The immediate cause of a rising tide of violence in the henequen zone between 1907 and 1911 can be put down to a sharp economic downturn that resulted in the owners of the estates cutting back on wages and conditions. A more fundamental cause of this resistance, however, was the relative absence of social control of the types and intensity adopted by the Aspíllagas—in short, the lack of any developed paternalistic relationship between planters and their workers; and the reason for this in Yucatán lay in absentee ownership by entrepreneurs who had no stake in the place but rather regarded their plantations as speculative pieces of real estate, which might at any moment be sold, traded, or gambled away. The absentee owners, moreover, were divided into warring factions and were themselves at odds with the national government. The workers meanwhile "turned elite feuds to their own advantage" and adopted strategies of "political conflict and popular insurgency" that took the region to the brink of generalized rebellion.

The question of paternalism as a means of control is central to the question of resistance and accommodation on plantations, but such a recognition has not always been so. In the mid-1960s, Eugene Genovese made the point that historians, in analyzing how slaves in the antebellum South had survived and fought back, had erred:

> The trouble is that we keep looking for overt rebellious actions—the strike, the revolt, the murder, the arson, the tool-breaking—and often fail to realise that, in given conditions and at particular times, the wisdom of a people and their experience in struggle dictates a different course and an emphasis on holding together individually and collectively.[76]

Genovese developed this theme further in *Roll, Jordan, Roll,* his famous book on the world the slaves made in order to counter the "elaborate network of paternalistic relationships" that structured their existence. Whereas masters sought possession of the hearts and minds of their slaves (or total cultural hegemony), the slaves' response to this paternalism did not normally involve hopeless acts of overt resistance that could not succeed. Instead, they redefined their masters' paternalism, and in doing so developed "a sense of moral worth" that involved no "surrender of psychological will." In this they "transformed their acquiescence in paternalism into a rejection of slavery itself, although their masters assumed acquiescence in the one to demonstrate acquiescence in the other."[77] That is to say, the slaves resisted within the parameters of accommodation, but it was not a case of adopting one to the exclusion of the other. As Genovese says elsewhere:

> We do know of enough instances of deliberate sabotage to permit us to speak of a strong undercurrent of dissatisfaction and hostility. . . . And we need to remember . . . that the pattern of behavior we call accommodation (for want of a better word) itself represented a struggle for cultural autonomy and unity and as such had its own positive value for black people beyond the terrible and basic question of staying alive.[78]

The various constraints on labor, and the scope for alternative strategies along the lines suggested by Genovese, goes far in explaining why overt and large-scale manifestations of resistance were rare.

Slave revolts, when viewed in the span of time, bear out this contention. Whether they did or did not occur can be related to specific structural conditions and political circumstances. In the Colony of the Cape of Good Hope, for example, a comprehensive "structure of domination" had the mutually reinforcing effects of precluding the development of a slave culture—least of all one that spanned generations—and with it the possibility of organized rebellion.[79] These slaves' predicament bears striking similarity to that of Indian plantation workers in Fiji (as described by Lal in this volume). To go yet further back in time is to reveal a broadly similar picture. In classical antiquity there were no more than three significant slave revolts, and these occurred only because circumstances permitted —namely, a breakdown of the social order and a commonality of language and identity among sufficient numbers of slaves, in conjunction with individuals of outstanding leadership among them.[80] The necessary preconditions also existed in Brazil and the Caribbean:

> An important ingredient in the development of revolts out of local disturbances was the division of whites into warring factions and the general weakness of state apparatus. Together with these conditions went the general influence of geography in relation to state power. Where suitable terrain was combined with a weak state, runaway slaves could and did found maroon colonies, which directly fomented revolts and kept alive a tradition of armed resistance. With minor qualifications, these conditions did not exist in the United States.[81]

Here, the major manifestation of armed slave resistance was the Nat Turner rebellion of 1831, yet this so-called "cataclysm" was a small-scale event involving about one hundred slaves and confined to Southhampton County, Virginia.[82] A rebellion of similar extent and duration (three days of actual fighting) would have passed almost unnoticed in Brazil or the Caribbean. Yet even in these revolt-prone places, the usual pattern was one of failure or limited success; the only *comprehensively* successful slave rebellion was at Haiti, a fact that reinforces the frequent observation that overt resistance was risky at best.

Indentured plantation workers faced a similar problem, resulting in strategies of resistance that fell well short of outright rebellion. To give an example—one among many—of Indian laborers in Natal, Maureen Tayal has observed: "Acts of resistance on the estates were usually individualistic, often of a type which did not require premeditation, and almost always ineffectual except from the point of view of providing momentary psychological release from an intolerable burden of frustration.''[83] This perhaps explains why acts of overt violence by plantation workers were usually confined to often spontaneous attacks on overseers and drivers, lashing out at the immediate cause of frustration. There is the story, perhaps apocryphal, that a European overseer in Fiji quit his job in humiliation after being overpowered by a group of Indian women laborers, who held him down while each took her turn at urinating upon him.[84] On another occasion, again in Fiji, an overseer confronted a group of female workers who ''took up their hoes. He retreated, pleading with the women not to hit him, [and] moving backwards he landed in a sewer pit. The women then threw shit at him.''[85] This was the risk that overseers and drivers drawn from the ranks of the workers were prepared to take for the opportunity of advancement through collaboration: in the realm of social control, there were positive inducements to conform, and these are important in the socioeconomic structure of plantations.[86] There were also attacks upon lesser- or nonauthority figures—namely each other, either in the field or in more domestic settings, which would often go undetected by management. Otherwise, the sheer hopelessness of overt action and the realization on the laborers' part that indenture was a temporary condition that would only be prolonged by active resistance was usually sufficient to stifle it (see also Lal in this volume). Writing about peasants, whose situations are comparable in many respects to indentured laborers, James C. Scott has also drawn attention to the relatively small numbers of rebellions against authority and repression, and stresses instead

> what we might call *everyday* forms of resistance: the prosaic but constant struggle between the peasantry and those who seek to extract labour, food, taxes, rents, and interest from them. Most of the forms this struggle takes stop well short of collective outright defiance. Here I have in mind the *ordinary* weapons of relatively powerless groups: footdragging, dissimulation, false-compliance, pilfering, feigned ignorance, slander, arson, sabotage, and so forth. These Brechtian forms of class struggle have certain features in common. They require little or no co-ordination or planning; they often represent a form of individual self-help; and they typically avoid any direct symbolic confrontation with authority or with elite norms. To understand these commonplace forms of resistance is to understand what much of the peasantry does ''between revolts'' to defend its interests as best it can.[87]

It is not surprising that the findings of his seminal work *Weapons of the Weak* have been applied to several of the case studies in this volume (see chapters on Queensland, Samoa, the Solomon Islands,

Yucatán, Peru, and Guatemala).[88] Indeed, Scott's conclusions are relevant to all the chapters, given the relative powerlessness of indentured laborers against the wider authority structures that confronted them. But the situation varied among the locales examined in this book, and any discussion of workers' resistance must go beyond the undifferentiated listing of the forms it might take (as does Ronald Takaki), or to distinguish in terms of polar opposites, for example overt/covert, individual/collective (as does Kay Saunders), or in seeing resistance as an ever-pervading phenomenon (as does Shaista Shameem) without pausing to ask why a particular form of resistance manifested itself, if at all, and under what circumstances this was possible and feasible. Conversely, one must ask why particular forms of resistance were less prevalent or, alternatively, why accommodation and acquiescence were the norm. As Alan Warde has pointed out, if in a very different context: "The specific structural conditions under which they labor and the concrete political situations in which they find themselves will affect their actions."[89]

The case studies in this volume highlight this very point. Strike action—an explicitly overt form of collective resistance—was rare to the point of being exceptional within the institution of indenture, as Brij Lal's case study makes clear. Quite apart from the fact that striking was a forbidden aspect of indentured service, although it did occur from time to time, the diverse cultural background of the Indian laborers hindered the development of common interests and values, and because they were young and generally of lowly status in their own communities, they lacked leadership and organizational skills. Being confined on widely scattered plantations only increased the difficulties of mounting concerted opposition to the plantation regime, and in any case many had no incentive to fight the system. Rather, there was pressure to save money and to return to India at an early date: these were the primary considerations in the sojourners' calculations, and they acted as powerful disincentives to any involvement in time-consuming and potentially costly struggles against the planters. Add to this the authority the planters had over their laborers and the influence they enjoyed with the colonial government, and it becomes apparent that collective, organized resistance, such as striking, was not a viable option, despite laborers having strong reasons for grievance. The silence of the oppressed and their lack of action, however, should not be taken to imply indifference to or satisfaction with the status quo. Even in Queensland, where many Melanesians would have had an incentive to work well within the system and gain a good reputation, thus earning higher wages later as time-expired workers, the response was often only one of temporary compliance. Once they had become time-expired workers, these same Melanesians often assumed a more intransigent posture, and if this resulted in frequent court appearances, it seemed not to worry them because they had the money to pay the fines and, in short, "the ability to confront the system" (see Moore in this volume).

Absconding and desertion were disruptive of the plantation work routine and a tempting alternative to laborers who could escape across national borders with little fear of being extradited, or who could join maroon communities that were beyond the effective reach of the authorities.[90] It could also have its drawbacks, and frequently the risks involved outweighed any rational hope of success. By imposing collective punishment on workers for the misdeeds of an individual, planters had a ready-made means of making desertion a risk that only a few would take.[91] Another form of collective punishment, possible when workers were drawn from surrounding communities, took place in the context of debt bondage in Latin America, where employers had the option of exerting pressure against the debtor's close kin; they in turn would put pressure on the debtor to meet his labor obligations to the creditor-employer.[92] In the present volume, the case study on Gilbertese plantation workers in Samoa demonstrates the quandary of workers who had no real means of escaping from an extremely oppressive plantation regime. Because they so frequently came in family groupings, absconding Gilbertese laborers were deserting their families as well as their masters, so there was a strong disincentive to flight. Moreover, the hostility generally evinced by indigenous peoples toward imported laborers usually worked to the advantage of planters, because laborers were often safer on the plantations than at large in the countryside. Those laborers foolhardy enough to sneak off into the night were severely punished if caught; the likelihood of being captured was a near certainty because local Samoans would return them for the reward money; in the unlikely event of evading capture they were even less likely to survive in the bush.[93] Absconding was only likely to become an option when imported laborers and local people had established some sort of rapport, however tenuous. Desertion in Samoa became prevalent only after the late 1880s, when the Samoans directed their enmity toward the DHPG rather than the laborers.[94] But there is more to it than that: the DHPG's work force by this time was overwhelmingly made up of young, unmarried Melanesian males, whose desertion would not involve abandoning their families.

Obvious parallels exist between the situation in Samoa and "the isolation mechanism" used by planters in Yucatán to mobilize and control the work force (see Wells and Joseph in this volume). This involved several "reinforcing mechanisms of isolation." On the haciendas, negative constraints such as debt and physical punishment with a leavening of positive inducements such as housing and medical care were deployed to bind workers to the workplace. Beyond the hacienda, the use of bounty hunters, the state militia, and the local authorities combined to discourage desertion. At another level, the encroachment of the expanding hacienda system on the physical boundaries of villages increasingly resulted in the villages offering a diminishing source of refuge to runaways, while the villagers, in turn, found themselves increasingly drawn to employment on haciendas as alternative means of a livelihood contracted. Add to this the ecological limitations of the surrounding countryside and the difficulty of traveling beyond the region to freedom, and the final components of "the isolation mechanism" are in place: faced with such constraints, most rural dwellers in Yucatán therefore "made an economically rational decision in opting to reside on the haciendas" and to stay there.

The ability of workers to strike back or to otherwise resist is situationally determined in other ways. A plantation of mature coconut palms, for example, offers far less scope for sabotage than a sugar plantation. In contrast to coconut palms, a sugar crop is highly flammable and can be torched with little risk of detection, and placing concealed rocks in the mill's rollers could have spectacular results in terms of damaged equipment. These are the "weapons of the weak" or "everyday forms of resistance," and they assume a covert guise because those involved are operating from a position of essential weakness.

Recourse to the law as a countervailing force was not usually feasible. But the provisions of the law could change to the advantage of the laborer. More often than not, changes to the legal framework were imposed upon a reluctant local legislature by a higher external authority. The prime example is the termination of indenture when Hawaii was annexed in 1900, since contract labor was repugnant to mainland law. The effect of outside pressure can be seen in other ways: as already mentioned, the adverse opinion in the growing nationalist movement in India toward overseas labor migration impelled planters to improve the working conditions of their Indian workers lest they be denied access to further supplies, whereas in previous decades they were unfettered by public criticism, whether from the local press, missionary bodies, or humanitarian organizations.[95] In Queensland, the legislation covering indentured

Melanesian laborers became increasingly protective. This meant that Solomon Islanders, who formed the bulk of the sugar industry's labor force after the mid-1880s, went to Queensland "when conditions on the plantations were at their best, and when the rewards for labour migration were at their highest."[96] New Hebrideans, by contrast, were mostly recruited to Queensland before these reforms and therefore had to bear the brunt of employer abuses whether they liked it or not. In none of these instances did the workers actually contribute to their improved circumstances. Rather, they were beneficiaries of a situation they did nothing to create.

On other occasions, however, workers actively sought the aid of a third party in their contest for a greater measure of control in the workplace. If planters often became more sophisticated in their methods of social control, so too the workers became more adroit in their resistance. Recognizing the futility of one course of action, they would often turn to another should the opportunity arise. Thus the Gilbertese laborers in Samoa working on Frank Cornwall's estates in the 1880s sought out as "protector" the local British consul, and in doing so they demonstrated a fine appreciation in identifying their only viable source of redress. When the DHPG's Gilbertese laborers were placed under British consular control a decade later, they extracted the last ounce of advantage from their new situation and showed such intransigence that the company arranged the early repatriation of its entire Gilbertese labor force, glad to be rid of them (see Munro and Firth in this volume). Given the vulnerability of indentured laborers and the multitude of disabilities with which they had to contend, it was critical that they had a "protector," whether official or unofficial—someone who could act on their behalf, articulate their grievances, ensure that their rights at law were upheld, publicize injustices, represent their case in court if need be, and serve their interests generally.[97] Frequently enough, official protectors of laborers were actively hindered or insufficiently supported by the very governments that employed them.[98] Those private individuals who assumed the role of "protector" were often harassed, and on one occasion in Hawaii even lynched.[99] But the situation was a variable one, as the Gilbertese in Samoa demonstrate: Cornwall's laborers showed shrewdness in identifying their only source of redress; while the DHPG's Gilbertese laborers manipulated the fact of British consular protection in a manner that went well beyond what the British authorities had originally envisaged. In Guatemala, by contrast, "by far the most common mode of resistance was the undramatic but often effective petition of rights and grievances, of which the indigenous population filed thousands . . . For the authorities who received these petitions, they presented no small problem and could rarely be ignored with impunity" (see McCreery in this volume).[100] The "weapons of the weak," if circumstances allowed, could have unexpected potency.

The final question to be addressed concerns the boundaries of, and between, resistance and accommodation. At a conceptual level, this brings to mind the less problematic distinction often made by social historians of agrarian life and early industrialization between outright crime and social protest, arguing that it is not the action but the motivation that matters; and the interminable debates over the distinguishing features of revolutions, coups, "internal wars," insurrections, *jacqueries*, and other uprisings against authority. In the case of resistance and accommodation, scholars by and large have taken the meanings of the two terms as given. One of the few attempts at actual definition has been provided by James Scott, for whom resistance means "*any* act(s) by member(s) of a subordinate class *intended* to either mitigate or deny claims . . . made on that class by superordinate classes . . . or to advance their own claims . . . vis-à-vis those in superordinate classes."[101] But it is not quite so straighforward; Scott also maintains that the "insistence [by some scholars] that acts of resistance must be *shown* to be intended" can create difficulties:

> [In] acts like theft . . . we encounter a combination of immediate individual gain and what *may* be resistance. How are we to judge which of the two purposes is uppermost or decisive? What is at stake is not a petty definitional matter but rather the interpretation of a whole range of actions that seem to me to lie historically at the core of everyday class relations. The English poacher in the eighteenth century *may* have been resisting the gentry's claim to property to wild game, but he was surely as interested in rabbit stew. . . . Which of these inextricably fused motives are we to take as paramount? Even if we were to ask the actors in question, and even if they could reply candidly, it is not at all clear that they would be able to make a clear determination.[102]

Part of the problem, as Scott stresses, is the seeming inarticulateness of the subordinate classes, and the fact that so little of the contemporary documentation emanates from them. The superordinate classes, whether these be employers or the state, not only create much of the archival source material but interpret events to suit themselves.[103] More contentiously, Scott also maintains that one

ought not to rule out theft, for example, as an act of resistance
unless proven to the contrary. The gist of Scott's argument is that
innumerable acts of petty resistance, which in their totality can be
enormously disruptive to the superordinate classes, are only successful if undetected, and if undetected they will find little or no
echo in the historical sources. Thus, observable resistance is much
like the tip of an iceberg—simply a small fraction of the totality.

But what starts off as an exhortation not to rule out resistance
unless otherwise demonstrated comes very close to saying that any
given action *may* be regarded as resistance until there is proof to
the contrary. This seems to be a case of off-loading the burden of
proof to the other side in ways that are contentious. Absolved of
the obligation to demonstrate intent, the likelihood is that resistance will be seen where it does not occur; this in turn encourages
the risk of broadening the spectrum of resistance to unsustainable
bounds. Indeed, there is always danger that "resistance" will be
pressed into service as the overarching explanatory device for
worker behavior. Often the term has been made to work too hard,
and has been indiscriminately applied to encompass all manner of
actions on the part of workers. A genuinely indisposed laborer is
self-evidently not resisting the authority or demands of an employer any more than he or she is accommodating. Neither are
large-scale brawls among Melanesian plantation workers a protest
against treatment and working conditions. Rather, they are a function of the tensions that typically prevail among strangers in
Melanesia and often also of drunkenness, and in fact, they serve to
suppress a sense of group solidarity that might otherwise have provided a springboard for resistance proper. Nor does uncooperative
behavior necessarily amount to resistance: "Malingering may have
reflected no more than a disinclination to work, especially when
the rewards were so meagre. Likewise, what is taken for sabotage
may have originated in apathy and indifference."[104] In the case of
the present volume, resistance is a response to plantation life and
labor, but clearly not every response can be said to constitute an act
of resistance.

It is also a mistake, as previously mentioned, to impose a dichotomous model and to view resistance and accommodation as polar
opposites. That the degree of convergence between the two can
sometimes be quite marked is illustrated by Rebecca Scott's study
of the transition from slavery to abolition in Cuba. Between 1880
and 1886, the institutional mechanism by which this was imper-
fectly accomplished was the *patronato,* which so altered the existing social relationships between slaves and their masters that the

distinction between resistance and accommodation becomes exceedingly blurred as the *patrocinados* (the slaves being eventually freed under this arrangement) sought to wring further concessions in the face of attempts by slaveowners to hold the line. As Rebecca Scott explains, "The initiatives of *patrocinados* thus emerge as a hybrid activity, and fit neither the category of accommodation or that of resistance."[105]

A further consideration concerns the extent to which culture retention and the expression of cultural norms can be equated with resistance. Is cultural retention a *form of resistance* or should it more properly be regarded as a *source for resistance,* and when is it neither? The maintenance of ethnic religion among Japanese interned in the United States during World War II clearly had elements of both: religion in this case was a form of resistance against efforts to Americanize them, and a source for resistance in that it preserved a sense of identity and solidarity and thus averted social disintegration.[106] In other cases the distinctions are not so straightforward. Referring to the Melanesian workers in the Mackay district of Queensland, Clive Moore puts the view (in this volume) that

> their lack of obvious active resistance or protest relates not only to the authoritarian character of the plantation system but also to their deliberate choice of methods of response. . . . [T]hey maintained Melanesian social equilibrium by depending on major inward-focusing Melanesian social mechanisms coming from their ethnic backgrounds to provide satisfactory balances in their lives. Their response was to mediate their working and private lives through their own beliefs, institutions, and forms of behavior, which were in partial operation in the colony.

Moore draws a fine line between the more conventional notion of worker "resistance" as a reaction against the unsatisfactory aspects of plantation life and labor, and "response," or the coping strategies of workers. The thrust of Genovese's eloquent portrayal of the world that slaves made in order to counter their masters' paternalism is that by creating this separate moral world, exempt from the insinuations of their masters' paternalism, slaves provided themselves with psychological elbow room and a sense of inner integrity.[107] It is a subtle argument reflecting how resistance can assume equally subtle guises and hence why resistance can sometimes be confused with purely neutral forms of behavior, such as the retention of a hairstyle.

Nevertheless, the point can still be made that what does not appear on the surface to be resistance may indeed be just that. Less controversially, it can be said that resistance and accommodation on plantations are located within specific structural situations, and the forms they take are culturally conditioned to a greater or lesser extent. The chapters that follow all relate to situations of oppression and constraint, which effectively rule out overt or organized forms of resistance, at least on a sustained basis, and encourage instead a quiescent, less confrontational response from the laborers. Given the authoritarian nature of the plantation and the means of enforcement at planters' disposal, it is surprising perhaps that so much resistance actually occurred, and all the contributors to this volume remark on the ingenuity of workers in finding ways to fight back. Without romanticizing, the point can still be made that even when resistance failed, it was nevertheless admirable in that an attempt had been made to beat the odds, which might indirectly result in later improvements. Slave resistance, for example, in the United States was "essential to abolitionist propaganda, even though the slaves were often far removed from the immediate circumstances."[108] Nevertheless, indentured plantation workers generally lacked the tactical power to do much beyond brandishing the "weapons of the weak," winning the occasional skirmish, perhaps a battle or two, but losing the war until the structural conditions that governed their conditions of life and labor were altered by whatever means. Until that time arrived, their response can perhaps be summed up by the Ethiopian proverb: "When the Great Lord passes, the wise peasant bows deeply and silently farts."[109]

Notes

Many debts were incurred during the preparation of this chapter. I am especially grateful to Brij Lal. He first aroused my interest in the issues of resistance and accommodation, and our discussions materially assisted in the writing of this chapter. Ralph Shlomowitz commented incisively on the first draft and insisted that I adopt a yet more comparative approach. The award of a Visiting Research Fellowship by The Flinders University of South Australia, October–December 1990, made this possible. I also wish to thank James C. Scott, Clive Moore, Geralyn Pye, Colin Newbury, David McCreery, Barry Carr, and Stephen R. Niblo for assistance of various kinds. The financial support of the Australian Research Council is also gratefully acknowledged.

1. Michael J. Gonzales, "Chinese Plantation Workers and Social Conflict in Peru in the Late Nineteenth Century," *Journal of Latin American Studies*, 21:3 (1989), 385.

2. George L. Beckford, *Persistent Poverty: Underdevelopment in plantation economies of the Third World* (Jamaica/London, Zed Books edn., 1983), xxii.

3. J. H. Bernardin de Saint Pierre (1773), quoted in Sidney W. Mintz, *Sweetness and Power: The place of sugar in modern history* (New York, 1985), frontispiece quotation.

4. See Michael Moynagh, *Brown or White?: A history of the Fiji sugar industry, 1873-1973* (Canberra, 1981), 9-11.

5. C. C. Goldthorpe, "A Definition and Typology of Plantation Agriculture," *Singapore Journal of Tropical Geography*, 8:1 (1987), 30.

6. Edgar T. Thompson, "The Plantation as a Social System," in Pan American Union, *Plantation Systems of the New World* (Social Sciences Monographs, 7, Washington, D.C., 1959), 26, quoted in Beckford, *Persistent Poverty*, 10-11. Thompson was writing from the unusual perspective of one born and raised on a United States plantation. The chapter in this volume on Yucatán (by Wells and Joseph) deals with haciendas rather than plantations; the authors draw the distinction between the two in the present volume. For a more detailed discussion, see Allen Wells, "From Hacienda to Plantation: The transformation of Santo Domingo Xcuyum," in Jeffrey Brannon and Gilbert M. Joseph (eds), *Land, Labor and Capital in Modern Yucatán: Essays in regional history and political economy* (Tuscaloosa/London, 1991), 112-142.

7. Beckford, *Persistent Poverty*, 12.

8. For estimates of the scale of intercontinental labor flows, see Stanley L. Engerman, "Servants to Slaves to Servants: Contract labour and European expansion," in P. C. Emmer (ed.), *Colonialism and Migration: Indentured labour before and after slavery* (Dordrecht, 1986), 271-274.

9. Brij V. Lal, *Girmitiyas: The origins of the Fiji Indians* (Canberra, 1983), 13; Stanley L. Engerman, "Contract Labour, Sugar, and Technology in the Nineteenth Century," *Journal of Economic History*, 43:3 (1983), 648.

10. Charles A. Price with Elizabeth Baker, "Origins of Pacific Island Labourers in Queensland, 1863-1904: A research note," *Journal of Pacific History*, 11:2 (1976), 110-111; David McCreery, personal communication, 30 November 1988.

11. Doug Munro, "The Origins of Labourers in the South Pacific: Commentary and statistics," in Clive Moore, Jacqueline Leckie, and Doug Munro (eds.), *Labour in the South Pacific* (Townsville, 1990), xxxix-li.

12. J. W. Davidson, "Problems of Pacific History," *Journal of Pacific History*, 1 (1966), 5-21; H. E. Maude, "Pacific History—past, present and future." *Journal of Pacific History*, 6 (1971), 3-24; K. R. Howe, "The Fate of the 'Savage' in Pacific Historiography," *New Zealand Journal of History*, 11:2 (1977), 137-154. For a recent "outsiders' appraisal" of the Canberra school, see Robert Borofsky and Alan Howard, "The Early Contact Period," in Alan Howard and Robert Borofsky (eds.), *Developments in Polynesian Ethnology* (Honolulu, 1989), 243-246.

13. Dorothy Shineberg, *They Came for Sandalwood: A study of the sandalwood trade in the South-West Pacific, 1830-1865* (Melbourne, 1967).

14. See especially Deryck Scarr, "Recruits and Recruiters: A portrait of the labour trade," *Journal of Pacific History;* 2 (1967), 5-24; Peter Corris, "Pacific Island Labour Migrants in Queensland," *Journal of Pacific History*, 5 (1970), 43-64; Peter Corris, *Passage, Port and Plantation: A history of Solomon Islands labour migration, 1870-1914* (Melbourne, 1973).

15. The academic writers include Adrian Graves, "The Nature and Origins of Pacific Islands Migration to Queensland, 1863–1906," in Shula Marks and Peter Richardson (eds.), *International Labour Migration: Historical perspectives* (London, 1984), 112–139; Adrian Graves, "Colonialism and Indentured Labour Migration in the Western Pacific, 1840–1915," in Emmer (ed.), *Colonialism and Migration*, 237–259; Kay Saunders, *Workers in Bondage: The origins and bases of unfree labour in Queensland, 1824–1916* (Brisbane, 1982). For appraisals, see Clive Moore, "Revising the Revisionists: The historiography of immigrant Melanesians in Australia," *Pacific Studies*, 15:2 (1992), 61–86; Doug Munro, "The Historiography of the Queensland Labor Trade," ms.

16. K. L. Gillion, *Fiji's Indian Migrants: A history to the end of indenture in 1920* (Melbourne, 1962), ch. 6; Brij V. Lal, "Labouring Men and Nothing More: Some problems of Indian indenture in Fiji," in Kay Saunders (ed.), *Indentured Labour in the British Empire, 1834–1920* (London/Canberra, 1984), 126–157; S. G. Firth, "German Recruitment and Employment of Labourers in the Western Pacific before the First World War," D.Phil. dissertation, Oxford University, 1973; Stewart Firth, *New Guinea under the Germans* (Melbourne, 1982), ch. 6; Michel Panoff, "Travailleurs, Recruteurs et Planteurs dans l'Archipel Bismarck de 1885 à 1914," *Journal de la Société des Océanistes*, 64 (1979), 159–173; Edward D. Beechert, *Working in Hawaii: A labor history* (Honolulu, 1985), chs. 3–5.

17. Arnold J. Bauer, "Rural Workers in Spanish America: Problems of peonage and oppression," *Hispanic American Historical Review*, 59:1 (1979), 34–63.

18. Clive Moore, *Kanaka: A history of Melanesian Mackay* (Boroko/Port Moresby, 1985).

19. In a celebrated article, Kerry Howe deplored the trend in Pacific historiography toward "finding out more and more about less and less" and called for a more synoptic orientation. K. R. Howe, "Pacific Islands History in the 1980s: New directions or monograph myopia?" *Pacific Studies*, 3:1 (1979), 81–90.

20. As expressed, for example, in O. W. Parnaby, *Britain and the Labor Trade in the Southwest Pacific* (Durham, N.C., 1964).

21. P. M. Mercer and C. R. Moore, "Melanesians in North Queensland: The retention of indigenous religious and magical practices," *Journal of Pacific History*, 11:1 (1976), 66–88.

22. Moore, *Kanaka*, ch. 7.

23. Colin Newbury, "The Melanesian Labor Reserve: Some reflections on Pacific labor markets in the nineteenth century," *Pacific Studies*, 4:1 (1980), 2–3.

24. This includes "Markets for Indentured and Time-expired Melanesian Labour in Queensland, 1863–1906: An economic analysis," *Journal of Pacific History*, 16:2 (1981), 70–91; "The Fiji Labor Trade in Comparative Perspective, 1864–1914," *Pacific Studies*, 9:3 (1986), 107–152; "Mortality and the Pacific Labour Trade," *Journal of Pacific History*, 22:1 (1987), 34–55; "Epidemiology and the Pacific Labor Trade," *Journal of Interdisciplinary History*, 19:4 (1989), 585–610, among others.

25. Hugh Tinker, "Into Servitude: Indian labour in the sugar industry,

1833–1970"; Donald Denoon, "The Political Economy of Labour Migration to Settler Societies: Australasia, southern Africa and southern South America between 1890 and 1914"; Colin Newbury, "The Imperial Workplace: Competitive and coerced labour in New Zealand, northern Nigeria, and Australian New Guinea," in Marks and Richardson, (eds.), *International Labour Migration*, 74–89, 186–205, 206–232. Other collections of essays dealing with indentured labor that have a Pacific content are Saunders (ed.), *Indentured Labour in the British Empire, 1835–1920* and Emmer (ed.), *Colonialism and Migration.*

26. Ronald Takaki, *Pau Hana: Plantation life and labor in Hawaii, 1835–1920* (Honolulu, 1985), chs. 3 and 5. See the illuminating review of this book by Noel J. Kent in *Pacific Studies*, 8:2 (1985), 135–140.

27. Kay Saunders, " 'Troublesome Servants': The strategies of resistance employed by Melanesian indentured labourers on plantations in colonial Queensland," *Journal of Pacific History*, 14:3 (1979), 168–183 (republished in *Workers in Bondage*, ch. 6).

28. Saunders, " 'Troublesome Servants,' " 174; Corris, *Port, Passage and Plantation*, 25.

29. Saunders, " 'Troublesome Servants,' " 173, 179.

30. Brij V. Lal, "Veil of Dishonour: Sexual jealousy and suicide on Fiji plantations," *Journal of Pacific History*, 20:3 (1985), 153–154; Gillion, *Fiji's Indian Migrants*, 127–128. The pressures that could lead to suicide are graphically delineated in Brij V. Lal and Barry Shineberg, " 'The Story of the Haunted Line': Totram Sanadhya recalls the labour lines in Fiji," *Journal of Pacific History*, 26:1 (1991), 107–112. For comparative material see William D. Pierson, "White Cannibals, Black Martyrs: Fear, depression and religious faith as causes of suicide among new slaves," *Journal of Negro History*, 62:2 (1977), 147–159.

31. Corris, *Passage, Port and Plantation*, 43.

32. Ralph Shlomowitz, "Time-Expired Melanesian Labour in Queensland: An investigation of job turnover, 1884–1906," *Pacific Studies*, 8:2 (1985), 25–44; Shlomowitz, "Indentured and Time-Expired Melanesian Labour in Queensland," *Journal of Pacific History* 10:1 (1981), 70–91; Moore, *Kanaka*, 160–164.

33. Thomas Flory, "Fugitive Slaves and Free Society: The case of Brazil," *Journal of Negro History*, 64:2 (1979), 116–130; Rebecca J. Scott, "Gradual Abolition and the Dynamics of Slave Emancipation in Cuba, 1868–1886," *Hispanic American Historical Review*, 63:3 (1983), 449–477.

34. " 'Troublesome Servants,' " 183. Compare with " 'The Black Scourge': Racial attitudes towards Melanesians in colonial Queensland," in Raymond Evans, Kay Saunders, and Kathryn Cronin, *Exclusion, Exploitation and Extermination: Race relations in colonial Queensland* (Sydney, 1975), 167–207; "The Pacific Islander Hospitals in Colonial Queensland: The failure of liberal principles," *Journal of Pacific History*, 11:1 (1976), 28–50; *Workers in Bondage*, chs. 4–5; "The Worker's Paradox: Indentured labour in the Queensland sugar industry to 1920," in Saunders (ed.), *Indentured Labour in the British Empire*, 213–259.

35. Shaista Shameem, "Girmitiya Women in Fiji: Work, resistance and survival," in Moore, Leckie, and Munro (eds.), *Labour in the South*

Pacific, 148–154. The overemphasis on resistance is a common enough feature of the wider literature. Terence Ranger, in discussing his book *The African Voice in Southern Rhodesia, 1898–1930* ten years after publication, noted that one of its weaknesses was the concentration on resistance without at the same time his reflecting on "adaptation and accommodation." See Ranger, "The Rural African Voice in Zimbabwe Rhodesia: Archaism and tradition," *Social Analysis*, 4 (1980), 100.

36. Brij V. Lal, "Kunti's Cry: Indentured women on Fiji plantations," *Indian Economic and Social History Review*, 22:1 (1985), 69–70.

37. Brij V. Lal, "Murmurs of Dissent: Non-resistance on Fiji plantations," *Hawaiian Journal of History*, 20 (1986), 188–214 (revised version published in this volume).

38. Stewart Firth, "Governors versus Settlers: The dispute over Chinese labour in German Samoa," *New Zealand Journal of History*, 11:1 (1977), 166.

39. I. C. Campbell, "Race Relations in the Pre-colonial Pacific Islands: A case of prejudice and pragmatism," *Pacific Studies*, 8:2 (1985), 76. There are obvious exceptions, such as the Maltese in the Queensland sugar industry (Barry York, "Sugar Labour: Queensland's Maltese experiment, 1881–84," *Journal of Australian Studies*, 25 [1989], 43–56) and the Portuguese, Norwegians, Germans, and Poles, among others, who labored in the sugar fields of Hawaii. European indentured laborers, however, resented the inferior status of indenture and the abusive conditions that went with it and generally sought to terminate their contracts at the first opportunity, the Norwegians in Hawaii being a prime example. See Beechert, *Working in Hawaii*, 86–87; Takaki, *Pau Hana*, 38–41. As if to underline the question of racism in Hawaii, the Portuguese were not accepted as being Caucasian, either by other Caucasians or by Hawaiians, and this may have been due to "their swarthy skin or perhaps because approximately three quarters of the Portuguese immigrants had been illiterate peasants." This matter is discussed by Laurence H. Fuchs, *Hawaii Pono: A social history* (New York, 1961), 53–59.

40. Gillion, *Fiji's Indian Migrants*, 129.

41. E. J. Hobsbawm, *Industry and Empire: An economic history of Britain since 1750* (London, 1969), 67; Takaki, *Pau Hana*, 91. See also E. P. Thompson, "Time, Work-Discipline and Industrial Capitalism," *Past & Present*, 38 (1967), 56–97; John Burnett (ed.), *Useful Toil: Autobiographies of working people from the 1820s to the 1920s* (Harmondsworth, 1977), 39–40; Eugene D. Genovese, *Roll, Jordan, Roll: The world the slaves made* (New York, Vintage Book edn., 1976), 285–294; Bauer, "Rural Workers in Spanish America," 57–58. A useful case study of the contradictions between the natural rhythms of task-oriented work and the unfamiliar routine of time-oriented work is in Takaki, *Pau Hana*, 6–12. Takaki's discussion also highlights the problems of maintaining plantation discipline in the absence of strong *in situ* coercive measures reinforced by the support of an external authority.

42. Kusha Haraksingh, "Control and Resistance among Overseas Indian Workers: A study of labour on the sugar plantations of Trinidad, 1875–1917," *Journal of Caribbean History*, 14 (1980), 3–4. A contrast is evident

in the antebellum United States, where the "driver" (slave foreman) often mediated between slaveowner and slaves in the interests of harmony and productivity. See Eugene D. Genovese, *In Red and Black: Marxian explorations in Southern and Afro-American history* (Knoxville, University of Tennessee, 1984), 125. Genovese also points out that "there has not been a single study of a driver" in the antebellum South (p.105) and this generally holds true for their counterparts in other locales involving the coercive use of plantation laborers. There are no studies of sirdars on Fiji plantations, for example, even though this line of approach would probably reveal a great deal about plantation life under indenture—only the posthumously published account of a European overseer. See Walter Gill, *Turn North-East at the Tombstone* (Adelaide, 1970). One of the very few accounts of the role of drivers is Kusha Haraksingh, "Indian Leadership in the Indenture Period," *Caribbean Issues*, 2:3 (1976), 241–247.

43. Stewart Firth and Doug Munro, "Compagnie et Consulat: Lois germaniques et emploi des travailleurs sur les plantations de Samoa, 1864–1914," *Journal de la Société des Océanistes*, 91 (1990), 124.

44. Robert Ross, *Cape of Torments: Slavery and resistance in South Africa* (London, 1983), 106–115. Galant's experiences form the basis of André Brink's novel *A Chain of Voices* (London, 1982).

45. Eric R. Wolf, "On Peasant Rebellions," in Teodor Shanin (ed.), *Peasants and Peasant Societies* (Harmondsworth, 1971), 268.

46. Eric Wolf, "Specific Aspects of Plantation Systems in the New World: Community sub-cultures and social classes," in Michael M. Horowitz (ed.), *Peoples and Culture of the Carribean* (New York, 1971), 164; Hugh Tinker, *Race, Conflict and the International Order* (London, 1977), 23–25; Beckford, *Persistent Poverty*.

47. Roger M. Keesing and Peter Corris, *Lightning Meets the West Wind: The Malaita massacre* (Melbourne, 1980), 34.

48. Judith A. Bennett, *Wealth of the Solomons: A history of a Pacific archipelago, 1800–1978* (Honolulu, 1987), 169–170. See also Kenneth M. Stampp, *The Peculiar Institution: Slavery and the ante-bellum South* (New York, Vintage Book edn., 1956), 105–106, for a discussion of tensions and war of nerves during the crucial settling-in period when a new overseer took control of a gang of slaves.

49. Peter Kolchin, "The Process of Confrontation: Patterns of resistance to bondage in nineteenth-century Russia and the United States," *Journal of Social History*, 11:4 (1978), 479. The notion that "an oppressed group or class develops an autonomous conception of their economic and social rights, essentially drawing a line across which the ruling class cannot legitimately step," is also developed by Alex Lichtenstein, " 'That Disposition to Theft, with which they have been Branded': Moral economy, slave management, and the law," *Journal of Social History*, 21:3 (1988), 413–440; Emilia Viotti da Costa, "From All According to their Abilities, to All According to their Needs: Day-to-day resitance and slaves' notions of rights in Guiana, 1823–1832." Presented at the Conference on Slavery and Freedom in Comparative Perspective, University of California at San Diego, 4–6 October 1991. See also Genovese, *In Red and Black*, 111; and *Roll, Jordan, Roll*, 45–47. See also Wells and Joseph in this volume.

50. Genovese, *In Red and Black,* 125.

51. Firth, "German Recruitment and Employment of Labourers," 65; Adrian Graves, "Truck and Gifts: Melanesian immigrants and the trade box system in colonial Queensland," *Past & Present,* 101 (1983), 106–118.

52. Beechert, *Working in Hawaii,* 119–120.

53. Michael J. Gonzales, "Capitalist Agriculture and Labour Contracting in Northern Peru, 1880–1905," *Journal of Latin American Studies,* 12:2 (1980), 291–315; David McCreery, "Debt Servitude in Rural Guatemala, 1876–1936," *Hispanic American Historical Review,* 63:4 (1983), 735–759; Allen Wells, *Yucatán's Gilded Age: Haciendas, henequen and International Harvester, 1860–1915,* (Albuquerque, 1985), ch. 6; Alan Knight, "Debt Bondage in Latin America," in Leonie J. Archer (ed.), *Slavery and Other Forms of Unfree Labour* (London/New York, 1988), 206–224.

54. Beechert, *Working in Hawaii,* ch. 3.

55. See also Vijay Naidu, *The Violence of Indenture in Fiji* (Suva, 1980), 52–54; Shiu Prasad, *Indian Indentured Workers in Fiji* (Suva, 1974), 23. A similar disparity in the cases brought by employers and employees and in the proportion of cases decided against complainants and defendants occurred in Mauritius, another area of employment of Indian plantation workers. See Panchanan Saha, *Emigration of Indian Labour (1834–1900),* (New Dehli, 1970), 120. The disabilities at court of Indian plantation workers in British Guiana are discussed by Basdeo Mangru, *Benevolent Neutrality: Indian government policy and labour migration to British Guiana, 1854–1884* (London, 1987), 158–162.

56. Doug Munro and Stewart Firth, "Company Strategies—Colonial Policies," in Moore, Leckie, and Munro (eds.), *Labour in the South Pacific,* 13–22.

57. Ralph Shlomowitz, personal communication, 3 April 1989. Because of this, employers frequently attempted to disguise the mortality rates of their first-indenture workers by not reporting deaths, by listing deaths as desertions, and by entering free laborers and time-expired laborers as first-indentured laborers so as to apparently decrease the percentage of mortality among the latter. See Ralph Shlomowitz and Lance Brennan, "Mortality and Migrant Labour in Assam, 1865–1921," *Indian Economic and Social History Review,* 27:1 (1990), 93–94.

58. An even more extreme case of coercive labor mobilization involved the Yaqui Indians of northeast Mexico. Being the largest and most intransigent Indian grouping, they posed a serious threat to state authority. As part of the eventual subjugation of the Yaquis during the first decade of the present century, the Mexican government systematically evicted the Yaquis from their ancestral homeland; most of those not killed were deported to Yucatán province as forced labor on the henequen plantations. See Evelyn Hu-DeHart, "Development and Rural Rebellion: Pacification of the Yaquis in the late *Porfiriato,*" *Hispanic American Historical Review,* 51:1 (1974), 72–93.

59. Stewart Firth, "German Firms in the Western Pacific, 1857–1914," *Journal of Pacific History,* 8 (1973), 15–19.

60. A. G. Lowndes (ed.), *South Pacific Enterprise: The Colonial Sugar Refinery Company Limited* (Sydney, 1956), 443.

61. See the discussion in Bruce Knapman, *Fiji's Economic History, 1874–1939: Studies of capitalist colonial development* (Canberra, 1987), 13–14; and more generally J. H. Galloway, *The Sugar Cane Industry: An historical geography from its origins to 1914* (Cambridge, 1989), chs. 6–7. The mechanization of the sugar industry receives comparative treatment in Edward D. Beechert, "Technology and the Plantation Labour Supply: The case of Queensland, Hawaii, Louisiana and Cuba," in Bill Albert and Adrian Graves (eds.), *The World Sugar Economy in War and Depression, 1914–40* (London/New York, 1988), 131–141; Geoff Burrows and Ralph Shlomowitz, "The Lag in the Mechanization of the Sugarcane Harvest: Some comparative perspectives," *Agricultural History*, 63:3 (1992), 61–75.

62. Geoffrey Barraclough, *An Introduction to Contemporary History* (Harmondsworth, 1967), 44. It is a truism but perhaps also an exaggeration to say that sugar was the *only* plantation crop to undergo a scientific revolution before World War I. Coffee also experienced a scientific revolution, though to a lesser extent. See T. J. Barron, "Science and the Nineteenth Century Ceylon Coffee Planters," *Journal of Imperial and Commonwealth History*, 16:1 (1987), 5–21. Copra presents a contrast. There was no "scientific revolution" in the commodity itself, but in its application. Coconut oil had originally been used in soap production, but the discovery of a new refining process at the turn of the century enabled the substitution of coconut oil for animal fats in the production of margarine. Further refining was necessary to overcome problems of using coconut oil to make margarine in hot climates. K. Buckley and K. Klugman, *The History of Burns Philp: The Australian company in the Pacific* (Sydney, 1981), 159; K. Buckley and K. Klugman, *"The Australian Presence in the Pacific": Burns Philp, 1914–1946* (Sydney, 1983), 46.

63. H. C. Brookfield, *Colonialism, Development and Independence: The case of the Melanesian islands in the South Pacific* (Cambridge, England, 1972), ch. 1; Munro and Firth, "Company Strategies—Colonial Policy," 3–29; Galloway, *The Sugar Cane Industry*, ch. 9.

64. Sometimes, however, there was a tension between economy and control. It was a saving for planters if they were not obliged to provide rations for their workers, as was the case in Fiji with respect to Indian laborers after the first six or twelve months of their five-year contracts of indenture. For specific details, see Gillion, *Fiji's Indian Migrants*, 105 and note. On henequen plantations in Yucatán, workers were initially encouraged to grow their own food on small plots of land provided for the purpose by the planter. But the cultivation of corn was strictly prohibited, precisely because it was basic to their diet. Planters reasoned that this prohibition would give them a greater measure of worker control, despite the added costs. See Allen Wells, "Yucatán: Violence and social control on henequen plantations," in Thomas Benjamin and William McNellie (eds.), *Other Mexicos: Essays on regional Mexican history, 1876–1911* (Albuquerque, 1984), 218. On the general question of plantation workers being compelled to produce some or all of their own subsistence, see the discussion in Ann Laura Stoler, "Plantation Politics and Resistance on Sumatra's East Coast," *Journal of Peasant Studies*, 13:2 (1986), 124–143.

65. Quoted in Gillion, *Fiji's Indian Migrants*, 78.

66. Paul M. Kennedy, *The Samoan Tangle: A study in Anglo-German-American relations, 1878–1900* (Dublin, 1974), 107; Doug Munro and Stewart Firth, "German Labour Policy and the Partition of the Western Pacific: The view from Samoa," *Journal of Pacific History*, 25:1 (1990), 86, 99.

67. Moore, *Kanaka*, 160–164, 167–168; Shlomowitz, "Markets for Indentured and Time-Expired Melanesian Labour in Queensland," 70–91; Mark Finanne and Clive Moore, "Kanaka Slaves or Willing Workers?: Melanesian workers and the Queensland criminal justice system in the 1890s," *Criminal Justice History*, 13 (1992), 141–160. Compare these comments with Adrian Graves, "Crisis and Change in the Queensland Sugar Industry, 1862–1906," in Albert and Graves (eds.), *Crisis and Change*, 269.

68. Gillion, *Fiji's Indian Migrants*, 170–177.

69. For a telling statement on rewards and bonuses see J. B. Hirst, *Convict Society and its Enemies: A history of early New South Wales* (Sydney, 1983), 36; and for a more extended discussion, Ralph Shlomowitz, "On Punishments and Rewards in Coercive Labour Systems: Comparative perspectives," *Slavery and Abolition*, 12:1 (1991), 97–102; Robert E. Elson, "The Mobilization and Control of Peasant Labour under the Early Cultivation System in Java," in R. J. May and William J. O'Malley (eds.), *Observing Change in Asia: Essays in honour of J. A. C. Mackie* (Bathurst, N.S.W., 1989), esp. 84 ff. See also Bennett in this volume.

70. Michael J. Gonzales, "Economic Crisis, Chinese Workers and Peruvian Sugar Planters, 1875–1900: A case study of labour and the national life," in Albert and Graves (eds.), *Crisis and Change*, 181–195; Gonzales, "Chinese Plantation Workers and Social Conflict in Peru," 406–413.

71. Ralph Shlomowitz and Doug Munro, "The Labour Trade of Ocean Island (Banaba) and Nauru, 1900–1940," *Journal de la Société des Océanistes*, 94 (1992), 103–117.

72. Geralyn Pye, "Hegemonic and Counter-Hegemonic Uses of Physical Culture in Cuba before 1959," *Flinders Journal of History and Politics*, 14 (1990), 3–4.

73. Beechert, *Working in Hawaii*, 85–86.

74. Ahmed Ali, *Plantation to Politics: Studies on Fiji Indians* (Suva, 1980), 43–106; K. L. Gillion, *The Fiji Indians: Challenge to European dominance, 1920–1946* (Canberra, 1977), ch. 2; Takaki, *Pau Hana*, ch. 6; Beechert, *Working in Hawaii*, chs. 8, 10; Kay Saunders, "Masters and Servants: The Queensland sugar workers' strike, 1911," in Ann Curthoys and Andrew Markus (eds.), *Who Are Our Enemies?: Racism and the working class in Australia*, (Sydney, 1978), 96–111; John Armstrong, "The Sugar Strike, 1911," in D. J. Murphy (ed.), *The Big Strikes, Queensland, 1889–1965* (Brisbane, 1983), 100–116. See also the case studies by Beechert and Lal in this volume.

75. Ross, *Cape of Torments*, 34.

76. Genovese, *In Red and Black*, 108. In particular, Genovese was referring to the the work of Kenneth Stampp (*The Peculiar Institution*, ch. 3),

who emphasized that the resistance of slaves made them "a most trouble-some property"; and Herbert Aptheker, *American Negro Slave Revolts* (New York, 1943, 1966), who exaggerates the number of slave rebellions. The issues are more complex than the bald summary in this note might indicate. Some of that complexity has been delineated in Robert William Fogel, "Cliometrics and Culture: Some recent developments in the histori-ography of slavery," *Journal of Social History*, 11:1 (1977), 34–51; David Brion Davis, *From Homicide to Slavery: Studies in American culture* (New York/Oxford, 1986), 187–206; Peter J. Parish, *Slavery: History and historians* (New York, 1989).

77. Genovese, *Roll, Jordan, Roll*, 658; Genovese, *In Red and Black*, 127.

78. Genovese, *In Red and Black*, 135.

79. Ross, *Cape of Torments*, ch. 8.

80. M. I. Finley, "Slavery," in *International Encyclopedia of the Social Sciences*, vol. 14 (New York, 1968), 311. See also Keith R. Bradley, *Slavery and Rebellion in the Roman World, 140 B.C.–70 B.C.* (Bloomington, 1989).

81. Genovese, *In Red and Black*, 132. See also Eugene D. Genovese, *From Rebellion to Revolution: Afro-American slave revolts in the making of the modern world* (Baton Rouge, 1979); Robert L. Paquette, "Slave Resistance and Social History," *Journal of Social History*, 24:1 (1991), 681–685. "It is remarkable that almost every one of the serious rebellions [in Jamaica] during the seventeenth and eighteenth centuries were insti-gated and carried out mainly by Akan slaves who came from a highly developed militaristic régime, skilled in jungle warfare." Orlando Patter-son, *The Sociology of Slavery* (London, 1967), 276.

82. Herbert Aptheker, *Nat Turner's Slave Rebellion* (New York, 1966). See also Genovese, *Roll, Jordan, Roll*, 592–597, and *In Red and Black*, 131.

83. Maureen Tayal, "Indian Indentured Labour in Natal, 1890–1911," *Indian Economic and Social History Review*, 14:4 (1977), 543. Similar comments abound in the literature. See Pieter Emmer, "The Importation of British Indians into Surinam (Dutch Guiana), 1873–1916," in Marks and Richardson (eds.), *International Labour Migration*, 106–610; José Arturo Güémez Pineda, "Everyday Forms of Mayan Resistance: Cattle rustling in northwest Yucatán," in Brannon and Joseph (eds.), *Land, Labor and Capi-tal in Modern Yucatán*, 18–50.

84. Gill, *Turn North-East at the Tombstone*, 34.

85. Ahmed Ali (ed.), *The Indenture Experience in Fiji* (Suva, 1979), 55.

86. Robert S. Starobin, "Privileged Bondsmen and the Process of Accommodation: The role of houseservants and drivers as seen through their own letters," *Journal of Social History*, 5:1 (1971), 46–70.

87. Jim Scott, "Everyday Forms of Peasant Resistance," *Journal of Peasant Studies*, 13:2 (1986), 6. See also James C. Scott, *Domination and the Arts of Resistance: Hidden transcripts* (New Haven, 1990). Scott's work on resistance has been discussed by such writers as Michael Adas, " 'Moral Economy' or 'Contested States'?: Elite demands and the origins of peasant protest in Southeast Asia," *Journal of Social History*, 13:4 (1980), 521–546; Gilbert M. Joseph, "On the Trail of Latin American Ban-dits: A reexamination of peasant resistance," *Latin American Research*

Review, 25:3 (1990), esp. 25–33 (republished in Jaime E. Rodríguez [ed.], Contested Patterns in Mexican History [Wilmington, Del., 1992]); Paquette, "Slave Resistance and Social History," 684.

88. James C. Scott, Weapons of the Weak: Everyday forms of peasant resistance (New Haven/London, 1985).

89. Alan Warde, "Conditions of Dependence: Working-class quiescence in Lancaster in the twentieth century," International Review of Social History, 35:1 (1990), 71.

90. Ronnie C. Tyler, "Fugitive Slaves in Mexico," Journal of Negro History, 57:1 (1972), 1–12; Stuart B. Schwartz, "The Mocambo: Slave resistance in colonial Bahia," Journal of Social History, 3:4 (1970), 313–333; and especially David M. Davidson, "Negro Slave Control and Resistance in Colonial Mexico, 1519–1650," Hispanic American Historical Review, 46:3 (1966), 235–253; Gad Heuman (ed.), Out of the House of Bondage: Runaways, resistance and maroonage in Africa and the New World (London, 1986).

91. Genovese, Roll, Jordan, Roll, 654. A notable application of the deterrent effect of collective punishment concerns the lead mines of Siberia, where German prisoners of war were set to work after 1945. Should an escaped worker be captured, he was forced to run a gauntlet formed by his companions, who by that stage were in an ugly mood, having been put on reduced rations. Since the Russian guards had obligingly provided strips of wood, trouser belts, pieces of iron—"anything to clout him with as he ran the gauntlet, he the trouble maker"—the escapee received a severe beating, the idea being to make absconding so unpopular that no one would try. See J. M. Bauer, As Far As My Feet Will Carry Me (London, Mayflower Book ed., 1966), 37–41. The use of a gauntlet is not unknown in the Pacific. As late as 1948, at Nauru, rioting Chinese phosphate workers, once subdued, were forced to run a gauntlet of baton-wielding police in a display of excessive force that resulted in two deaths, several injuries, and a subsequent inquiry. See Maslyn Williams and Barrie Macdonald, The Phosphateers: A history of the British Phosphate Commissioners and the Christmas Island Phosphate Commission (Melbourne, 1985), 380.

92. Tom Brass, "The Latin American Enganche System: Some revisionist reinterpretations revisited," Slavery & Abolition, 11:1 (1990), 78–79.

93. Local hostility, of course, extended to management as well as the laborers and frequently enough led to attacks on plantations. In frontier settings, such as pre-cession Fiji, this often resulted in planters and their laborers becoming dependent on one another for mutual protection against outside attack. See John Young, Adventurous Spirits: Australian migrant society in pre-cession Fiji (Brisbane, 1984), 143, 286–287.

94. Some remarks about the complex relationship between Samoan villagers and imported Melanesian plantation workers are provided by Malama Meleisea, O Tama Uli: Melanesians in Samoa (Suva, 1980), 18–19.

95. Gillion, Fiji's Indian Migrants, 93–94.

96. Corris, Passage, Port and Plantation, 43, 97.

97. For discussions on the role of "protector," see Doug Munro, "Planter versus Protector: Frank Cornwall's employment of Gilbertese plantation workers in Samoa, 1877–1881," New Zealand Journal of His-

tory, 23:2 (1989), 173–182; Sandra J. Rennie, "Contract Laborers under a Protector: The Gilbertese laborers and Hiram Bingham, Jr., in Hawaii, 1878–1903," *Pacific Studies*, 11:1 (1987), 81–106; and an incomparable essay by Hugh Tinker, "Odd Man Out: The loneliness of the Indian colonial politician—the career of Manilal Doctor," *Journal of Imperial and Commonwealth History*, 2:2 (1974), 226–243.

98. A. A. Graves, "Pacific Island Labour in the Queensland Sugar Industry, 1862–1906," D.Phil. dissertation, Oxford University, 1979, 250–256; Moore, *Kanaka*, 133–337; Gillion, *Fiji's Indian Migrants*, 80–95; see also Lal in this volume.

99. Beechert, *Working in Hawaii*, 115.

100. Petitions and letters were a common device for redress among plantation workers from many parts of Latin American and the Caribbean. Puerto Rican plantation workers in Hawaii in the early years of the present century also sent letters of grievance to government authorities, resulting in an official investigation but not in appreciable improvements in working conditions. See Carmelo Rosario Natal, *Exodo Puertorriqueño: Las emigraciones al Caribe y Hawaii, 1900–1915* (San Juan, P.R., 1983), 88–100.

101. James C. Scott, "Everyday Forms of Resistance," in Forrest D. Colburn (ed.), *Everyday Forms of Peasant Resistance* (Armonk, N.Y./London, 1989), 7; Scott, *Weapons of the Weak*, 290.

102. Scott, *Weapons of the Weak*, 291. See the discussion in Joseph, "On the Trail of Latin American Bandits," 27–29.

103. The inarticulateness of subaltern classes and the indifference of their articulate contemporaries create well-known and almost insuperable problems for later historians. E. J. Hobsbawm and George Rudé, *Captain Swing* (London, 1969), 11–12, strikingly evoke the difficulties in recreating the world of the nineteenth-century English farm laborer. A similar situation applies for plantation workers in the Pacific Islands. Of the almost 61,000 Indian laborers in Fiji, only one is known to have written (and published) accounts of his experiences. See Totaram Sanadhya, *My Twenty-One Years in Fiji and The Story of the Haunted Line*, ed. and trans. by J. D. Kelly and U. K. Singh (Suva, 1991). A worthwhile family history by the descendant of a Kanaka in Queensland is Noel Fatnowna, *Fragments of a Lost Heritage*, ed. Roger Keesing (Sydney, 1990). The quantity of contemporary narratives by slaves in the United States, by contrast, is exceptional.

104. George M. Frederickson and Christopher Lasch, "Resistance to Slavery," *Civil War History*, 13:4 (1967), 318.

105. Rebecca J. Scott, "Gradual Abolition and the Dynamics of Slave Emancipation," 475.

106. Gary Y. Okihiro, "Religion and Resistance in America's Concentration Camps," *Phylon*, 45:3 (1984), 220–33.

107. Genovese, *Roll, Jordan, Roll*, 658.

108. James Oakes, "The Political Significance of Slave Resistance," *History Workshop Journal*, 22 (1986), 94.

109. I am grateful to Professor James Scott for drawing my attention to this.

2

Patterns of Resistance and the Social Relations of Production in Hawaii

Edward D. Beechert

The history of sugar production throughout the world goes to prove that to make a success of the business, it is absolutely necessary not only to have cheap labor, but to have absolute control over this labor. Plantation history has been dominated by the basic character of its labor supply. Sugar plantations have traditionally used workers racially distinct from both the indigenous population and the plantation power structure, because adequate local supplies were not generally available, and "race was a convenient means of controlling labor."[1] The forms of labor control have ranged from slavery to indentured labor. The range of repression varied greatly according to location and historical circumstances. In the seventeenth century, for example, Cuba probably had the most extreme form of chattel slavery. A similarly wide range of repression characterized the institution of indentured labor.[2]

Most of the plantation economies generated racial ideas, which entered the political structure to create superior and inferior racial classifications. The distinction to be made between race and ethnicity is largely one of acculturation. Groups identified as racial become, over time, ethnic groups, that is, groups distinguished by common traits or customs that set them apart from other groups. When systems of control are put in place, the defining term is usually race, conveying a sense of separateness.[3] In Hawaii, the term *race* was coupled to expressions of unassimilability, depravity, and uneducability. Ironically, the planter group in Hawaii often argued that a particular race had those characteristics needed for a docile, servile labor force. The opponents of immigrant labor inverted these arguments to describe an undesirable population.[4] But briefly, the notion of ethnicity and race as applied to plantation systems tended to follow the lines of racist concepts so familiar in American history.

The problem of labor control and worker response, however, cannot be dealt with in simple sociological terms. The circumstances affecting the social relations of production in plantation systems are so diverse and change so dramatically over relatively short periods of time that one is tempted into sweeping generalizations. The point was well made by Alan Dawley: "Trouble begins when the whole fluid historical process is squeezed to fit simple sociological models."[5] The fluidity of the Hawaiian labor situation, evolving under three distinct forms of government and law, is a good example of this problem. Andrew Lind identified the Hawaiian characteristics of the term *race* as being defined by "criteria chiefly social and cultural in nature which can therefore appear and be widely recognized, or in some cases also disappear within a single generation. Consequently barriers and distances separating racial groups in Hawaii have shifted markedly from time to time and place to place."[6] The plantation was a dynamic institution, responding to changes in the surrounding political economy, rapidly changing technological conditions of cane sugar production, and changing international conditions that dramatically altered the supply of replacement labor in the late nineteenth and early twentieth centuries. Edgar Thompson characterized the plantation as a "race-making situation," that is, one in which a labor problem was eventually defined as a race problem.[7] These rapid changes in the political economy can best be seen in the relations of production. The struggle for job control, which is present in the productive process under capitalism, is the most dialectical of all of the processes of class.[8] The issue of control over working conditions takes forms other than those of conflict. More subtle forms of resistance to exploitation and control are more common or typical of worker responses. Two considerations affected the unfree worker of the indenture contract: penal sanctions and expulsion—that is, loss of job and place. For those not indentured, the possibilities of response were somewhat greater but still limited. The job and a place in the plantation community had more importance than many observers have attached to such employment. For the worker, to be out of work was to court disaster. Unemployment is the ultimate worker disease. The patterns of capital investment create a dialectic that produces changes in the attitudes and responses of the worker. Using the rubric of the development of the sugar industry in the nineteenth and twentieth centuries in Hawaii, we will examine the workers' response to that industrial expansion.

Westerners were struck by two important facts on arriving in early nineteenth-century Hawaii. The abundance of vacant land and the benign climate suggested the great wealth that this combination could produce. Of the many crops under consideration and experimentation, sugar was one of the frontrunners. Whatever the crop, labor was a necessary ingredient, and the indigenous Hawaiian was thought to be in need of a gainful occupation to replace the "immorally" high level of idleness. Happily, profit and salvation could be combined. Between 1826 and 1850, vigorous attempts were made to convert the Hawaiian commoner into an appropriate, western-oriented labor force to accompany the conversion of the Hawaiian communal land system into a fee simple, private property status, suitable for capitalist development. The chiefs proclaimed idleness to be a vice and an offense under the rapidly evolving western-style political structure. The crumbling Hawaiian political authority attempted to transform the traditional power of the ruling chiefs, based upon a communal society,[9] into one of wealth accumulation, based upon the labor of the Hawaiian commoner. The results were indifferent, to say the least.

A survey conducted by the newly installed western-style government in 1846 revealed the degree of failure of the efforts to convert the commoner to a compliant, wage-working status. Reports from all areas of the islands demonstrated the Hawaiians' refusal to work for low wages. They could be attracted into wage labor for varying periods of time when the offer was attractive enough to persuade them to leave their subsistence activities.[10] It was this refusal to submit to low wages and the various legal efforts to compel the Hawaiian to work for wages that played a major role in the enactment of an indentured labor system in 1850, duplicating the experience of other sugar-producing areas of the world. The Masters and Servants Act provided for the signing of labor contracts, enforced with penal sanctions. The act, almost incidentally, provided for the importation of indentured workers.[11]

After an initial flurry of sugar planting and crude efforts to refine it, the activity slowly dwindled between 1836 and 1861, as the lack of capital and an adequate market forced a majority of planters out of the business. Spurred by the sudden appearance of a market created by the American Civil War, sugar production expanded rapidly after 1861, from 572 tons in 1861 to 8,865 tons in 1864. This market continued with the growth of the Pacific Coast population and became a virtual monopoly market for Hawaiian sugar. Planters had considerable difficulty in financing the development of land and mills. The semi-arid nature of the available land dictated the building of expensive and massive irrigation systems to ensure the expansion of production. Rapidly changing sugar refining technology was also expensive. Hawaiian sugar could not meet

the Pacific Coast market with the old, low-grade cake-sugar of earlier years. It was at this point that Honolulu merchant capital
entered the picture to finance the expansion of production. Under
this influence, plantations were consolidated and improved. A
greater uniformity in sugar production processes accompanied the
rising level of efficiency.[12]

Up to 1875, labor demands in sugar had been met largely with
Hawaiian labor. At that point some two-thirds of the sugar workers
were Hawaiian. As the old, traditional economy became more difficult to sustain, with restricted access to the extended land areas
necessary to the old Hawaiian system of subsistence, the people's
scale of relative values changed accordingly. Wage employment
had become a necessity of survival.[13] Although there had been
much discussion of the "Labor Question" before 1875, much of it
was empty rhetoric, designed to ward off demands for higher
wages. With the signing of the Reciprocity Agreement in 1875, the
pattern changed dramatically. Admission duty free to the American
sugar market meant, in effect, a subsidy of approximately two-and-
a-half cents per pound above the market price.[14]

The evidence of Hawaiian reluctance to accept low wages, which
was clear in the 1848 report, was still operative in succeeding
years. While the number of Hawaiians working on sugar plantations
was remarkably constant in the period 1850–1876, the Hawaiians
clearly maintained their selective attitudes, much to the dismay
and disgust of those perennial experts on labor—the newspaper editors. The Hawaiian refusal to accept low wages was uniformly seen
as evidence of "indolence" and ethnic inferiority. Typical of the
complaint was the observation of the leading newspaper: "But
with rare exceptions, they [Hawaiians] refuse to work on plantations and laugh at the idea. Few are tempted by the wages made in
Honolulu—from $2 to $2.50 per day about the docks."[15] Against
these views are the facts that the Hawaiians made up more than 50
percent of the plantation work force and that the predominantly
Hawaiian longshore force had conducted no fewer than four
recorded strikes in the years between 1867 and 1880. In each case,
the newspapers optimistically urged the replacement of the Hawaiians: "the natives will find when too late they have killed the goose
that laid the golden egg."[16] This apparently was not sufficient evidence to warrant a change in the stereotype.

The tendency of the Chinese to work out defensive mechanisms
was held to be one of their "undesirable" traits. Since Britain and
China had combined to prevent the signing of reembarkation contracts, the Chinese were arriving as free immigrants.[17] "John

Chinaman thoroughly understands the principles of trade unions, and the Chinese are not wanting in organizations among themselves to maintain the present price of labor."[18] The ability of the Chinese immigrant to choose alternatives to the penal contract was an effective weapon against the tools of control over the worker.

The 1880s dilemma of low prices for sugar, coupled with difficulties in the volume of the labor supply, created an upward pressure on wages. One answer to the dilemma was for the planters to organize. Forming the Kauai Planters' Association in 1883, they resolved to "fix a standard rate of wages of $17 per month for Chinese day laborers, *shipped [indentured] labor remaining at about the same amount*" (emphasis added). Hoping to regain control of wages, the new association proposed to institute a system of "certificates," which would detail the employment history of the worker. This was a recurring notion, taking the form of legislation during the period of the Republic, but to no avail. The continuing shortage of labor and the hope of expanding production to take advantage of the United States subsidy effectively neutralized such a tactic.[19] The extent to which the Hawaiian government was willing to go to maintain a supply of labor is seen in the instructions to the Hawaiian consul in Hong Kong in 1881: "This Government will do all that lies in its power to prevent any undue influence being brought to bear upon the Chinese immigrants, and to give them every opportunity of finding employment wherever it may best suit them, and at the best wages obtainable." The point of the letter was to assure the Chinese government that all efforts were being made to put a stop to charges of a "coolie" trade. The Minister of Foreign Affairs assured the consul that the government had been in communication with the Chinese government and had been assured that the Chinese minister would make every effort to "get the Chinese to bring their families with them. . . . It will be the duty of this Government to reciprocate by seeing that the Chinese when they arrive are kindly and fairly treated in every respect."[20] Objections from both the United States and the Honolulu urban community to the continued importation of Chinese workers led to a search for alternative labor supplies. The ideal was to "Europeanize" the work force. As had other sugar producers before them, the Hawaii planters turned to the Atlantic Islands to bring in Portuguese sugar workers, drawing on both the experience of Azores and Madeira people and their persistent poverty. The high cost of Portuguese importation was discouraging, as was the tendency for Portuguese to depart Hawaii soon after arriving.

The use of the indenture contract proved to be a continued point of opposition to the importation of European workers. Taking advantage of a depression in Norway, planters brought a shipload of Norwegian craftsmen and urban workers to Hawaii in 1880 under the indenture contract. A literate group of skilled craftsmen, they complained vociferously about the poor conditions and jobs to which they were subjected. Their letters of grievance reached the *San Francisco Chronicle,* which promptly put the letters to work in its campaign against the Hawaiian Reciprocity Treaty. Before long, European newspapers were running headlines about the "slave labor conditions in the Sandwich Islands."[21] The fiasco that resulted from the importation of Norwegian workers in 1880 produced a blast of unfavorable publicity on the mainland and in Europe. The reluctance of European workers to accept the conditions attaching to Hawaiian sugar work, whether Portuguese, Norwegian, German, or Spanish, placed a severe restriction on the available numbers of workers. The consequence was an increased pressure to find a sufficient labor supply, regardless of origin. Asia was the only practical prospect.

The fear of the Hawaiian aristocracy about the continuing threat of a loss of sovereignty from the decline of the commoner population coupled with the growing need for cheap and hopefully docile labor produced a solution that would solve both problems. The solution was to find a "cognate race" that could merge with the Hawaiian population and produce a new "native" population and at the same time produce an adequate labor supply to relieve the upward pressure on wages. The two goals were often contradictory. For example, the vision of four million Negroes being freed by the Civil War seemed to offer an opportunity to expand the labor supply by as much as 20,000. "We could perhaps admit with advantage to ourselves, say 20,000 freed Negroes, pay them the wages and give them the treatment of free men." The king, however, was adamant in insisting on "an amalgamation with the kindred races of Polynesia, or with industrious farmers, mechanics and laborers from the U.S. and Europe." The Minister of the Interior concluded his dispatch with this observation: "One thing I look forward to as a very serious impending fact and it is this—unless we take opportunely effective measures to secure a supply of not over-dear labor, all our best agricultural enterprises will fail."[22]

After experimenting with various Melanesian and Micronesian imports,[23] the search for labor turned to Japan—the only possible source of significant supplies, given that China and India were either too difficult or off limits. Conflicting with this management viewpoint was the view of the Hawaiian legislature, which was vigorously opposed to the importation of labor. In 1880, a severely restrictive Chinese immigration bill was enacted that prohibited the importation of Chinese workers except under an indentured contract and required the expulsion of any Chinese worker not under a current contract. The bill was vetoed by the king as contradicting the government's policy of repopulating the country with a "cognate race" to protect Hawaiian sovereignty.[24]

While the king could block anti-Chinese legislation, and did, he could not control the obvious resentment of the Hawaiian working class toward imported labor. Chinese workers found it more profitable and congenial to work on the growing number of rice plantations or to go into urban work than to work on the sugar plantations. Coming as free workers, they could choose among the opportunities. The Minister of the Interior complained of the slowness of the sugar planters to "advance wages and so forced large numbers of Chinese workers into rice culture. This is to be regretted as sugar is more profitable to the nation than rice." Government plans to import five hundred Chinese workers in 1877 were deemed to be useless: "unless more comprehensive and practical views are adopted [by the planters], the laborers will exchange the cultivation of sugar under task-masters, for the more independent and profitable rice culture."[25] The increasing difficulties of recruiting in China, and local anti-Chinese attitudes, turned the search for a labor supply to the only remaining possibility—Japan. Despite an early unfortunate experience in 1868, the Japanese government was persuaded in 1885 to initiate a system of exporting "surplus" population to Hawaii. The Japanese government set up a quite rigorous and formal system to control the exodus of Japanese workers under a three-year indenture contract.[26]

The introduction of Japanese workers brought to the industry a new element of control. Unlike the Chinese workers, the Japanese arrived with the indenture contract signed in Japan under the supervision of the prefecture officials who selected the emigrants. The agreement attempted to surround the importation of the Japanese workers with safeguards against the well-known abuses connected with plantation and indentured labor.[27] Japanese physicians were to be stationed throughout the kingdom. A Japanese Inspector of Immigrants was to protect against possible abuses. The Japanese consul, of course, had the primary responsibility for safeguarding the immigrant workers. By contrast, Chinese workers had no effective representation in Hawaii.

Evidence suggests that in practice such arrangements left much

to be desired. Consuls and inspectors of whatever nationality
tended to find indentured workers to be deficient and ungrateful.
The inspectors identified with and sided primarily with the employer. Perhaps the greatest difficulty facing the worker was the
lack of opportunity to file complaints with the proper authority.
Leaving work to register a complaint with the local magistrate was
an offense in itself, if undertaken without permission. The consul
and inspector were in distant Honolulu. These factors must be
weighed against the restraining influence of official sanctions for
abuse of workers. Given the constraints of available alternative
sources of labor, growers had to accept limitations on their power
lest they be cut off from additional supplies. In a time of rapid
expansion of planting, a shortage of labor meant heavy losses. The
letters of planters between 1885 and 1900 reveal chronic labor
shortages, with difficulties in both replacing departing workers and
obtaining needed labor for expansion.

In theory the law also required the employer to meet all of the
conditions of the contract and forbade the use of corporal punishment, debt peonage, unilateral extension of the contract, reduction
in wages, and failure to provide housing and/or food if specified.
Needless to say, in such a system the worker was at a serious disadvantage. Workers brought to the magistrate on charges of "refusing
bound service" were subject to jail terms and/or fines if convicted.
Until such fines and trial costs were paid, the worker was confined
to prison. In theory, if not in practice, the only limit to the length of
such terms was life. In practice, such penalties were confined to
more modest terms. The persistent shortage of labor placed limits
on the utility of long jail terms. Given the high levels of arrest and
punishment, it is clear that such punishment did not operate effectively. Data on desertion suggest that the threat of imprisonment
was not an effective means of control (see Table 1).

One reason for the ineffectiveness of penal threats was the labor
shortage. Planters were quite willing to add to their work force
from whatever source. Letters to the factors show constant complaints about the willingness of plantations to hire runaways. This
led to repeated demands to enact an internal passport that would
effectively prevent such hiring. Unlike many other sugar-producing
areas, in Hawaii workers always had the option of working as
free day laborers, whether immigrant or citizen. After repeated
demands for an internal passport, a bill was enacted in 1892 but
was vetoed. The "free Japanese and Chinese," as the attorney general termed them, continued to plague the job control effort.[28] As
free day workers, a more normal employee-employer relationship

Table 1. Cases of Deserting and Refusing Bound Service

Year	No.	Work Force %	Year	No.	Work Force %
1876	2,099	—	1888	2,830	18.2
1878	2,478	—	1890	3,095	17.3
1880	4,476	—	1892	3,992	19.4
1882	3,454	33.7	1894	3,403	18.0
1884	3,164	—	1898	5,876	20.6
1886	2,955	20.6	1900	4,335	11.0[a]

SOURCES: Reports of the Chief Justice, Hawaiian Kingdom, 1876–1900, Summary of Civil Cases in District Court. The work force data are in Romanzo Adams, *A Statistical Study of the Races of Hawaii* (Honolulu, 1925), 22–23.

[a] six months only

was assumed, allowing the worker the option of withdrawing his labor power. Given the ongoing expansion of production and the chronic shortage of labor, maintaining a skilled, experienced work force was a matter of considerable importance and a source of frequent competition among planters for workers.

Chinese workers, imported to meet the demand of expansion after 1875, quickly took advantage of this option by organizing themselves into contracting companies through which they undertook various tasks, dividing the work and payment according to their own rules. The government newspaper complained that a shipment of 114 Chinese would have little impact on wages. "They are free immigrants who are under no contract to labor, but are ready to engage in any thing. They come in gangs or companies of eight to seventeen, each having its own head or chief."[29] This form of labor contracting tended to remove the worker from immediate supervision of the field superintendents. Historically, the low-level supervisor has been responsible for the majority of abuse associated with plantation labor.

A situation was thus created whereby three distinct labor forms could be found at any given time: indentured labor under penal compulsion, free day labor able to withdraw at any time for any reason or to be discharged for any reason, and a self-organized gang labor system contracting their services. Some planters combined all three forms; a handful of plantations refused to use indentured workers.

Given the pressure to expand after 1875 and the frequent uncertainty as to labor supply, the plantations had far less flexibility than is frequently assumed in the degree of authority they could exert, and they certainly were limited in the extent of physical abuse that might have otherwise been permissible. Worker response to the conditions on the Hawaii plantations was, in turn, limited to individual reactions to personal abuse and sometimes to spontaneous violence or group turmoil. The organization of the work did not offer the possibility of worker organization.

Trying to balance the negative aspects of indentured plantation labor with the positive factors of secure, cash-wage labor is difficult. It is clear from the behavior of the Japanese workers that they frequently complained and even rebelled against abusive treatment and expressed their discontent in the rapid turnover and desertion rate characteristic of the period 1885–1900. Desertions after 1894 were substantially different from those occurring before that date. At issue after 1894 was the sum of money withheld from the worker's wage to ensure his return to Japan. If the desertion occurred within three months of employment, the importing company had to refund its fee, in addition to the wages withheld. The declining number of indentured workers after 1894 lessened the importance of the desertion rate. The other side of the coin indicates that the workers were remaining in Hawaii as free sugar workers and sending to Japan startling sums of money. Between 1885 and 1894, the volume of money sent to Japan is estimated at two-and-a-half million yen per year.[30]

There is little or no evidence of any sense of cohesion or group identity among the different groups of workers and types of workers. Spontaneous protest and individual response were the primary reaction of the worker. The varying records of plantation managers suggest that many were keenly aware of the negative effects of abusive or coercive methods on production. Others clearly had no understanding of the problem and firmly believed in force and a show of authority. Those plantations were generally associated with a poor record of production and profit. Against the successes of Makee and Grove Farm plantations on Kauai must be set the dismal performance of Makawao and Lydgate on Maui and Hawaii in the 1880s.

In response to the rapidly growing sentiment against Chinese importation, efforts were made to devise a system whereby Chinese plantation workers would be confined to agricultural work and subject to immediate deportation should they leave such employment. A considerable struggle ensued between those desiring a total ban on Asian labor and those in sugar who needed a continuous supply of cheap labor. The popular sentiment was expressed by one of the leaders of the Hawaiian Anti-Asiatic Union:

"We do not say as elsewhere that the Chinese must go, we simply say that Chinese and Japanese must be stopped from coming here any more."[31] Facing a shortage of labor in 1890, the sugar industry was successful in securing legislation that would permit the importation of what amounted to Chinese serfs, bound to remain at labor on plantations under the threat of instant deportation. Wages were subject to a charge to guarantee the cost of the deportation should the Chinese worker leave agriculture. The constitutional provision that had permitted immigrant workers to remain in Hawaii was thus abrogated. The Hawaii Supreme Court found the provision for instant deportation to be a violation of the Hawaii Constitution, since it presumed guilt before the fact by forcing the worker to pay for his deportation without having violated any law. The constitution was promptly amended to allow such a provision. The overthrow of the monarchy made such concern over constitutional niceties unnecessary. Despite the changes in the law, there were no deportations during the period of the Republic, 1893–1898.

The annexation of Hawaii in 1898 and the abolition of the penal contract drastically altered the terms of the labor supply. Torn between a desire to exert greater control over the work force and the importance of gaining permanent entry to the American sugar market, the planters had a difficult choice to make. The flood of Chinese workers imported during the period of the Republic had attracted severe criticism from the mainland, where anti-Asian sentiment was reaching new heights. The importance of the United States subsidy was never better illustrated than in the decision to cut off the attractions of indentured labor in return for the certainty of the subsidy. The argument was offered, somewhat with hindsight, that more than half of the work force in 1898 was free labor and that the indenture form was no longer essential. Henceforth only two forms of labor would be available—day labor and short- and long-term contracting. The third, unfree element, was no longer available to restrain the responses of recalcitrant workers.

Reviewing the generalizations made about labor in the period 1842–1898, one can see quite clearly the racism that pervaded the Hawaiian situation. From the reluctance of the Hawaiian to accept substandard conditions was derived the notion of the "innate" indolence of the Hawaiian. Such ethnic stereotypes quickly entered the conventional wisdom of the community and were accepted without question. Each group of workers in turn was hailed as the "solution" to the problem of an adequate, low-cost, docile labor supply. Following the Hawaiian, the Chinese worker was the magic worker, to be followed by the Japanese, hailed as a paragon of the docile worker accepting authority with unquestioning devotion to the master. Each of the European groups was presumed to have cultural or ethnic traits that would make them amenable to strict control and efficient production. Each group in turn was found to be wanting, failing in some respect to meet their employers' expectations. An example of this stereotyping is found in a conclusion drawn by the Commissioner of Labor: "The Chinaman was the more steady and reliable but less energetic [than the Japanese]. . . . The Japanese represents the radical, the Chinaman the conservative side of oriental civilization. . . . His white employers consider him [the Japanese] mercurial, superficial and inquisitors [sic] in business matters."[32] Within a few years, another United States official reporting to Congress added to the list of stereotypes by including the two ultimate groups to be imported to Hawaii: Puerto Ricans and Filipinos. He said, "The Porto Rican [sic] was considered very much inferior to all the others until the Filipino was brought in, and it is conceded by all, that the latter is the poorest specimen of man that was ever introduced into the Islands."[33] The circle was now complete—each of the ethnic groups, without exception, whether Asian or European, had been tested and found wanting.

As the Hawaii sugar planters confronted the new situation of free unbound labor, one must compare their position with that of other sugar producers. Sugar plantations around the world had displayed remarkably similar attitudes toward workers and systems of control. In almost all cases, workers were from Asia, Africa, the Pacific Islands, and to a limited extent, from the poverty areas of southern Europe, such as Spain and Portugal. In all cases, the labor force was either racially distinct or ethnically identifiable from the managerial class, which was drawn largely from the ruling class or their direct representatives.

One of the more significant ideas of labor control was that of exploiting racial differences. A corollary was to justify rigid control on the basis of racial "inferiorities." At its optimum, the exploitation of racial differences implies the ability to lay one group of workers against another to enable the threat of substitution to act as the coercive device for each of the groups involved. This presumes that the work force can be easily replaced at any given time. The fact is that at no time in Hawaii's sugar history was this true. Nor, for that matter, was it true anywhere else. The disruption of production and the loss of skills was too prohibitive to contemplate. Even in Queensland, where the overwhelming political demand for a White Australia Policy demanded the end of Melanesian labor in sugar, numerous exceptions had to be made as sugar producers found reasons why "their" workers could not be deported.[34]

The realities of labor supply and the costs of importing labor required making do with whatever was available. Efforts by the plantations in the Hawaii strikes of 1909, 1920, and 1924 proved the use of strikebreakers an expensive, largely futile exercise. There were never sufficient numbers of racially different skilled workers available to replace the massive numbers involved in these strikes.[35]

The global recitation of ethnic deficiencies was a part of the worldwide psychological rationalization of labor exploitation. From the sixteenth century to the present, such rationalizations of racial slavery are important in justifying the exploitation of those who "look different." An example of the extent to which such rationalization exercises could be carried is seen in Max Weber, the eminent sociologist, who justified the poor treatment of Polish workers in Prussian sugar beets: "It is not possible for two nationalities with different bodily constitutions, stomachs of different construction, to quite freely compete in one and the same areas. It is not possible for our [Prussian] workers to compete with the Poles."[36]

New possibilities opened up for Hawaii's sugar workers with the beginning of United States control. The most immediate gain was the abrogation of all indenture contracts as of 14 June 1900. Little else was gained. The United States Supreme Court soon ruled that the Constitution did not follow the flag. Protections such as the 5th, 6th, and 7th Amendments were not available in Hawaii.[37] In reality, the protection of United States legal institutions was not significant, given the difficulty of worker access to such remedies. However, the absence of any penal recourse made the coercive efforts of the planters more difficult and less certain.

The almost immediate use of organized strikes by Japanese workers, now the dominant work force, presented a new challenge to the planters. The planters organized into a largely informal organization, the Hawaii Sugar Planters' Association (HSPA), and were soon forced to adopt a more formal and disciplined approach to deal with the disconcerting militancy of the Japanese. Strikes in 1903, 1904, and 1905 demonstrated an unsuspected ability on the part of the Japanese to organize and to take concerted action.

The uncertainty about immigration policy after annexation sharply cut into Japanese immigration. In the first year after annexation, there was a net loss of 3,490 Japanese workers, largely to California agriculture.[38] The militancy of the workers climaxed in 1905, when the abuse of a worker by a field superintendent at

Lahaina produced a riot and an attack by armed deputies, resulting in the death of a worker. Alarmed by the militancy of the workers, the governor called out the militia, and the United States Army sent a squad of Signal Corps troops, even though the manager had made several important concessions, including the dismissal of the foreman.[39]

The unity of the employers was solidified largely through strike actions such as those in 1909 and 1920. These strikes were concerned not only with economic issues, but important community and social issues of dignity as well. Issues of camp sanitation, decent housing, and the quality of supervision were as important as the demands for the abolition of racial pay scales and more adequate wages. On all of these issues, the workers made major gains and forced significant concessions from the planters—this despite the use of the old American technique of jailing the strike leadership on a variety of charges revolving around "sedition." The planters firmly believed that the plantation worker was incapable of acting on his own initiative and without leadership would be easy prey.

The strike of 1920 objectively transformed the organization and control of the sugar industry. The increasing financial control by factors over the individual plantations converted the trustees of the HSPA into a board wearing two hats—one as the trustee for the industry and one for the individual factors, now representing their wholly owned sugar companies. The gap between the two was narrowing and would eventually vanish in the strike of 1924.

Despite a deep belief by management that an ethnically diverse work force would work to prevent collaboration, the two major elements of the work force demonstrated the ability to come together in the strike of 1920. With all of the difficulties imposed by language and culture and the intensive countereffort of the HSPA, the Filipino and Japanese labor organizations forged an elementary unity. Despite a highly coordinated espionage and agents' provocateur effort, the two disparate organizations held out for six months, until meager funds ran out.[40]

That the style of organization differed radically between the two groups demonstrated only the differences in cultural approach. Expressed in different modes, both groups displayed a high degree of working-class unity. The degree of cooperation was inhibited only by the fact that labor organization had no legal sanction in Territorial or American law at that time, and the worker organizations possessed severely limited financial resources. They had to struggle not only against the employer, who was united and well financed, but against the political and judicial system. The specter of racial cooperation shocked the industry into massive changes and reorganization in order to maintain labor control. As in the period of indenture, when the pressure to produce had limited the severity of the employers' reaction, so in this later period the lure of the subsidy dictated some adjustment in the labor system that would provide a maximum of control over the wage level without affecting the labor supply or harming production.

Borrowing a leaf from the notebook of mainland industry, the Hawaii planters organized a program of welfare capitalism following the 1920 strike. The success of the workers in resisting the combined attack of the planters and the territorial government was an object lesson to the industry. Massive changes were put into effect. Shifting from closely supervised day labor to short- and long-term contracts for cultivating, irrigating, and harvesting, the industry hoped to reduce the group cohesiveness, and, they hoped, the militancy displayed in the dual ethnic strike. The results were mixed. The Japanese were effectively dealt with by converting them to contractors and by their steady exodus from the industry. With none coming as replacements, the Japanese were quickly displaced as the dominant element in the work force. The planters met their needs during the continuing expansion of the 1920s with a sharp increase in the number of Filipinos. For these mostly single workers, the improvements in camp welfare and housing were minimal. What was largely overlooked was the fact that the number of married Filipinos was increasing sharply. Further, even the single workers expected more community facilities, rather than the barebones camps. As it had with the Japanese, the altered perspective of the Filipino worker generated new pressures on the work situation.

The 1924 strike came as a serious blow to the elaborate welfare programs undertaken after the 1920 strike. Despite the disorganization of the Filipinos, the strike had a greater impact on the industry than any of the earlier organized strikes. The 1924 strike came closer to being industrywide than any of the previous strikes. It was the first strike to move throughout the islands, lasting over seven months.[41] There is some evidence to suggest that the Japanese, although not participating in any overt manner, did supply money and food to various groups of strikers. The Japanese of Kohala and Kona contributed significantly to the Filipino effort on the Island of Hawaii. Similar efforts seem to have been made on Maui. The greatest impact of the strike is seen in the organization of the HSPA. The employer offensive was, unlike the earlier strikes, conducted en-

tirely by the central organization, with a massive effort made to use
police and judicial powers to break the strike at the earliest point.
The frustration of the HSPA executives is apparent in their memos,
demanding that plantations furnish the central office with advance
warning of meetings of the Filipinos so that they could be arrested.
One agent wrote to the manager of Olaa Sugar Company about the
situation in the following terms:

> The police at this time are prepared to jail a large number of men
> and are arresting the Filipinos on the slightest indications of
> wrong doing, and that is why they are so anxious to have every-
> thing reported to them which occurs on the plantations so that
> they can act promptly. The police have forbidden any of the strik-
> ers to hold meetings in town or elsewhere, and if they can catch
> any of these men in the act of holding meetings they are anxious
> to do so, so as to arrest them.[42]

The shock generated by the failure of the elaborate program of welfare capitalism pushed the industry to examine its operation. A. H.
Young, Inc., was hired to conduct an intensive examination of the
industry with a view to eliminating the conditions that had produced two severe strikes in four years.[43]

The report noted the inefficiency of the ten- and twelve-hour
days, recommending eight-hour shifts in the mill. The poor quality
of field supervision came in for a major share of the blame for labor
disputes. In general, the report found a contradiction between the
welfare programs and the day-to-day operation of the plantations.
The centralized control operated only in time of crisis. For the daily
conduct of the plantation, the individual manager still had a great
deal of leeway. This made for great variations in the quality of the
work experience. Most of the suggestions were shrugged off as
unrealistic and too radical; expanding production and improving
efficiency were considered more important.

Workers had more subtle forms of resistance in the struggle over
control of the work process. As the industry entered the decade of
the 1920s, both the methods of production and efforts to control
the work force became much more sophisticated. A few examples
will illustrate the process.

In the strike of 1920, workers on the outer islands voted to continue working and to contribute to the support of the Japanese
workers on Oahu. In at least one case, this resulted in a sharp
change in the employment policies of one plantation. The Japanese
workers at Grove Farm approached the manager with the demand
that a list of Japanese workers who had agreed to contribute a por-

tion of their pay to the strikers be fired from the plantation because they had failed to pay their assessment. If the defaulting workers were not fired, all of the Japanese workers would quit. The manager reluctantly informed the owner that he had no choice, fearing to lose so many valuable workers. To replace the fired workers, the plantation was forced to break its boycott of Filipino workers.

A less dramatic example concerns the rapid mechanization of plantation tasks, particularly in cultivating, which accelerated in the early 1920s. A disagreeable task was the spreading of guano fertilizer on young cane. Traditionally, this was done by hand casting. Ewa plantation, encountering difficulty in finding workers who would accept the assignment, devised a mechanical spreader. The manager at Kohala plantation questioned the need to invest in such machinery, pointing out they had no trouble obtaining workers to do the hand spreading. The manager at Kohala observed that citizen workers at Ewa had many alternatives to employment at Ewa, whereas workers at Kohala had few viable options. Before long, mechanical spreading was the common practice. The point to note here is that more than a question of saving labor costs was involved. Clearly, mechanical spreading involved the expenditure of capital and was not shown to be a saving in labor costs. It eased a labor shortage which took the form of resistance to the task.

The ability of the worker to resist surrendering control over the work process is illustrated in a variety of ways. Many are subtle and nonconfrontational. They nonetheless limit the ability of the employer to exert full control over the work process. One such example is found in the drive to meet several problems in cane harvesting. Cutting and loading cane were both onerous and labor-intensive processes. A long search, extending from the 1880s, for mechanical means of dealing with these two tasks had occupied sugar growers in many areas other than Hawaii.[44] In the 1920s, Hawaii succeeded in developing mechanical loading devices to replace the *hapai-ko* work, which had been performed primarily by husband and wife teams, often assisted by their children. In addition to the loss in earnings, there was a resentment of displacement by machines. Spurring that search was the realization that Japanese laborers were no longer available. Fearing the Filipino replacements were too small physically to accomplish the work, the industry accelerated the drive to develop mechanical devices. The first machines bundled the cane in preparation for its being lifted by cranes. The loaders were put to this much lighter physical task. To the dismay of the Hawaii growers, the loaders frequently refused to accept this task. Not until the pay scale was adjusted to produce income similar to loading under piecework rates was the objection overcome.

This type of resistance has often been noted as an obstacle to mechanization. The social construction of skill has generally been ignored in discussions of mechanization. Opposition is seen as an example of Luddism—simpleminded fear of machines that leads to smashing.[45] The question of whether or not a job or task is skilled is frequently a matter of management decision and can be seen as a means of attempting to cope with worker resistance. Piecemeal mechanization leads to many complications in the production process. The supporting mechanisms are slower to develop than the actual machine process. At the heart of the problem of introducing mechanical solutions to cultivating and harvesting problems is the difficulty of adjusting the sequential process of sugar production to changes in any one area. Worker fear of displacement as well as resentment of loss of status have been major obstacles to be overcome.[46]

A part of the problem in describing these work processes and worker reaction stems from the fact that work is seldom defined in terms other than pay, hours, difficulties, and the like. Little or no attention is given to the consideration of work as a social process. The evidence in folk literature clearly indicates that work has an important social worth, and societies generally place a central importance on the necessity of work. Modern social science has long since departed from the path laid out by Smith, Ricardo, and Marx, whose studies focused on the central role of work. "Economists have disregarded the notion of work as a cultural and social value."[47]

The depression of 1930 brought the final changes to the work situation on Hawaii's plantations. Federal regulations governing the payment of subsidies and quotas for sugar production required compliance with a wide variety of federal statutes such as the National Labor Relations Act, the Fair Labor Standards Act, and Social Security. Henceforth, questions of control over the work process had to be considered in the context of these controlling statutes. What had been a dual situation of employee-management relationship became a tripartite affair. The anticipation of the federal government reached into the most minute areas of the work situation. Housing and perquisites, which had long since been converted from a necessity occasioned by the isolation of the plantation into a means of worker control, were now regulated by a variety of statutes, not the least of which was the tax liability incurred by the inclusion of perquisites in the wage. As early as 1939, the industry was seeking to establish a job classification that would simplify the complex job structure of the industry. There were no less than 357 different job descriptions in use in Hawaii's industry.[48] The pressure of union organization and the development of industrywide collective bargaining transformed the historical work relationships into a much more bureaucratic process over which the worker now exercised job control through a voice in the collective bargaining process.

Conclusion

From the indentured worker in 1850 who ran away from the penal contract, to the union member who files a grievance under the sugar contract, the worker has not been without some means of bringing pressure to bear on the work situation. The noisy complaints of the planter objecting to the behavior of workers who demand higher wages and better working conditions are sufficient evidence of the effectiveness of the resistance of workers to abusive conditions. Given the circumstances of Hawaiian production, workers there achieved a high degree of control over the conditions of work, within the limitation of not leaving the job. The laborer turnover of approximately one-third annually suggests both that the opportunity to escape into other occupations was available and that this turnover exerted on the planter a constant pressure to conserve his work force. Perhaps one of the most effective tools to control abuse of the worker is to be found in the complaint of one planter who advocated "the introduction of East India laborers in large numbers, if it can be done without too much expense and without political considerations on the ground that this class of immigrant will prove an offset to Chinese labor in this country."[49] When the historical conditions permitted it, Hawaiian sugar workers turned immediately to classic labor organizing, albeit along ethnic lines. Given the level of communication available to them, their lack of experience in trade unions, and the degree of political and economic control by the planter class, the levels of achievement were remarkable. The ultimate transformation of sugar agriculture in Hawaii into an industrial operation brought the benefits of mass organization to a unique class of agricultural workers. That success is now threatened by circumstances from abroad and beyond the political or economic power of either the sugar companies or the worker organization—the chaotic conditions of the world sugar market and the political pressure of industrial sugar users to avail themselves of cheap world prices.

Abbreviations

F.O. & Ex. File   Archives of Hawaii, Hawaiian Kingdom, Foreign Office
and Executive File.

F.O. Letter Book Archives of Hawaii, Foreign Office Letter Books.

HSPA   Hawaiian Sugar Planters' Association.

PCA   *Pacific Commercial Advertiser.*

Notes

1. George Beckford, *Persistent Poverty: Underdevelopment in plantation economies in the Third World* (New York, 1972), 67.

2. See Manuel Moreno Fraginals, *The Sugar Mill: The socio-economic
complex of sugar in Cuba* (New York, 1976); Hugh Tinker, *A New System
of Slavery: The export of Indian labour overseas, 1830-1920* (London,
1974), for Indian indenture experiences.

3. Edgar Thompson, *Plantation Societies, Race Relations and the South:
The regimentation of populations* (Durham, N.C., 1975), 115. See also Jay
Mandle, *The Roots of Black Poverty* (Durham, N.C., 1978), 10; D. M. Benn,
"The Theory of Plantation Economy and Society: A methodological cri-
tique," *Journal of Commonwealth and Comparative Politics*, 12 (1974),
249-260; Sidney Mintz, "From Plantations to Peasantries," in Sidney
Mintz and Sally Price (eds.), *Caribbean Contours* (Baltimore, 1985); David
Montgomery, *Worker Control in America* (New York, 1971); Adrian
Graves and Peter Richardson, "Plantations in the Political Economy of
Colonial Sugar Production: Natal and Queensland, 1860-1914," *Journal of
Southern African Studies*, 6:2 (1980), 214-229; Alexander Saxton, *The
Indispensable Enemy* (Berkeley, 1971).

4. Edward D. Beechert, *Working in Hawaii: A labor history* (Honolulu,
1985), 77-78.

5. Alan Dawley, "American Workers/Workers' America: Recent works
by Brody and Green," *International Labor and Working Class History*, 23
(1983), 40.

6. Andrew Lind, "Race and Ethnic Relations: An overview," *Social Process*, 29 (1982), 130-150.

7. Thompson, *Plantation Societies, Race Relations and the South*, 115.

8. Karl Marx, *The Eighteenth Brumaire of Louis Bonaparte* (New York,
1936), 13; Karl Marx, *The German Ideology*, C. J. Arthur (ed.) (New York,
1981), 57-59; Montgomery, *Worker Control in America*, 4, 14.

9. The term *communal* is used here to indicate the relationship of the
commoner in the Hawaiian community. The chiefly system depended upon
a distribution of the land to subordinate chiefs, who in turn made it available to commoners. The commoners' principal attachment was to his family and a particular *'ohana* (extended family). See E. S. Craighill Handy
and Elizabeth Green Handy, *Native Planters in Old Hawaii* (Honolulu,
1972), 287.

10. Willey, Minister of Foreign Affairs, Annual Report, 1848, "Answers

to Questions . . . Addressed to all Missionaries in the Hawaiian Islands. May 1846."

11. Kingdom of Hawaii, Penal Code, 1850, Sections 1–21, reenacted in Civil Code, 1859, Sections 1417–1426.

12. William Taylor, "The Hawaiian Sugar Industry," Ph.D. dissertation, University of California (Berkeley, 1935).

13. *Planters' Monthly*, 4 (1887), 499, 539.

14. Taylor, "The Hawaiian Sugar Industry," 16, 65–66; George McClellan, *The Handbook of the Sugar Industry of the Hawaiian Islands* (Honolulu, 1899), 8.

15. *PCA*, 19 April 1879.

16. *PCA*, 4 May 1867; 17 July 1869; 19 April 1879; 31 March 1880.

17. Hawaiian Kingdom, Foreign Office, Letter Book 57:311–315, 14 February 1879.

18. *Saturday Press*, 11 December 1880.

19. *PCA*, 1 September 1883.

20. F.O. & Ex. File, Box 87, 2 July 1881.

21. Hawaiian Kingdom, Cabinet Council Minute Book, 11 January 1882; F.O. Letter Book 58, 14 January 1882. For the Swedish reaction see F.O. & Ex. File, Box 91, Consul General, Sweden and Norway, 18 January 1882; Great Britain, F.O. & Ex. File, Box 92, 24 January 1882; 25 January 1882.

22. R. C. Wyllie to Edward Everett, Boston, 13 September 1862, F.O. Letter Book 42.

23. J. A. Bennett, "Immigration, 'Blackbirding,' Labour Recruiting: The Hawaiian experience, 1877–1887," *Journal of Pacific History*, 11:1 (1976), 3–27; Sandra J. Rennie, "Contract Labor under a Protector: The Gilbertese and Hiram Bingham, Jr., in Hawaii, 1878–1903," *Pacific Studies*, 11:1 (1987), 81–106.

24. *Hawaiian Gazette*, 24 August 1880.

25. Carter to Allen, 20 April 1877, F.O. Letter Book 55, file 77; *Hawaiian Gazette*, 1 March 1877.

26. Beechert, *Working in Hawaii*, 96–97; Hilary Conroy, "The Japanese Expansion into Hawaii, 1868–1898," Ph.D. dissertation, University of California (Berkeley, 1949), 120; Hawaii, Bureau of Immigration, Report, 1886: 283, p. 36

27. Masters and Servants Act, 21 June 1850. Penal Code, 1850:170–176, amended in 1859, 1869, 1872, 1876, 1880, 1882, 1886, 1892, 1894, and invalidated by annexation on 14 June 1900.

28. Beechert, *Working in Hawaii*, 112.

29. *Hawaiian Gazette*, 2 June 1875.

30. James Okahata, *The History of The Japanese in Hawaii* (Honolulu, 1971), vol. II:26.

31. Much of the impetus for a complete ban on immigrant labor came from the Portuguese urban community, supported by some of the urban press. Ralph S. Kuykendall, *The Hawaiian Kingdom* (Honolulu, 1967), III: 176–178.

32. U.S. Bureau of Labor Statistics, Report on Hawaii, 1905:34–35.

33. U.S. Immigration Commission, Industrial Conditions in Hawaii, 1911: 3–4.

34. Peter Corris, " 'White Australia' in Action: The repatriation of Pacific Islanders from Queensland," *Historical Studies*, 15 (1972), 237-250; Clive Moore, *Kanaka: A history of Melanesian Mackay* (Boroko/Port Moresby, 1985), 274-285.

35. The plantations created a loss-sharing formula in the 1909 strike whereby struck plantations were paid for the lost production by a tax on the tonnage of nonstruck plantations. The losses in each case outweighed the production of the strikebreakers. Nonstruck plantations were unable to obtain sufficient labor to effectively harvest crops, as labor was diverted to the struck plantations. See Beechert, *Working in Hawaii*, 173-174.

36. Max Weber, "Die Landliche Arbeiter Verfassung," *Schriften des Veriens für Sozial Politik* (Berlin, 1893), 75.

37. *Hawaii v. Mankichi*, 190 U.S. 197 (1903). Neither the Newlands Resolution of annexation nor the Organic Act of 1900 "had seemed to incorporate the Hawaiian Islands in the United States." The Bill of Rights was not applicable to Hawaii.

38. The Governor reported that "all plantation stocks have fallen owing to the uncertainty of the labor supply." Archives of Hawaii, Report of the Governor, 1901:63.

39. Ernest Wakuwaku, *A History of the Japanese in Hawaii* (Honolulu, 1938), 133-134; Governor, Report, 1905:64-65.

40. Beechert, *Working in Hawaii*, 201-208; John E. Reinecke, *Feigned Necessity: Hawaii's attempt to obtain Chinese contract labor, 1921-1923* (San Francisco, 1979), 87-136.

41. John E. Reinecke, "The Filipino Piecemeal Sugar Strike of 1924," unpubl. ms., 1963, Hawaii-Pacific Collection, University of Hawaii, is the most thorough documentary history of the strike available.

42. C. Brewer & Co., Hawaiian Agricultural Co., HSPA Memos, 28 January 1925 (microfilm, University of Hawaii Library).

43. Beechert, *Working in Hawaii*, 244-245. The survey surfaced during the strike of 1937. The executive director of the HSPA was impressed by the fact that the demands of the workers were almost identical with many of the recommendations of the 1926 survey. He observed: "What seemed too radical in 1926 has now become common place and in many cases required by law."

44. Mechanization of field work became a major goal of the HSPA during the 1920s and became an imperative after 1934, when the Philippines were cut off as a source of labor. See Edward D. Beechert, "Technology and the Plantation Labour Supply: The case of Queensland, Hawaii, Louisiana, and Cuba," in Bill Albert and Adrian Graves (eds.), *The World Sugar Economy in War and Depression, 1914-1940* (London/New York, 1988), 131 ff.

45. E. A. Wilson, "Introduction" to Stephen Wood (ed.), *The Degradation of Work: Skill, deskilling and the labour process* (London, 1982), 17.

46. Nathan Rosenberg, "The Direction of Technological Change: Inducement mechanisms and focussing devices," *Economic Development and Cultural Change*, (October 1969), 18; Nathan Rosenberg, *Inside the Black Box: Technology and economics* (New York, 1982).

47. Sandra Wallman (ed.), *The Social Anthropology of Work* (New York, 1979), 367. See also Stephen Wood (ed.), *The Degradation of Work* (London, 1982).

48. HSPA, Proceedings, 1939:12–13.

49. *Hawaiian Gazette,* letter from S. T. Alexander, sugar planter, 8 December 1880.

3

The Counterculture of Survival: Melanesians in the Mackay District of Queensland, 1865–1906

Clive Moore

Labor migration was central to Pacific colonialism. It has also been an important theme in the development of the new Pacific historiography, which depicts Pacific Islanders as active rather than passive agents in dealing with the process of European expansion.[1] The colonial state, the indenture process, and colonial society contained powerful mechanisms that controlled the working and private lives of the laborers. While always stressing this, Pacific historians have also tried to study the response of the laborers, but most have failed to deal sufficiently clearly with the complex relationship between resistance and accommodation; and what occurred when laborers from circular-migratory systems became immigrants in colonial societies, such as with Asians in Hawaii and Fiji and Melanesians in Queensland. In particular, previous writers on the resistance of Melanesian plantation workers in Queensland created a contradiction by arguing that successful resistance was carried out within a successfully coercive regime (see Chapter 1 in this volume).[2]

The first Pacific historian to recognize the continuum between resistance and accommodation on Pacific plantations, and to take cultural differences sufficiently into account in the various responses employed by plantation laborers, was Brij V. Lal, writing on Indo-Fijians (republished in this volume). Lal argued that accommodation was one of their strategies for survival, and that their diverse social and cultural background reduced their potential for collective action.[3] The present study develops Lal's argument but suggests that in the case of Melanesians in Queensland, although their diverse origins may have prevented collective industrial action, they were not an impediment to collective political action when it was needed in the face of the White Australia Policy in the 1900s.

Methods of Response

This chapter is a study of Melanesian methods of response to their lives in Queensland. In excess of 62,000 contracts of indenture were entered into by Pacific Islanders, predominantly male and mainly from Melanesia, recruited to work in the colony from 1863 until 1904.[4] Conditions on the plantations and farms were often difficult, and given that the laborers came from precapitalist societies and had little prior contact with the world beyond their islands, the first years of their indenture contracts could be quite traumatic. Nevertheless, many of the Melanesian indentured laborers renewed their contracts; considerable numbers chose to stay in the colony for ten and even twenty years, and many fought to remain in Australia in the face of the mass deportation proposed in 1901. The first indentured laborers fresh from the islands were always worst off, and must often have found their working conditions to be extremely regimented and oppressive. Yet none of the Melanesians were ever the servile, docile employees that the colonists desired.

Strikes were illegal within the institution of indenture, and although there were occasional cases of strike action, almost all in the late 1880s and 1890s, there is no evidence of any large-scale strikes by Melanesians anywhere in Queensland.[5] A fair indication of the success of the indenture system in keeping the work force compliant is the fact that the large strikes in the Pacific—beginning in Hawaii in 1909—occurred only after the local abolition of indenture.[6] The only large-scale disturbances in Queensland were caused by excessive consumption of alcohol,[7] or because of aggression between different language or island groups, not protests to employers over work conditions.

This seems logical in light of the diverse origins of the laborers: circular-migrants on short contracts, coming from more than seventy different islands of the southwest Pacific, speaking numerous languages, and having no sense of unity other than that imposed upon them by the regime under which they worked. Their main methods of resistance, all small-scale, sporadic, and spontaneous, were illustrative of the relative powerlessness of indentured laborers against authority structures. This approach is typical of methods used by plantation laborers the world over: what James C. Scott called the "weapons of the weak," everyday forms of resistance stopping well short of collective outright defiance.[8] Yet in 1901 the Melanesians went beyond the "weapons of the weak" by forming the Pacific Islanders' Association, the only large organized response to the harsh mass deportation ordered by the new Commonwealth of Australia. With headquarters in Mackay and branches in several other coastal towns, the association was at the time the largest European-style political movement organized by Melanesians anywhere.[9] The association was the concrete expression of a determination to resist deportation against a wider background of no previous basis among Melanesians in Queensland for organized resistance and protest.

This chapter seeks to explain this apparent anomaly and the methods of response to their new environment employed by Melanesians in colonial Queensland; it uses as a case study Melanesians in the Mackay district from 1865, when the first Melanesians arrived in the district, through the early 1900s. In the argument presented here, Melanesian resistance has been divided into two types: overt, meaning unconcealed, when the Islanders directly confronted the imposed authority structures that controlled their working lives, and covert, meaning secret or disguised, when the Islanders' methods of response owed more to inward-focusing social adjustment mechanisms flowing from their own societies than to control mechanisms imposed by the state and colonial society. The chapter examines the legal controls and conditions of employment to which the indentured laborers were subjected, and argues that their lack of obvious active resistance or protest relates not only to the authoritarian character of the plantation system but also to their deliberate choice of methods of response. In short, they maintained Melanesian social equilibrium by depending on major inward-focusing Melanesian social mechanisms coming from their ethnic backgrounds to provide satisfactory balances in their lives. Their response was to mediate their working and private lives through their own beliefs, institutions, and forms of behavior, which were in partial operation in the colony.

Allied to this accommodation to colonial society was the development of a Melanesian counterculture of survival. This was integral to the emergence of a pan-Melanesian society in Queensland, created around the different categories of labor. The Melanesians were never a totally separate unit within colonial society, never a group that remained apart for forty years and then disappeared, but an immigrant group that interacted and eventually merged with wider Queensland society. The inward-focusing struggles and adjustments of the Islanders as an immigrant group are central to any analysis of resistance and protest. A similar interpretive thrust has been proposed by Adrian Graves, who views Melanesian resistance on Queensland plantations in terms of personal, ethnic, and

class integrity (what he terms "ethnicity") as part of his wider
argument that the Islanders were undergoing partial proletarianiza-
tion within a system of capitalist tropical industrial agriculture. The
main categories that he examined are forms of ethnic expression,
crime and violence, actions undermining production, manipulation
of the labor market, and finally the protests against deportation.
Even though he flirts with "ethnicity" as a motivating force of
Melanesian actions, he never gets beyond seeing the laborers as
pawns of peripheral capitalism, and never comes to terms with the
complexity of Melanesian methods of response.[10]

The present discussion takes a somewhat different tack and
argues that the inward-focusing social mechanisms and their accommodation in developing a pan-Melanesian society strengthened
and unified the Melanesian community at the same time as it
became an established sector of the working class in colonial
Queensland. This is in line with attempts by Adrian Graves, Mark
Finnane, and myself to examine the development of proletarianiza-
tion of the immigrant Melanesians, but at odds with Graves' attacks
on Pacific historians over their interpretation of indigenous cultural change and adaptation within colonial society.[11]

Drawing together their various methods of response, we can
explain their ability to confront the colonial regime and to operate a
European-style political movement at the turn of the century,
directly resisting the actions of the fledgling Commonwealth of
Australia.

The Mackay District Sugar Industry

The Mackay township is the major center of the 80-kilometer-long
Pioneer Valley. First viewed by Europeans in 1860, sugar cultivation commenced soon after in 1865. The valley is wide and amply
supplied with water courses; its bottom half consists of broad allu-
vial plains very suitable for growing sugarcane. By 1885, the district had produced 21,604 tons of raw sugar: there were thirty plantation mills operating, the majority growing most of their own cane
but also accepting cane from surrounding small-farm cultivators,
and two growing very little of their own cane, operating primarily
as central mills. For the first two decades, the industry was dominated by plantations: large farming units with vertical integration
of the farming and milling process and extensive human capital
investment using mainly indentured Melanesian, but also some
European and Asian, labor. The whole process was personally
supervised by frontier agricultural entrepreneurs or their manag-

ers. The planters of the Pioneer Valley, although socially and politically prominent at the time, lasted a single generation and soon gave way to large companies and farmer-operated central mills.

The industry boomed in the early 1880s. Speculation was rife. The acreage under sugarcane increased dramatically, and trade in Melanesian laborers reached its peak. But Queensland's sugar industry was profitable only because the high price of sugar on the world market covered its inefficiency, low productivity, and rising labor costs. The world sugar price, declining since 1881, fell by a third in 1884, and continued to fall until a slight recovery in 1889–1890. It was another three decades before the 1883 price was again equaled, and by that time the Queensland sugar industry had been completely restructured.[12]

The first farmer-controlled central mills in Queensland were completed at Mackay in 1888, financed by the government. By 1900, five farmers' mills had been built in the district and in that year for the first time outproduced the remaining seven plantation mills. While the farmers were predominant by the end of the century, small-scale farmers existed right from the 1860s. Farms and freehold land encircled the plantations clustered around the river and nearby creeks. Melanesian laborers were employed on these farms as well as on the plantations, and by the 1890s they were more likely to be working on small farms than on plantations.

Leadership among the Melanesian Workers

Most of the Melanesian laborers who came to Queensland served out their three-year indenture contracts and returned to their islands, wiser about the world and richer in material possessions. These were called first-indenture laborers. Although they were always the most numerous, other categories existed as well, considerably varying the pattern.

Not all laborers were enlisting for the first time. Reenlisting occurred in the islands from the late 1860s, and by the 1890s about one-quarter of all Melanesians entering the colony had previously worked in Queensland, Samoa, or Fiji. Those who completed one three-year agreement but opted to remain in Queensland and complete another work contract were called "time-expired" workers. The proportion of time-expired laborers to those working under first indentures increased steadily from the 1860s, but no register of their contracts was kept until 1883: in that year there were 1,476 time-expired Melanesians in Queensland, about one-quarter of the total number. At Mackay between 1888 and 1904, 84 percent of the total number of agreements were made by time-expired laborers. Already trained, less likely to fall ill, opting for shorter contracts, and mobile within and between districts along the Queensland coast, they worked primarily for small-farm operators, preferring the more casual life-style of the family farms to that of the regimented plantations.[13]

The final category of Melanesians was an elite but constantly diminishing group known as the ticket-holders. In 1884, the Queensland government placed occupational restrictions on all immigrant Melanesians in the colony, except for 835 who had been resident since before September 1879. Expressed as a proportion of the overall Melanesian population of Queensland, from 1885 to 1906 in any one year ticket-holders constituted between 7 and 11 percent of the Melanesian population. They had their own farms, seldom worked as indentured laborers, and occupied a variety of other occupations from boardinghouse keeper to market gardener. By the 1880s, time-expired and ticket-holding Islanders made up 40 to 60 percent of all Melanesians in Queensland.[14]

The 1880s were as pivotal to the Melanesian work force as they were for the sugar industry as a whole. Reenlisting, time-expired, and ticket-holding Melanesians constituted an ever-increasing proportion of the sugar industry work force in the 1880s, 1890s, and early 1900s. They were comfortable in their colonial surroundings, quite skillful in their negotiations with employers, and seem also to have fared best in handling prosecutions brought under the Masters and Servants Act. They were the major leaders within the Melanesian community, controlling resistance to work situations on the plantations and farms, and controlling the inward-focusing social mechanisms that regulated Melanesian behavior and intersected with social control mechanisms employed by the owners of the means of production.

When it came to overt political action at the time of the mass deportation ordered by the Commonwealth government in 1901, the Islanders were able to organize the Pacific Islanders' Association. There is no evidence of European participation in the association. Its leaders were literate and articulate, mission-educated, time-expired, and ticket-holding men, all of long residence in the colony. They rose to the occasion and confronted the new state and federal governments. That they would lose the battle was inevitable, given the general fervor for a White Australia. But they partially succeeded: in collaboration with various humanitarian groups, they managed to moderate the draconian plans so that about 2,500 Islanders were eventually able to remain in Australia.[15]

Conditions of Employment

Control of Melanesians working in Queensland was achieved through the mechanisms of control on the plantations, with the backing of the coercive apparatus of the state, in particular the legal system, the police, and the special acts and regulations.[16] Illiterate, pagan Melanesians were held to legal agreements that they signed by touching a pen or placing their mark on a contract. Even the literate Christians among the laborers must seldom have understood the full ramifications of their contracts. Until 1868, the only basis for indentured labor agreements in Queensland was the 1861 Masters and Servants Act. Abuses in the Pacific labor trade during the 1860s and 1870s led to the introduction of the 1868 Polynesian Labourers Act in Queensland and an 1872 British antikidnapping act (Pacific Islanders' Protection Act), amended in 1875. The 1868 act, substantially amended in the 1870s and 1880s, with the Masters and Servants Act, were the major legal powers governing relations between Melanesian laborers and their employers for the remainder of the century, and the first few years of the next. Under the terms of the Masters and Servants Act, servants could not strike, negotiate for higher wages during a contract, form trade unions, or leave indentured service before their agreements expired. The 1872 Pacific Island Labourers Act stipulated three-year contracts for first-indenture laborers, minimum wages of £6 per year, minimal levels of health care, clothing, and sustenance, and repatriation at the employer's expense at the end of the contract if the laborer chose not to reengage.

Administration of this vast circular-migration involving over 800 voyages, more than 62,000 initial indenture agreements, thousands of transfers of agreements, and tens of thousands of new and renewed contracts was handled by the Pacific Island Branch of the Queensland Immigration Department. The branch was headed by the Immigration Agent in Brisbane, supervising Government Agents placed on all recruiting vessels from 1871, assistant immigration agents (known as Inspectors of Pacific Islanders) in each of the major coastal sugar towns, and police magistrates who also served as assistant imigration agents in smaller centers. The Inspectors and their deputies supervised the arrival of all Melanesian immigrants and administered the acts and numerous regulations. Local government medical officers completed health inspections of newly arrived laborers, and from the 1880s there were segregated hospitals for the Melanesians, with resident and consultant doctors and wardsmen, also under the control of the local inspector.

Overt Resistance

The data on which much of this chapter is based are expressed in Table 1, which summarizes the 1,598 offenses charged against Melanesians in the Mackay district between 1871, when the first case was recorded, and 1907. These statistics have been compiled primarily from court reports in newspapers, especially the *Mackay Mercury*, as reliable government statistics, for Mackay's courts only cover the years 1873–1881.[17] As indicated in Table 1, more than 45 percent (729 of the 1,598 cases) of the charges recorded relate primarily to the Melanesians' working lives: common assault, using abusive language, offenses in breach of the Masters and Servants Act and the Pacific Island Labourers Act, failure to obey a court order, and vagrancy. Plantation owners, managers, overseers, and farmers brought charges under the Masters and Servants Act. These were heard before the district magistrate or justices of the peace, themselves often owners and managers of plantations or pastoral stations employing Melanesians. Punishments varied considerably, from being dismissed with a caution, to a fine of from 5 to 20 shillings, or a jail sentence of from twenty-four hours to a maximum three months. Sentences of more than thirty days were given only for refusing to obey the bench's order to return to work, and were served in Rockhampton jail, not the Mackay district lockups.[18]

Prosecutions under the act were more frequent from the 1890s, when the number of Melanesians in the district was declining, but 40 to 60 percent of them were time-expired or ticket-holders, who preferred to work on small farms rather than plantations. These post-1890 cases differ in several ways from the earlier ones. Most of the cases of breach of contract were not brought by planters or their overseers, but by small-farm operators. Heavier sentences were awarded for absconding than for disobedience, but fines were more severe, from a minimum of 5 shillings to about £5. Fewer Melanesians served time in the lockup. Most either paid their own fines, got friends to pay, or arranged to have the amount deducted from wages due to them. Many obstinately resisted European authority, challenging orders from field overseers. It seems likely that most of these cases were brought against time-expired or ticket-holding Islanders, or perhaps laborers returning to the colony for a second or third time, or those who had previously worked in Fiji or Samoa. These categories of Melanesians had the money to pay fines and enough experience to confront the system, so the court's punishments held less fear for them. The paradox is that the time-expired laborers had no particular need to engage in direct and potentially costly confrontations. Being wage laborers on short-term contracts that were constantly renegotiated, they had options; they could choose for whom they worked and became extremely knowledgeable about the labor market and the value of their skills, negotiating over wages and conditions.

On paper, Melanesians in Queensland seemingly had equal rights with their employer before the law. A feature of the Queensland Masters and Servants Act, contrary to the act in neighboring New South Wales, was that it enabled employees to lay criminal charges against their employers. But they seldom laid charges: only twenty-two charges are known to have been made in the Mackay district between 1871 and 1907 against almost 1,598 charges brought against them by employers and police (see Table 2). This imbalance indicates both the effectiveness of employers in controlling the legal system and the incidence of disputes in the fields and mills going unreported when the aggrieved party was a Melanesian.

The low number of cases brought by Melanesians (22 compared with 1,598 brought against Melanesians) is not any indication of the frequency of offenses against Melanesians. This is illustrated by the infrequency with which assaults by Europeans on Melanesians were reported. Only eight cases reached the courts, and with one exception the defendant was a plantation manager or overseer. Three of the charges were dismissed, the outcome of one was not reported, and in the three that were successful, the fines imposed were between 10 and 50 shillings.[19] For a comparable offense against a European, Melanesians usually received a jail sentence of from seven to fourteen days. The remaining case contained an element of the ridiculous. It concerned Mrs. I. J. Atherton, the wife of a leading pastoralist, who pleaded guilty to striking her Melanesian general servant; the court ordered her to pay his wages and cancelled the agreement, but no sentence was imposed.[20] Of the eleven other cases brought by Melanesians against an employer, one was for theft of money. The circumstances seemed to support the accusation, yet the case was dismissed.[21] Ten other cases involved employers who had breached the terms of the Masters and Servants Act: six were wage claims, five of which were dismissed. Another charge, for illegal detention of property, was upheld. The quality of justice that Melanesians could expect, at least in the 1870s, is graphically indicated by the case of two Islanders who had been arrested on suspicion of murdering another. They were marched at

Table 1. All Known Charges Brought against Melanesians in the

	1871	1872	1873	1874	1875	1876	1877	1878	1879	1880	1881	1882	1883	1884	1885	1886	18
Homicide																	
1. Murder			3									2					
2. Attempted murder																	
3. Unlawful wounding					1												
4. Assault	2			1	3		3	5	1			2	6	5			
5. Assault and robbery																	
Offenses Against Sexual Mortality																	
6. Rape of a female														8			
7. Indecent asssault of a female														1			
8. Indecent exposure																	
Petty Offenses: Disturbing the Peace																	
9. Drunkenness					2	3	3	4	3					1	1	14	1
10. Unlawfully supplying alcohol																	
11. Traffic offense	1																
12. Resisting police									1								
13. Inciting prisoner to resist																	
14. Abusive language										1				1	2		
15. Disturbing the peace/disorderly conduct											1	2			3		
16. Carrying weapons with intent																	
17. Possession of firearms								4		5							
18. Breach of the M. & S. Act/disobeying lawful order	2			2								1		1	3	6	8
19. Disobey Court order																	
20. Abscond or be absent from hired service	2	10	23	6	1	7	5	1	2					2		6	

Mackay District, 1871–1907

1888	1889	1890	1891	1892	1893	1894	1895	1896	1897	1898	1899	1900	1901	1902	1903	1904	1905	1906	1907	Total
1		5	1	12		9	7	8					2	2	2		1	5		60
						1	1	1	2											5
6		5		6			4	4	1			3				1			3	35
4	5	11	11	12	14	3	4	13	12	23	5	24	3	2		2	14	4	4	205
				5						1		1								7
																				312
	2			6				1												17
					1				1											3
						1									1					2
																				22
21	10	29	32	19	21	8	18	37	9	17	23	17	15	9	5	4	15	13	28	391
							1					2			2					5
																				1
4		1						1												7
											1									1
1			1	1		3			1								2			13
7	3	7	5	2	3		3	3	2	3	6	2	32	1		2	1	1		89
			1		1															2
1			1								2		1							14
4	8	2	9	5	3	7	6	11	18	26	35	32	40	6	2	4				236
	1										2									3
8		14	3	10	6	10	3	9	12	11	24	20	16	7	2	3	10	2		236

Continued

Table 1. continued

	1871	1872	1873	1874	1875	1876	1877	1878	1879	1880	1881	1882	1883	1884	1885	1886	1887
21. Vagrancy										1				4			
22. In a common gambling house																	
Offenses Against Property																	
23. Arson														1			
24. Willful destruction of property																	
25. Burglary																	
26. Larceny								1	2	1	3	1		7	3	1	
27. False pretenses																	
28. Illegally on the premises					2					1					1	1	
Other																	
29. Of unsound mind								1									1
30. Small Debts Court																	
31. Other offenses														1			
Total	5	0	10	28	10	8	15	15	10	17	5	2	0	20	24	39	2

SOURCE: Clive Moore, *Kanaka: A History of Melanesian Mackay* (Boroko/Port Moresby, 1985), 362–363.

a forced pace, and witnesses saw the police beat them with a whip when they fell down. The *Mackay Mercury* was outraged, but the case against the police was dismissed with a caution.[22]

In other respects too the coercive apparatus of the state had the effect of discouraging organized resistance. Under the Masters and Servants Act, indentured servants were not entitled to go on strike. The occasional strike action did occur, but these were isolated events involving only individuals or small groups.[23] The vagrancy laws, moreover, were increasingly used to hold unemployed Melanesians who were considered to be creating a nuisance by wandering about the district, or were suspected of being guilty of more serious crimes. It is significant that most vagrancy charges occurred after 1896, when the government passed regulations intended to force Melanesians to reengage as indentured laborers within a month of the end of a contract, or to return to their islands.[24]

With the legal system operating against their interests, Melane-

1888	1889	1890	1891	1892	1893	1894	1895	1896	1897	1898	1899	1900	1901	1902	1903	1904	1905	1906	1907	Total
									5		2	2	6				16			36
												15								15
																				1049
				2			1		1											5
	1	6	1	1										2						11
		3																		3
1	1	3	3	9	1	1	1			2	2	2	2	1	2	1		2	1	55
								1												1
1	1							1	1			1		1		2				13
																				88
	1								3	2	4	1	3	3	3	2	1	2	2	29
						1	1		78			3	1	4	2					90
					1				1	1	2	1			1					8
																				127
59	33	86	67	91	50	47	50	93	146	90	108	121	127	38	14	21	47	43	36	1596

sians tended to take matters into their own hands. Physical violence and recalcitrant behavior are typical Melanesian reactions to aggravation. In 1879, for example, thirty Melanesians attempted to rescue one of their number who had been arrested in town for drunken deportment; five additional police had to come to the help of the arresting officer.[25] On plantations, such behavior was an expression of grievance, dissatisfaction, or just plain exasperation —and Mrs. Atherton was lucky not to have been violently assaulted for her lack of self-control. The vast majority (one report suggests 90 percent) of cases of Melanesians refusing to obey orders or assaulting an employer were never reported, but there are still many reports of overseers and employers being assaulted or threatened with murder or assault by Melanesians brandishing guns and knives. John Spiller was the first man to plant cane in the district in 1865 and employed Melanesian laborers from 1867, priding himself that he got on well with them. But this was not always the case; in 1872, a little drunk, Spiller tried to interfere and show a gang of Islanders how to cut cane. They attacked him and screamed abuse,

Table 2. The Total and Melanesian Population of the Mackay District Compared with Charges Brought in the District's Magistrates Courts

	District Population	Persons Taken into Custody in the District	District Melanesian Population	Charges Brought against Melanesians	Charges Brought by Melanesians	Charges Brought by Non-Melanesians Concerning Melanesians
1871	1,400		700	5	1	
1872			700			
1873			900	10		
1874			1,200	28	1	
1875		311	1,700	10		
1876	3,766		1,600	8	6	2
1877			1,400	15	3	
1878				15	1	
1879				10		1
1880		370		17	2	
1881	5,787		2,087	5	1	3
1882				2		
1883						
1884				20		3
1885		600		24	2	
1886	11,082		2,686	39	1	3
1887			2,077	23		4
1888			1,911	59		8
1889			2,121	33	1	
1890		542	2,816	86	1	8
1891	10,538		2,450	67	1	14
1892			2,102	91		11
1893			2,081	50		14
1894			2,168	47		4
1895		692	2,167	50		5
1896			1,941	93		13
1897			1,780	146		14
1898			2,074	91		13
1899			1,684	107		8
1900		559	1,775	121		4
1901	11,144		1,475	127		8
1902			1,708	38		4
1903			1,301	14		2
1904			1,260	21	1	4
1905				47		11
1906		887	959	43		5
1907			400±	36		3

SOURCE: Adapted from Moore, *Kanaka*, 363.

causing their employer to vault onto his horse and gallop home, much chagrined. His employees, victorious, quietly returned to work. On Inverness plantation in December 1879, thirty Islanders, all armed with cane knives, ganged up on an overseer and threatened to kill him. On Te Kowai plantation in 1887, one overseer got more reaction than he expected when he chastised an Islander for throwing cane onto the ground instead of handing it to him. The Islander approached the overseer, grabbed him in an embrace, and bit him on the chest. On Nindaroo plantation in 1884, Tallygooner attempted to murder an overseer named Hynes, but he was not convicted because it was clear that Hynes had provoked him. And in another case, two laborers were working on Ashburton plantation one morning in the winter of 1896. Occasionally Mackay can become very cold in winter. Garry and Harry were feeling cold, so they stopped work and lit a fire. They refused to obey the overseer's order to put the fire out and get back to work, wounding him in the ensuing fight.[26] There would undoubtedly have been more such attacks had overseers not taken precautionary measures. Melanesian oral testimony reveals that Islanders knew that it was pointless to sneak up on overseers at night, because the latter routinely kept a dog and a gun for their deterrent effect.[27]

The use of abusive language precipitated many other attacks on Europeans. Melanesians regard swearing as a much greater offense than most Europeans would. On the central Solomon Island of Malaita, the home of large numbers of the laborers, swearing at anyone or their ancestors can cause grave offense and warrant violent retaliation, which explains the murder of Morris Summers in 1901. Summers objected to the amount of noise that Oufough and other Malaitans were making and swore at them. Oufough flew into a rage and smashed in Summer's skull with an iron bar.[28] The provocations behind instances of assault were more varied, but the majority of the 205 cases recorded were Melanesian attacks on European overseers and other European employees.

There is only one documented report, from 1897, of a plantation overseer who was probably murdered by his laborers, and even then the police were uncertain.[29] Noel Fatnowna, a prominent Mackay Pacific Islander of Malaitan descent, had stories of secret retaliations by Melanesians. One overseer was supposed to have been killed out in the fields, his body hidden in a heap of cane trash and subsequently burned. Another tale was recounted to Fatnowna by George Farquhar, whose father had been a plowman on Nindaroo. Some Malaitan men were working in the fields. One caught a rat, which he skinned ready to cook and eat. The overseer, a much hated man, came and watched. The Malaitan pushed a stick through the rat and meaningfully told the overseer that if he ever set foot on Malaita he could expect the same treatment. The overseer hurried away, vowing never to visit the island.[30]

Most attacks on Europeans were for the obvious reason that overseers frequently assaulted, threatened, or swore at their Melanesian employees, and occasionally killed them.[31] Use of the whip on recalcitrant Melanesians was not de rigueur, but common enough for "Bondsmen and stripes" to write to the *Queenslander* in 1877, advocating legalization of the use of the whip, giving as reasons that the practice was already widespread and that it would be better to have it regulated.[32] Some overseers wore revolvers, which did not go down well with the laborers. In August 1895, Islanders working at Alexandra, part of Palms plantation, threatened to shoot the overseer because he was wearing a revolver, which he said he carried because they refused to obey orders. Although most such problems arose on plantations, some incidents did occur on farms. In 1897, two Malaitans assaulted their employers, because one always carried a revolver when working with them. And in 1900, a Melanesian employee on Solway's farm, for reasons unknown, armed himself with a rifle and threatened to shoot his employer.[33]

The frequency of attacks on Europeans is a reflection of the absence of other avenues of redress. To put it another way, the small number of cases brought by Melanesians is no indication of the frequency of the offenses by Europeans that gave rise to them. The technicalities of British justice were also a powerful deterrent. Melanesians probably never fully understood the charges against them, or brought by them. When Melanesians laid charges, it was usually the work of the police magistrate, the Inspector of Pacific Islanders, and sometimes even the Immigration Agent, acting on their behalf. More often than not it was on the basis of a complaint laid by a Melanesian who had walked off the plantation or farm to report the matter.[34] There were even instances where Melanesians and Europeans raced each other to the inspector. In 1892, for example, two Melanesians walked the short distance from Meadowlands plantation into Mackay township to lodge a complaint against W. H. Hyne, one of the proprietors. Getting wind of what had happened, Hyne rode ahead on horseback and charged the pair with absconding. At their trial the two "absconders" were represented by the inspector, who remonstrated vigorously against Hyne's unfairness, and the case was dismissed.[35]

Most of the time, however, court decisions went the other way.

The system of adversary justice inherent in British law was alien to Melanesian practice. As well, Melanesian evidence was often rejected by the courts, as most were not Christian and could not be placed on oath. They had little chance of success before a bench made up of justices of the peace who were all leading citizens and usually themselves employers of indentured servants. There can be no doubt that Melanesians had little difficulty in assessing their chances of winning a case, using an alien system of justice, and chose alternative means of redress. Melanesians in Queensland seldom confronted their employers with direct organized industrial action. Instead, they found other methods, such as negotiation and covert retaliation, and finally, accommodation more effective.

Covert Resistance and Inward-Focused Social Adjustment Mechanisms

Much of the literature on resistance and protest by agricultural workers looks at the mechanisms by which employers subordinate and exploit workers, and stimulate productivity. It is as if plantation workers had no control of their own lives and were merely units of labor employed by the owners of the means of production. While this is true in terms of the indenture agreements and the courts' rulings on breaches under the acts, this is an exceedingly mechanistic approach and cannot totally explain Melanesian behavior within colonial society. Although they seldom chose to confront their employers directly through the legal system, or by strike action, there is evidence that they worked on a much wider punitive front using the "weapons of the weak," and more ethnically Melanesian methods of response.

Absconding is resistance to work, but it is also a classic Melanesian evasive tactic that is better dealt with as covert response. There were 236 charges of absconding or being absent from hired service. Some absconders managed to live for months in the bush around the district, terrorizing isolated settlers, stealing from gardens and poultry yards, being seen at a distance naked and armed. There are two cases of Melanesians who absconded and tried to walk back to their islands overland down the Queensland coast. In the mid-1870s, a Pentecost Island woman named Vadim was sexually molested and watched her husband kill two men for the offense. She set out walking south and was returned from Rockhampton, 400 kilometers away. Soon after, she set off again and was found near Port Curtis (Gladstone), 500 kilometers south. And in 1896, an Islander who was thought to be insane was shipped south to Maryborough to board a vessel going back to his island. He made his way back to Mackay overland, about 650 kilometers, arriving seminaked and quite deranged.[36]

Malingering and working slowly were other covert means by which Melanesians resisted their work environment. Once out in the fields, if they could escape the beady eye of the overseer, they could take a dip in a nearby creek or have quick naps among the cane rows. Pretending to be ill was common, but this could have unfortunate repercussions. The standard cure-all in the nineteenth century was a large dose of castor oil,[37] and if malingering was suspected, a laborer might find himself on the receiving end of an overseer's boot. One of the best examples of Melanesian oral testimony concerning malingering was provided by Malaitan Henry Bobongie of Mackay. His relative Tom Robins was working at Mossman in the north in the 1890s. One morning he stayed in bed, too sick to go to work. The overseer arrived and poked Robins with a stick to force him to get up. Robins did as requested but punched the overseer with enough force to shoot him back through the doorway. The result was a trip to court for Robins.[38] Unfortunately, Melanesians suffering from tuberculosis or gastrointestinal diseases were often accused of malingering when they were actually quite ill.

Sabotage in the fields and mills was common and is another covert response. Melanesians hid rocks in the cane going to the mills, with disastrous effects on machinery.[39] Acres of cane mysteriously caught fire and mill sheds also went up in smoke. Sometimes this was accidental, but often laborers were suspected, particularly when the fires started from the center of the field and burned outward, or when mysterious fires occurred in quick succession.[40] Much of the cane on the north side of the river was grown on steep slopes. Winding trails were built around the hills, enabling horses and drays to reach the top. The early drays had no brakes, so the load of cane was positioned to drag on the ground at the back as a brake. Later drays had brakes fitted, and a favorite caper of the laborers was to let the brakes off deliberately, allowing a dray to run out of control down a hill, mangling both horse and dray in the process. Cane was also brought down the slopes by "flying-fox," a stretched wire rope to which bundles of cane were attached. These were sabotaged so that the load fell apart halfway down, scattering the cane; although they had to pick it up again, there was satisfaction in frustrating the overseers and the general good order of the plantations and farms.[41]

Other forms of covert resistance were less bound to the immediate circumstances of the workplace but rather are explicitly Melanesian methods of response. The cultures and cosmologies of the Melanesian immigrants, though disrupted and truncated, were successfully transplanted to Queensland. In Melanesia, the physical and spiritual worlds intermingled. Sickness and death, even accidents, were believed to have a supernatural cause. An examination of sickness and death provides a perspective for establishing a more holistic approach to resistance and control.

Traditionally, when individual Melanesians fall sick it is because they have broken some societal rule: the sickness is a direct punishment for misconduct, inflicted by ancestors whom they must appease in order to survive. The Melanesian attitude toward death is similar: there is no such thing as natural death in the sense that Europeans would understand the term. Illness and death were not the natural corollaries of disease, epidemics, or old age. Epidemics were thought to be caused by antagonistic spirits, ever circulating in the sky, waiting for a chance to commit a malevolent act. Sickness was viewed as a direct punishment for misconduct, inflicted by ancestors who had to be appeased if the ailing individual were to survive. Death had two explanations—either from illness caused by ancestors or murder by living people. People murdered, either by sorcery or by some physically violent means, had to be avenged to maintain societal equilibrium.[42]

There were vast conceptual differences between European and Melanesian views of sickness and health. Given the cultural and linguistic gap, there was little that Melanesians could do to explain their reactions and feelings. Nor did it help that nineteenth-century European scientific knowledge was extremely limited, leaving little chance that settlers in colonial Queensland could understand the problems of health and cultural adjustment faced by the migratory Melanesians. Melanesians, moreover, were epidemiologically unsuited for long-distance labor migration. The more isolated a human community, the more specialized and individual its disease environment is likely to become.[43] The Pacific Islands, with their small, isolated communities, were virgin soil populations, sheltered from the outside world. As a result their populations were unable to acquire the most significant immunities and remained susceptible to infection.

The interaction between Melanesians in Queensland, as well as between Melanesians, Aborigines, Europeans, and Asians, was devastating for newly arrived recruits. The three-year length of the circular-migration exacerbated the problem with diseases, constantly exposing a fresh supply of laborers to the new disease environment.

Ralph Shlomowitz has calculated that the estimated crude death rate of Melanesians in the first year of their indentures (81 per 1,000) was over three times as great as the estimated crude death rate for the rest of the Melanesian population (26 per 1,000). But if a Melanesian survived the first three years in Queensland, he or she had every chance of living to old age.[44]

Between 4,000 and 5,000 Melanesians died in the Mackay district between 1867 and 1907.[45] These were mainly young men in the prime of life, aged sixteen to thirty-five years and judged medically fit when they first arrived. Over the four decades of the Queensland labor trade, the average annual death rate was around 55 per 1,000. At its height, in 1884, the Melanesian mortality rate was 147 per 1,000, unacceptable by humane standards anywhere, and at its lowest bound it was still many times higher than the mortality rate for Europeans in Queensland.[46] The major killers were bacillary dysentery and respiratory tract infections such as tuberculosis, pneumonia, bronchitis, and pleurisy. The next most important causes of death were other diseases of the gastrointestinal tract, infectious diseases (particularly measles), and fevers.

What is of concern here is the Melanesian reaction to death, sickness, and injury. Viewed in this light, the incidence of murders and assaults by Melanesians becomes more intelligible. At a bare minimum, Melanesians killed 100 people in the Mackay district between 1867 and 1907 (see Table 1); and if we take into account deaths from sorcery, the number would probably double or triple. Given that 4,000 to 5,000 died in the district, and Melanesian attitudes toward death and illness, one could even conclude that they were quite restrained in not killing more people.[47] Following the argument presented in the introductory section of this chapter, one should also consider what other inward-focusing social mechanisms were used to compensate for the high death rate.

In seeking deaths in compensation for deaths, Melanesians did not feel obliged to kill the person they thought directly responsible. If possible, they would locate and deal directly with the murderer; otherwise, they were satisfied with any death. Murders by Melanesians were usually brutal, but they could be reasonably explained as retaliation for an act supposed to have been committed by the victim or someone close to them. This accounts for the seemingly irrational murders of innocent European children.[48] Swaggies (the Australian expression for hobos) were more vulnerable still, as they had no ties to the district and might not even be missed.[49] Folk memory at Mackay records that Europeans had an easy way of knowing if a murder had been committed by a Melanesian: if the victim's head was smashed or missing, then the culprit was sure to be Melanesian. Iron bars and lumps of lead on ropes, whirled around to add momentum before making contact, were favorite methods of dispatch, as well as the ubiquitous axes and cane knives, spears, clubs, bows and arrows, and occasionally a gun.[50] Those involved often fled the district, using their rights under English law to be presumed innocent. The *only* constant factor is the extreme violence of the incidents.

The Queensland police often kept suspects under surveillance for years, until there was sufficient evidence to press charges. Given the covert nature of many of the killings, the police sometimes apprehended the wrong victim as (probably) in 1894, when Johnny Assina from Malaita was convicted of the murder of Sarah Baumgarten of Double Peak near North Eton. Assina had arrived at Mackay on the schooner *Fearless* in 1891, worked his first contract on Habana plantation, then accepted a one-year contract with a farmer near North Eton. Sarah Baumgarten, a farmer's wife, took in washing to supplement the family income. Earlier she had assisted at the Melanesian Mission (Anglican), while it was based at the local township of Marian. She accepted Assina's washing, but there was a disagreement over the filthy state of his clothes. She was shot one night, and Assina was found in possession of the revolver. He was tried and sentenced to death, although the evidence was circumstantial. When statements were later made in support of Assina's innocence, his sentence was commuted.[51]

At the time of his conviction, the *Mackay Mercury* editorialized on the tension existing between Europeans and Melanesians in the district:

> In the sentence of death the first blow is struck in the district to free it from the terrorism of the Malayta [Malaita] Islanders. The attitude of these Kanakas deserves something more than passing notice. At the present time 50 per cent of the Polynesians in the district are natives of Malayta, while 10 years ago the proportion was only about 15 per cent. The habits of the Malayta boys are of particular interest. Their thirst for blood is notorious. . . . Of recent years the aggressiveness of these boys in the district has become unbearable. Apart from any graver offences of which Malayta Islanders may be suspected, it is notorious that they boast of the immunity with which they can defy their employers and commit crime. White men who have lived in the district for years, now sleep with revolvers near at hand, and admit the ter-

rorism inspired by this class of boys. . . . [I]t is the boast of the
Islanders that the white man will not hang them. It is argued also
that, if we hang Kanakas, reprisals will follow in the Island trade,
and to this we reply that hanging or no hanging, the Malayta boys
would cut off the first ship they could, and would murder any
white man they could get a show at.[52]

Not all antagonisms were vented on non-Melanesians, and it
seems likely that many intertribal and interisland fights may have
been methods of letting off aggressive energy perhaps originally
motivated by work or interracial tensions. Violent, large-scale
fights involving hundreds of men were reasonably common. Severe
injuries and sometimes fatalities were sustained.[53] More often,
however, the fighting seems to have been Sunday sport, full of ges-
ticulations, shouting, and threats, but with little real action. Similar
patterns of aggression are quite common in the islands, where the
intention is to draw blood but not to cause serious hurt or death.
Plantation overseers did try to defuse mobs of agitated Islanders,
but usually Europeans facing large-scale violence between groups
of Melanesians either got out of the way and let the battle rage or
called the police. In one incident in the 1870s, the overseer on a
plantation heard yells from the laborers' quarters and hurried to
investigate: ". . . as he rushed along yell succeeded yell, and the
cries and screams grew worse. There he found the boys in an
extreme delirium of excitement and fury. They were armed with
bows and arrows, guns, rifles, tomahawks, cane knives and clubs,
dressed in war paint, without any dress at all, stark naked, and
ready as it seemed to murder anyone and everyone."[54] In another
typical incident in 1883, the staff of Cassada plantation were having a quiet Sunday when shots were heard from the direction of
Walkerston, the nearest town. Then, suddenly, the ridges were
alive with Melanesians armed with bows and arrows and guns.
They commenced an attack on the Cassada laborers, who were
ready to repel their assailants.[55]

Just as interisland and intertribal fighting must be considered as
inward-focusing mechanisms of social adjustment, so must sorcery,
a completely immeasurable dimension, invisible but devastatingly
potent for those who believe in its power. Oral evidence indicates
that during the lifetimes of the original immigrant Melanesians, religious and magical practices figured prominently in the Queensland
Melanesian community.[56] Reputed sorcerers were regarded with
fear and awe, their reputations outlasting their lives by decades.
The last reasonably verifiable case of death by sorcery in the
Mackay district occurred in the 1940s.[57] But how can we ever

attempt to rate as effective social mechanisms the power of sorcery and the ability to conjure up the spirits of ancestors to come to one's aid in times of need?

The oral evidence suggests that there were two main categories of magical beliefs and practices among Melanesian immigrants in Australia: socially approved magic for protective and productive purposes, and destructive magic. Socially approved magic was not the property of specialists. Such magic was used in gardens and in fishing, in healing and for sexual procurement. Many of the original Melanesian migrants owned, and often carried around with them, magical stones, powders, amulets, and concoctions tied up in little packages, all said to have magical powers. In the darkest corners of their huts there were old baskets or coconut shells, blackened with the smoke of years, containing some magical treasure that no one else would dare touch or even go near. Informants report that in the early decades of this century, when the old Islanders went out visiting, they would hang their bags, containing these magical items, on trees or bushes well away from the house, never bring them inside, and never allow anyone to go near them. Love magic was a great favorite in the largely male community and many a young girl is supposed to have been lured irresistibly to the waiting embrace of an impatient lover.

To Melanesians, as previously mentioned, there were no natural causes of illness or death, and sorcery was often suspected. Destructive magic and protective magic was used constantly. Magic stones, with the power to cripple or kill any trespasser, were strategically placed around gardens, houses, and graves. Taboo markers were used to warn off the unwanted. Sorcerers are reputed to have crept around the outsides of houses at night looking for food scraps and other refuse or belongings to steal away for use in spells. To prevent this, the ground around houses was kept bare, cleanly swept each evening and closely inspected for suspicious footprints the next morning. Sorcerers were also reputed to be able to become invisible, to walk through walls, to fly, and to turn themselves into animals, reptiles, and birds.[58]

Oral accounts leave no doubt that belief in sorcery was widespread and must be taken into account in assessing Melanesian society in Queensland just as it would be in the islands, but is there any contemporary documentary evidence of the phenomenon? Dead Melanesians just turned up every now and again; no one ever seemed to have any explanation of how they had died. Although with most reports there is no indication of the age of the dead person, it can be presumed that few would have died of old age, given that the average age was sixteen to thirty-five. They may well have died from disease, but it seems likely that at least some were poisoned or the victims of sorcery. Take the case of Outapina from the New Hebrides, who died on River plantation in 1876. He had just completed his contract, had his wages, and was living in a boarding-house in Mackay. He became ill and died in two days: the inquest was inconclusive, and the diagnosis was "serious apoplexy."[59] In another case in 1881, Terrai from Foulden plantation was found dead on neighboring River plantation. The inquest reported death from natural causes but noted that Terrai's only abnormality was a slight throat infection. Perhaps unrelated to sorcery, but nevertheless interpretable as such, are the decomposed bodies of Melanesians found around the district. One was found on Te Kowai plantation in 1891 and another in a creek near Habana in 1893. In the Habana case, the man had absconded from a nearby farm eleven days before and had probably died about the same time.[60] Joseph Antoney found a decomposed body of a Melanesian, dead for about a fortnight, on his property near Eton in 1896; there were no suspicious marks on the body. And in 1899 a Melanesian died in strange circumstances on board a ship between Rockhampton and Mackay. At the beginning of the voyage he appeared normal, but he jumped overboard midvoyage and had to be rescued. Afterward he became extremely ill and died before the ship reached port.[61]

Melanesians also committed suicide, usually by hanging, the ultimate in inner-directed aggression. It is evident that a few laborers were mentally disturbed when they enlisted, and life in the colony drove them to greater depression, violence, deeper insanity, and suicide. Consumption of alcohol and subsequent depression may also have been responsible for some of the suicides. But if we examine certain of the cases closely, another interpretation is possible. For example, in 1877 there was an elaborate double suicide on Pioneer plantation, but it could just as easily have been ritual murder or the end product of sorcery. Charlie Lorbacco and his wife, Mary Narisse, both from Tongoa in the New Hebrides, had been in Queensland since 1869. Lorbacco was in charge of a small farm on the plantation. During the night of 13 August, he went to the house of a European laborer on the plantation, pleading to be taken in until morning because he feared sickness at his own house. He was sent home with promises of help in the morning, but the next day the couple were found shot dead. Their bodies were arrayed in all of the clothing they possessed, as well as beads, feathers, and plaited human hair, and they were wrapped in dozens of yards of calico recently stained with red and yellow ocher. All other belongings had been broken or destroyed by fire.[62] In another case, from 1891, George and Geor were working on Homebush plantation; the couple had a four-month-old baby. One morning Geor, distraught, told her husband that a "devil-devil" was chasing her and would kill her. She then hanged herself after attempting to strangle her child.[63] Was this postnatal depression, was she attacked by a malevolent spirit, or was this the result of sorcery?

While these deaths seem bizarre, they would not puzzle Melanesians in Queensland or in the islands even today. Melanesian Mackay last century abounded with spirits and sorcerers, haunted trees, magic stones and love potions, plants used to ward off evil, and fearsome tales of cannibal monsters and ghoulish creatures. Melanesian society was violent and rumbustious but self-contained. Melanesians coexisted closely with the Aboriginal population of the valley, wooing the women, and somewhere along the line acquiring their knowledge of local edible plants, medicinal herbs and poisons, and an appreciation of their life-style and spirit world.[64] But, except for a handful of European women who married Melanesians, Europeans never became closely enough involved with Melanesians to ever understand the complexities of their behavior.

Their accommodation to life in Queensland was on their own terms. Today in suburbs in the cane towns of Queensland and northern New South Wales, a few of the more traditionally minded of their descendants claim that they are still in communication with their ancestors and that the spirit world and the physical world still melt together. They certainly do not discuss this with their European neighbors, who they know would put it down to quaint superstition and laugh. In the nineteenth century, their forebears were even more reticent to explain or discuss sorcery and their spirit world. Yet how can we hope to appreciate their reactions to colonial society if we do not take their beliefs into account? Covert resistance and inward-focusing means of social adjustment were more important to them than the Masters and Servants Act ever was.

Conclusion

Eugene Genovese reminds us that radical historians too frequently make it their responsibility to provide the masses with a "glorious tradition of resistance to oppression, and to portray them as having been implacably hostile to the social order in which they have been held."[65] He goes on to say that it is often considered sacrilege, bordering on racism, to suggest that Black slaves in America may have lived in harmony as well as antagonism with Whites. Genovese suggests that there is little evidence to show that Blacks in America mounted large-scale organized opposition to the regime or developed a revolutionary tradition: "Our main problem is to discover the reasons for the widespread accommodation and, perhaps more important, the long-term effects both of the accommodation and of that resistance which did occur."[66]

Equally we must be careful not to provide Australia's immigrant Melanesians with a "glorious tradition of resistance" when they were merely using the "weapons of the weak" against their employers, behaving in ethnically Melanesian ways, accommodating themselves to colonial society, or just quietly going about their allotted tasks in the cane fields. The price of intransigence was too high, especially for first-indentured laborers: Marroko, a laborer on Nindaroo plantation, refused to obey an order, provoked the others working with him to do likewise, and was fined £3 sterling (or six months' wages).[67] Following the argument that resistance and accommodation are two aspects of the same process, Melanesian accommodation must be seen as part of their survival strategy in difficult circumstances. Part of this accommodation was the development of a pan-Melanesian society—an amalgam of elements from dozens of island societies; and part was their adaptation to life in colonial and twentieth-century Australia as they became absorbed into the wider working class. This was just as much a part of their strength as direct resistance or their continued dependence on their ethnic origins. Their methods of response to colonial society must be viewed in this complex way if we are to appreciate their survival as an Australian immigrant group.

The formation of the Pacific Islanders' Association came at the end of forty years of Melanesian migration to Queensland. It could not have operated successfully at an earlier time. Its success depended on the existence of pidgin English as a lingua franca, and a group of quite well-educated, financially secure Islanders who were able to move between and around sugar districts, and could stand their ground when negotiating with white Australians.[68] The association was elite controlled and certainly not a mass movement, but its members ran meetings, petitioned the British and Australian governments, and sent deputations to present their case to the governor-general and the prime minister. Although the 1906 Sugar Industry Labour Royal Commission was a product of much wider economic and humanitarian concerns, the Islanders certainly played their part in applying political pressure to negotiate better terms for their deportation.

Why then in decades of negotiating over industrial conditions did they choose the "weapons of the weak"? The reason is clear: initially there was no other choice. It was possible to focus the intent of a Queensland-adapted Melanesian elite for a few years on such a life-and-death issue as deportation. But collective negotiation over wages and conditions of employment by individuals, or at the most small ethnic groups, living in scattered communities along a developing agricultural frontier between 1863 and 1904 was not possible. What is seen with hindsight as an aggregate of 62,000 contracts of indenture amounted on an annual basis to no more than a few hundred, or at most a few thousand, often antagonistic individuals in each district, many of whom were in the colony for only three years. Their methods of response could only be through the "weapons of the weak" until accommodation finally developed to an extent that enabled political resistance in the 1900s.

Abbreviations

BOHC   Black Oral History Collection, Section B (Pacific Islanders). Department of History and Politics, James Cook University of Northern Queensland, Townsville.

COL   Records of the Colonial Secretary's Office. Queensland State Archives, Brisbane.

CPS   Records of the Court of Petty Sessions. Queensland State Archives, Brisbane.

DC   Delamothe Collection. James Cook University of North Queensland Library, Townsville.

JUS   Records of the Justice Department. Queensland State Archives, Brisbane.

Notes

1. See, especially, Dorothy Shineberg, *They Came for Sandalwood: A study of the sandalwood trade in the South-West Pacific, 1830–1865* (Melbourne, 1967); Peter Corris, *Passage, Port and Plantation: A history of Solomon Islands labour migration, 1870–1914* (Melbourne, 1973).

2. Kay Saunders, "The Black Scourge," in Raymond Evans, Kay Saunders, and Kathryn Cronin, *Exclusion, Exploitation and Extermination: Race relations in colonial Queensland* (Sydney, 1975), 192–234, and " 'Troublesome Servants': The strategies of resistance employed by Melanesian indentured labourers on plantations in colonial Queensland," *Journal of Pacific History*, 14:3 (1979), 168–183; A. A. Graves, "Pacific Island Labour in the Queensland Sugar Industry, 1862–1906," D.Phil. dissertation, Oxford University, 1979, chs. 9–11.

3. Brij. V. Lal. "Murmurs of Dissent: Non-resistance on Fiji plantations," *Hawaiian Journal of History*, 20 (1986), 188–214.

4. Charles A. Price with Elizabeth Baker, "Origins of Pacific Island Labourers in Queensland, 1863-1904: A research note," *Journal of Pacific History*, 11:2 (1976), 106-121.

5. See Kay Saunders, *Workers in Bondage: The origins and bases of unfree labour in Queensland, 1824-1916* (Brisbane, 1982), 74, 129, 140; Graves, "Pacific Island Labour," 315-317.

6. Ronald Takaki, *Pau Hana: Plantation life and labor in Hawaii, 1835-1920* (Honolulu, 1983), 153-164; Edward D. Beechert, *Working in Hawaii: A labor history* (Honolulu, 1985), 169-176.

7. Clive Moore, " 'Me Blind Drunk': Alcohol and Melanesians in the Mackay district, Queensland, 1867-1907," in Roy MacLeod and Donald Denoon (eds.), *Health and Healing in Tropical Australia and Papua New Guinea* (Townsville, 1991), 103-122; Graves, "Pacific Island Labour," 285-291; Saunders, *Workers in Bondage*, 113-115.

8. James C. Scott, "Everyday Forms of Peasant Resistance," *Journal of Peasant Studies*, 13:2 (1986), 5-35; and *Weapons of the Weak: Everyday forms of peasant resistance* (New Haven, 1985).

9. Peter Corris, " 'White Australia' in Action: The repatriation of Pacific Islanders from Queensland," *Historical Studies*, 15:58 (1972), 237-250; and *Passage, Port and Plantation*, 127-129; Clive Moore, *Kanaka: A history of Melanesian Mackay* (Boroko/Port Moresby, 1985), 274-285.

10. Graves, "Pacific Island Labour," 308-339 (ch. 9, "Resistance").

11. Mark Finnane and Clive Moore, "Kanaka Slaves or Willing Workers? Melanesian workers and the Queensland criminal justice system in the 1980s," *Criminal Justice History*, 13 (1992), 141-160. See also Adrian Graves, "Colonialism and Indentured Labour Migration in the Western Pacific, 1840-1915," in P. C. Emmer (ed.), *Colonialism and Migration: Indentured labour before and after slavery* (Dordrecht, 1986), 249 ff.

12. A. G. Lowndes (ed.), *South Pacific Enterprise: The Colonial Sugar Refining Company Limited* (Sydney, 1956), 442-443; Ralph Shlomowitz, "The Search for Institutional Equilibrium in Queensland's Sugar Industry, 1884-1913," *Australian Economic History Review*, 19:2 (1979), 91-122; Adrian Graves, "Crisis and Change in the Queensland Sugar Industry, 1862-1906," in Bill Albert and Adrian Graves (eds.), *Crisis and Change in the International Sugar Economy, 1860-1914* (Norwich/Edinburgh, 1984), 261-279, 361-369.

13. Ralph Shlomowitz, "Markets for Indentured and Time-Expired Melanesian Labour in Queensland, 1863-1906: An economic analysis," *Journal of Pacific History*, 16:2 (1981), 70-91, and "Time-Expired Melanesian Labour in Queensland: An investigation of job turnover, 1884-1906," *Pacific Studies*, 8:2 (1985), 25-44.

14. Shlomowitz, "Markets for Indentured and Time-Expired Melanesian Labour," 73; Moore, *Kanaka*, 164-168.

15. I am indebted to Dr. Patricia Mercer for providing this revised figure. It was previously thought that only 1,500 Melanesians had remained in Queensland.

16. Refer to Moore, *Kanaka*, 200-331, and to Adrian Graves' analysis of the wider mechanisms of state and social control in his dissertation, "Pacific Island Labour," chs. 8-10.

17. Police Magistrate's Letter Book of Mackay, 2 September 1873 to 2 May 1881, CPS 10B/G1. Because the cases listed in Table 1 are largely abstracted from the local newspaper, it is not always possible to differentiate between offenses under the Masters and Servants Act and those relating to laborers disobeying a court order, although it would appear that the latter offense was the more frequent.

18. Mackay's lockups and prison were small and always overcrowded, so longer-term prisoners were accommodated in Rockhampton. Clive Moore, "The Mackay Prison, 1888–1908," *Journal of the Royal Historical Society of Queensland*, 13:9 (1989), 329–330.

19. *Mackay Mercury*, 21 October 1871; 28 November 1874; 17 October 1877; 28 October 1881; 23 October 1886; 6 April 1887; Goodall to Atherton, n.d., CPS 10B/G1.

20. *Mackay Mercury*, 3 February 1877.

21. *Mackay Mercury*, 19 January 1878.

22. *Mackay Mercury*, 3 October 1874.

23. *Mackay Mercury*, 1 September 1889; 9 November 1889; 10 June 1890; 14 November 1891; *The Boomerang*, 16 November 1889; evidence by J. V. Chataway to Northern Territory Commission, *South Australian Proceedings of Parliament*, 1895, II:148; Philip Kirwan, "Mackay Sugar Industry," unpublished manuscript, 20 July 1954, held by Mackay Cane Growers' Executive.

24. Moore, *Kanaka*, 164.

25. *Mackay Mercury*, 31 December 1879.

26. *Mackay Mercury*, 1912 Jubilee Edition, 25; 6 December 1879; 23 October 1884; 25 October 1887; 12 November 1896.

27. BOHC, 51Bb.

28. *Mackay Mercury*, 26, 28 September; 2 November 1901.

29. *Mackay Mercury*, 6 March 1897.

30. Noel Fatnowna, *Fragments of a Lost Heritage*, ed. Roger Keesing (Sydney, 1989), 111. See my obituary to Noel Fatnowna in the *Daily Mercury* (Mackay), 14 March 1991.

31. All murders of laborers by overseers occurred in the 1870s. JUS/N34, 1872/200; *Mackay Mercury*, 19 February 1877; Goodall to Immigration Agent, 5 February 1877, CPS 10B/G1 (quoted in Evans, Cronin, and Saunders, *Exclusion, Exploitation and Extermination*, 197); JUS/N59 1878/207; Goodall to Attorney General, 31 July 1878, Goodall to W. Paxton, 2, 5 August 1878, CPS 10B1/G1; *Mackay Mercury*, 31 July 1878.

32. *Queenslander*, 10 March 1877.

33. *Mackay Mercury*, 6 December 1879; M. Rankin, Assistant Inspector of Pacific Islanders to Immigration Agent, 13 August 1895, in letter 10053 of 1895, COL/A792; *Mackay Mercury*, 12 June 1897; 28 April 1900.

34. Goodall to Atherton, n.d., CPS 10B/G1; Goodall to Avery, 19 November 1879, CPS 10B/G1; Mackay Planters' and Farmers' Minute Book, 1 January 1883, DC.

35. *Mackay Mercury*, 1 September 1892.

36. H. L. Roth, *Sketches and Reminiscences from Queensland, Russia and Elsewhere* (n.p., 1919), 13 (reprinted from the *Halifax Courier*, September 1915 to May 1916); *Mackay Mercury*, 25 September 1897.

37. Fatnowna, *Fragments of a Lost Heritage*, 98.

38. BOHC, Tape 41Ba.

39. *Mackay Mercury*, 26 June 1894.

40. *Mackay Mercury*, 18 March 1876; 1, 11 November 1890; 27 August 1895; 30 August 1900; 18 September 1902.

41. *Mackay Mercury*, 30 July 1975; and information from Noel Fatnowna, Mackay. Also refer to Fatnowna, *Fragments of a Lost Heritage*, 127.

42. Douglas L. Oliver, *Oceania: The native cultures of Australia and the Pacific Islands* (Honolulu, 1989) I:748–785, particularly I:768–769; Michele Stephen (ed.), *Sorcerer and Witch in Melanesia* (Melbourne, 1987); Daniel de Coppet, "The Life-giving Death," in S. C. Humphreys and H. King (eds.), *Mortality and Immortality: The anthropology and archaeology of death* (London, 1981), 175–204.

43. Philip D. Curtin, "Epidemiology and the Slave Trade," *Political Science Quarterly*, 83 (1968), 190–216; Ralph Shlomowitz, "Mortality and Workers," in Clive Moore, Jacqueline Leckie, and Doug Munro (eds.), *Labour in the South Pacific* (Townsville, 1990), 124–127.

44. Ralph Shlomowitz, "Epidemiology and the Pacific Labor Trade," *Journal of Interdisciplinary History*, 19:4, (1989), 585–610, and "Differential Mortality of Asians and Pacific Islanders in the Pacific Labour Trade," *Journal of the Australian Population Association*, 7:2 (1990), 116–127.

45. Moore, *Kanaka*, 244.

46. Ralph Shlomowitz calculated an average annual death rate of 54.44 per 1,000 from 1870 to 1906. Statistics are not as reliable for 1863–1869: the number of Melanesian immigrants was low during the 1860s (there were 2,607 in the colony in 1871), but the death rate was probably high, as labor recruiting was just beginning and only a small proportion of the recruits would have had any previous exposure to new diseases. The death rate among Europeans in Queensland over the same period—for all ages—was 15 per 1,000. The death rate for European males of similar age to the Melanesians in the colony was around 9 or 10 per 1,000. These rates were average for Australia. Shlomowitz, "Mortality and the Pacific Labour Trade," *Journal of Pacific History*, 20:1 (1987), 50; Moore, *Kanaka*, 244, 247.

47. Many of the ambushes, attempted murders, murders, and lootings of recruiting and trading ships in the islands were probably motivated by the need to compensate for deaths in the colonies. See Moore, *Kanaka*, 346–356; Judith A. Bennett, *Wealth of the Solomons: A history of a Pacific archipelago, 1800–1978* (Honolulu, 1987), 390–396; Roger M. Keesing, "The *Young Dick* Attack: Oral and documentary history on the colonial frontier," *Ethnohistory*, 33:3 (1986), 268–292.

48. For examples, see *Mackay Mercury*, 5 March, 9 April, 14 May 1884; 27 January 1906.

49. *Mackay Mercury*, 6, 8, 11, 13, 25 October 1892; 23 July, 8 August 1895; 19, 31 March, 4 April 1896.

50. Diary of John Ewan Davidson, 27 November 1867, DC; Goodall to Crown Prosecutor, Northern District Court, Bowen, 11 November 1874,

CPS 10B/G1; *Mackay Mercury*, 14 November 1874; 11 January 1875; 10 December 1881; 21 October 1885; 29 June 1886; 28 July, 11 December 1888; 18 December 1890; 7 March, 16 June, 24 December 1891; 25 October 1892; 28 December 1893; 9 January 1894; 5 September 1895; 4 March 1905.

51. Nancy Roberts, "Sarah and Johnny: Some sociological aspects of Queensland's colonial history during the 1880s and 1890s," *Journal of the Royal Historical Society of Queensland*, 10:4 (1978–1979), 74–163.

52. *Mackay Mercury*, 22 November 1894.

53. *Mackay Mercury*, 28 July 1888; 21 Febuary, 16 July 1891; 22 September 1892; 3 October 1893; 29 December 1896.

54. Roth, *Sketches and Reminiscences*, 13.

55. *Mackay Mercury*, 23 April 1884.

56. BOHC abounds with references to sorcery and magic.

57. Moore, *Kanaka*, 271.

58. P. M. Mercer and C. R. Moore, "Melanesians in North Queensland: The retention of indigenous religious and magical practices," *Journal of Pacific History*, 11:1 (1976), 66–88; C. R. Moore (ed.), *The Forgotten People: A history of the Australian South Sea Island community* (Sydney, 1979), 49–64; Fatnowna, *Fragments of a Lost Heritage*, 27–28, 79, 104.

59. JUS/N50 1876/281.

60. *Mackay Mercury*, 17 March 1891; 12 October 1893.

61. *Mackay Mercury*, 31 October 1896; 11 November 1899.

62. *Queenslander*, 22 September 1877.

63. *Mackay Mercury*, 9 June 1891. For other cases of suicide or attempted suicide, see JUS/N78 1881/297; *Mackay Mercury*, 3 May 1882; 26 March 1891; 16, 19 September 1893; 7 June 1895; 13 July 1895.

64. Fatnowna, *Fragments of a Lost Heritage*, 125–126.

65. Eugene D. Genovese, *In Red and Black: Marxian explorations in Southern and Afro-American history* (Knoxville, University of Tennessee Press ed., 1984), 130.

66. Genovese, *In Red and Black*, 130.

67. *Mackay Mercury*, 19 August 1886.

68. Refer to Moore, *Kanaka*, 300–331 on pidgin English, Christianity, literacy, and power and authority.

4

Samoan Plantations: The Gilbertese Laborers' Experience, 1867–1896

Doug Munro and Stewart Firth

No fewer than 12,500 immigrant Pacific Islanders and another 3,800 Chinese worked as contract laborers on plantations in Samoa before World War I.[1] Of the Pacific Islanders, approximately 2,500 were from the Gilbert Islands (now part of the Republic of Kiribati), a chain of coral atolls and reef islands to the northwest of Samoa. Nearly all those Gilbertese worked for the German firm of J. C. Godeffroy & Sohn (later the Deutsche Handels- und Plantagen-Gesellschaft, or DHPG), which dominated plantation enterprise in Samoa. The remainder, numbering little over 200, mostly worked on the estates of Frank Cornwall, the proprietor of the only sizable British plantations.[2] Over 80 percent of Gilbertese labor migrants to Samoa arrived between 1867 and 1880, and during this period they formed the overwhelming mass of Godeffroys' plantation work force (see Table 1). From 1881 on, Gilbertese were intermittently recruited, the last two contingents arriving in 1894.

Gilbertese plantation workers in Samoa, as elsewhere, were not by inclination the docile and submissive laborers that their employers might have wished. The paradox is that they mostly acquiesced. The regimentation of plantation life and the pace of work were much to their distaste, while the conditions of work and the liberal use of corporal punishment affronted their sense of dignity and self-respect. But power lay so one-sidedly with the employers that compliance was the only option. Initially the Gilbertese fought back, but the dismal failure of strike action during the 1870s destroyed their resolve to carry out further acts of collective protest. In 1894, however, the framework of German company rule that governed their working conditions gave way to a system of British consular oversight. Only then did the DHPG's Gilbertese laborers show their true colors. Taking full advantage of the new dispensation, they confronted their startled masters with such defi-

Table 1. Origins of Laborers Recruited

Year	Total	Gilberts[a]	Carolines	New Hebrides	Solomons	New Britain/ New Ireland
1867	81	81	—	—	—	—
1868	115	115	—	—	—	—
1869	40	40	—	—	—	—
1870	69	69	—	—	—	—
1871	48	48	—	—	—	—
1872	15	—	15	—	—	—
1873	438	358	80	—	—	—
1874	140	140	—	—	—	—
1875	280	280	—	—	—	—
1876	101	101	—	—	—	—
1877	251	251	—	—	—	—
1878	272	189	—	83	—	—
1879	718	115	—	570	—	33
1880	535	300	—	—	226	9
1881	378	—	—	179	199	—
1882	264	8	—	153	—	103
1883	355	2	—	29	37	287
1884	245	29	—	—	—	216
mid-1885	512	124	—	187	156	45
Total	4,857	2,250 (46%)	95 (2%)	1,201 (25%)	618 (13%)	693 (14%)

SOURCE: Stuebel to Bismarck, 27 January 1886, RKA 2316:51.

[a]This column includes some Marshall Islanders who were indiscriminately lumped together with the Gilbertese.

ance and intransigence that within two years the firm repatriated its entire Gilbertese labor force, glad to be rid of it.

Plantation development in Samoa grew out of European commercial activity centered in and around the port town of Apia. The early development of Apia was based on provisioning whaling vessels, which first called in 1836, by a multiplicity of small and often short-lived mercantile ventures. Among their proprietors were J. C. Williams, the United States Commercial Agent who pioneered the Samoan coconut oil trade in the early 1840s. But with production largely in the hands of Samoans, the annual export of coconut oil was limited to 300 tons at most. The obvious alternative was plantation agriculture, because it left a greater say about production in the hands of Europeans rather than Samoans. The latter, however, were loath to sell their land in sufficient quantities and would only engage in plantation labor at high wages and on a casual basis. This was the situation confronting Godeffroys when they established an agency in Apia in 1857.[3]

Godeffroys' initial aim was to use Apia as a center for trading in the archipelagoes beyond Samoa. With the massive capital backing of the parent company, a trading network stretching from Tahiti to the Marianas was established within ten years.[4] Godeffroys also turned an eye to plantations, tempted by the high cotton prices accompanying the American Civil War. For the next few years, high prices coincided with an abundance of laborers, as hungry Samoans were forced onto plantations by a long drought, a hurricane, and a pest plague that ravaged their gardens. Godeffroys took advantage of the Samoans' temporary distress in the mid-1860s to buy land from them, which became their first proper plantation, Mulifanua. From an insignificant 12 acres in 1863, Godeffroys owned 2,500 acres by 1868.[5] But when good times returned at the end of the 1860s, few Samoans wanted to pick cotton, most European planters went out of business, and Godeffroys was left as the sole company with sufficient funds for large plantations.

This capital backing enabled the firm to overcome the two besetting obstacles to plantation development—land and labor. During the Samoan wars of 1869–1873, when the Samoans' desire for guns and ammunition led them to sell their own and each others' lands, Godeffroys added further broad acres to their estates.[6] The question of labor was less easily solved, and occasionally Samoan chiefs would pass laws that prohibited their subjects from plantation labor at less than specified wages, and even tried banning labor from other islands.[7] Forced to look elsewhere, Godeffroys recruited their first 81 Gilbertese in 1867.[8] The group had several attractions as a labor pool. Its location between Samoa and the firm's trading stations in the Marshall Islands meant that company vessels could double as recruiters when passing through the group on the return voyage to Apia. Above all, the population, though scattered, was sizable and probably numbered 16,400–18,000 in the southern islands alone during the 1860s.[9]

The Gilbert Islands lie astride the equator in the central Pacific Ocean on the southeast extremity of the Micronesian cultural area. Stretching over 425 nautical miles of ocean, these sixteen low-lying islands have limited potential for supporting human populations. Freshwater supplies are restricted to shallow subsurface lenses, the soils are marginally fertile, and so the range of plant species that

can survive is severely restricted. The ultimate limiting factor to
subsistence agriculture, however, is the sparse and unreliable rain-
fall, which tends to decrease with proximity to the equator. Life on
the northern islands is less of a struggle for bare existence, but on
all the islands a delicate balance between population and resources
existed, particularly in the barren islands to the south, which are
prone to severe droughts. The southernmost island of Tamana, for
example, has experienced at least thirteen droughts since 1863,
often of several years' duration.[10] The unforgiving nature of these
visitations has been graphically depicted by Harry Maude, a former
Resident Commissioner of the Gilbert and Ellice Islands Colony.

> The Gilbertese drought can be a fearsome thing. I was in one on
> Nonouti during, I think, 1938 and it seemed as if the lagoon eco-
> system was determined to combine with the terrestrial in ousting
> man from the scene; for at the height of the drought, when the
> flora was dead or dying, the prolific fish population deserted the
> lagoon. Over-fished, you will say, but the Gilbertese felt rather
> that the food of the fish was no longer there. Historical evidence
> shows us that even in normal times many of the human inhabit-
> ants of such islands as Nonouti and Tabiteuea were habitually
> hungry, despite the efficient use of the ecosystem and often dras-
> tic population controls, the resources of the island being insuffi-
> cient to maintain the density of population.[11]

In traditional times, the populations were kept in tune with
resources by abortion, infanticide, and the more indirect method of
interisland warfare. But there was no answer to a severe drought,
which could lead to hundreds starving to death on one island alone,
until labor-recruiting vessels came to take them away.

Labor recruiting in the Gilberts began in the 1820s with the arrival
of whaling vessels in search of crewmen as well as whales. "The
numbers involved were not large," writes Macdonald, "perhaps no
more than a hundred or two over a thirty-year period, but they
were the first participants in a labour migration that was to
become, and remain, a dominant feature of life" in the Gilberts.[12]
During the remainder of the nineteenth century, at least another
9,400 Gilbertese were engaged in overseas labor migration to places
as far apart as Réunion in the Indian Ocean and Guatemala in Central America.[13] The earlier recruiting episodes were happenstance
events—twenty-two men left in 1847 to work on the northern New
South Wales properties of the entrepreneur and adventurer Benjamin Boyd; another fifty-one departed ten years later for Réunion;
and a further 312 were kidnapped by Peruvian slavers in 1863.[14]

Systematic labor recruiting began four years later, in 1867, with the arrival of Godeffroys' vessels, but the firm did not have the Gilberts to themselves. Within a year, recruiting vessels for plantations in Tahiti and Fiji, also beneficiaries of the American Civil War, arrived on the scene. The competition for labor hurt Godeffroys most—only a quarter of the 1,400 Gilbertese recruited to 1870 went to Samoa (see Table 2). Moreover, Godeffroys depended almost entirely on a Gilbertese plantation force, whereas Gilbertese supplemented Chinese coolie laborers in Tahiti and New Hebrideans in Fiji.[15] The level of competition also encouraged fraudulent recruiting practices and without doubt the majority of Gilbertese recruited during the late 1860s were taken by force or deception. Those bound for Samoa were carried in Godeffroy vessels that had no facilities for passengers: either they shared the hold with smelly bags of copra or endured the discomforts of an unsheltered deck.[16]

The prevalence of abuses resulted in a significant decrease in the level of recruiting by the early 1870s, and the flow of recruits was restored only by a severe drought afflicting the southern islands. "It seemed as though a fire had scorched and destroyed everything," wrote the Reverend S. H. Davies of the London Missionary Society as he approached Arorae in 1873. "We realized that the news received in Samoa concerning the drought had not been exaggerated. . . . The natives are literally starving. . . . I have never before in any place seen so many attenuated sickly-looking people."[17] Seeing overseas plantation work as the only escape from

Table 2. Labor Migration from the Gilbert Islands, 1867–1875

Year	Samoa	Fiji	Tahiti
1867	81	399	0
1868	115	135	0
1869	40	136	0
1870	69	224	192
1871	48	54	0
1872	69	427	92
1873	358	34	0
1874	140	85	0
1875	280	66	0
Total	1,200	1,560	284

SOURCES: Stuebel to Bismarck, 27 January 1886, RKA 2316:51; Jeff Siegel, "Origins of Pacific Islands Labourers in Fiji," *Journal of Pacific History*, 20:1 (1985), 46; C. W. Newbury, "Aspects of French Policy in the Pacific, 1853–1906," *Pacific Historical Review*, 27:1 (1958), 49.

their "land of heat and famine," Gilbert Islanders now recruited in
their hundreds rather than face starvation. It was a timely occurrence for Godeffroys, which needed extra laborers following the
land purchases of the early 1870s. Between 1873 and 1880, Godeffroys recruited over 1,700 Gilbertese compared with 353 from
1867–1872 (see Table 1). The recruits, moreover, went typically in
family groups rather than as individuals, which was to Godeffroys'
advantage, because the women and children were better than the
men at picking and weeding yet worked for lower rates of pay.[18]
And whereas planters in Fiji found Gilbertese laborers "sulky and
revengeful, and quite unfit for the work required of them,"[19]
Godeffroys in Samoa overcame such difficulties by imposing a stern
regime of discipline upon its overwhelmingly Gilbertese work
force.

Godeffroy und Sohn were indeed harsh taskmasters. Pressed for
labor, thanks to competition by recruiters from Fiji and Tahiti,
Godeffroys were prone to overtasking their work force, enforcing
their demands with a strict work discipline. The firm could do this
because it was a law unto itself in Samoa. The Godeffroy Samoan
manager doubled as the local consul for the North German Confe-
deration, as it then was, so there were no restrictions on the way in
which the firm obtained and used labor. He was "Licenser of the
ships that import the men, owner of the Ships; employer; and the
only person to whom the men can appeal for redress—Consul,
Judge and jury all in one—what can be expected?"[20]
    Although company records do not survive to tell the tale, the
abuses of the Godeffroy plantation system were observed by J. C.
Williams, the local British consul. In 1870, he reported that the
Gilbertese had been brought to Samoa under false pretenses and
once there, were being insufficiently paid and "beaten and
flogged."[21] The Gilbertese were not indifferent to their plight nor
were they submissive by inclination. Given their resentment of ver-
bal and physical abuse, their tendency to retaliate when provoked,
and their concentration in the one spot, it is at first sight remarkable that there was no rioting or even a general uprising.
    This never came about for two basic reasons. First, the severity
and pervasiveness of plantation discipline discouraged overt expressions of disobedience and resistance. Deserters could expect a
flogging, perhaps imprisonment and a deduction from their wages
as well, and possibly even an extension of their period of indenture.
In late 1869, for example, seven Gilbertese absconded in a company
whaleboat only to be caught and "flogged very unmercifully. They

were tied flat on their stomachs by the neck & feet on a board
raised about 2 to 3 feet above the ground the arms being round the
board underneath with shackles or Irons on & then flogged. . . .
They are sick of the treatment Work & pay they receive—pay at the
rate of one dollar per Month in Tobacco."[22] Theft was also dealt
with severely. One group of offenders were placed in the plantation
lockup with their hands secured around a post in such a fashion as
to prevent their lying down on the rough and uneven floor, and left
to rot. When they were released ten weeks later, they were too
weak to be flogged as intended, "& yet Mr Weber [the Godeffroy
manager] has told me that these people are happy," remarked the
incredulous Williams.[23]

Reinforcing the strong hand of plantation discipline was the realization by the Gilbertese that absconding was futile. To escape also
meant deserting one's kin, since the Gilbertese frequently came in
family groups. Besides, there was nowhere really to go. Upolu was
a small island, and the Samoans, who barely reconciled themselves
to the presence of plantation laborers, would never tolerate escap-
ees raiding their gardens. They were more likely to capture absconders and return them to Godeffroys for the reward money. For such
reasons individual absconders had no future as renegades, and
maroonage never developed.[24] Nor was it easy to work a passage
home on a ship, since most of the vessels bound for the Gilberts
were from the Godeffroy fleet.

Second, Godeffroys maintained discipline by virtue of the nature
of Gilbertese social organization on the plantations. Far from being
a cohesive group, the Gilbertese were socially fragmented with
their overwhelming loyalty lying with their kin, or perhaps with
friends from the same village if they came as individuals. In the
southern Gilberts, government was by the rule of the old men (or
unimane). But even when traditional leaders were present on the
plantations, they could not, in the circumstances of social fragmen-
tation, provide a focus to mobilize a strategy of resistance against
the employers and overseers. Given the overarching nature of plantation authority, it was difficult to even contemplate using what
James C. Scott terms "the weapons of the weak," those surreptitious day-to-day forms of resistance such as malingering and foot
dragging that often go undetected because of their overt nature.[25]

Nevertheless, strike action did take place, even if riots did not, in
1871 and again in 1875. Contributing to the latter event was the
emergence of a new type of leadership represented by one acquainted with European ways. The suppressed plantation work
force, in other words, spawned a leader more worldly-wise than his

fellows, called Holy Joe, whose name reflected a missionary train-
ing.[26] Under his prompting, the entire work force at the Mulifanua
plantation, 210 people in all and overwhelmingly Gilbertese, went
on strike against their pay and conditions, trekked the 30 miles to
Mulinu'u, near Apia, where the Samoans had formed a central government, and threw themselves on its mercy. The head of that government took the unprecedented step of appointing a special commission to investigate, which included his premier (the American
adventurer Colonel Albert B. Steinberger), three Samoan chiefs,
and representatives of the German, British, and United States consuls.[27]

The laborers had specific complaints—they were subject to corporal punishment; they were often compelled to work on Sundays
and after-hours; they were ill-treated when sick, badly housed, and
insufficiently fed. Other irregularities surfaced during the investigation—some laborers had been tricked into bonded servitude on
the pretense that they were being given passage to another island
in the Gilberts; others did not know the nature of their contract, it
being in the German language and not explained to them at the
time; some were assured they would receive $5 per month in
wages, but this was reduced to $2 when the Godeffroy manage-
ment heard what laborers in Fiji received; their wages had been
withheld in recent months (probably the factor triggering the
strike); they were not fed before the morning's work; and medical
treatment on the plantation was woefully inadequate. The most
serious single complaint concerned two sick women, guilty of
repeated absconding, who had died within a fortnight of being
flogged. In the opinion of James Lyle Young, the British consul's
representative, "The people can have no shadow of protection
under such a system unless a strong arm is interposed on their
behalf, and it is high time it was done here."[28]

The findings of the commission were at variance with the evidence. The commission proceeded from the premise that the laborers had engaged in "wilful misrepresentation" against their employer and dismissed their complaints about housing, clothing, and
food, though the state of the "so-called hospital" drew criticism.
The evidence presented about food is unambiguous, but the commission misrepresented the situation, saying that the laborers got
enough to eat and left it at that. In fact, only one laborer cross-
examined had complained about insufficient food: the gist of the
laborers' complaints was the unvarying diet of corn, and the cook
himself testified that he never prepared fish or peas, and beans
only on Sunday. The commission was not happy that laborers'

wages were being withheld and recommended that half their earnings be kept until the expiry of their contracts and the other half paid at the end of each month. The question of numbers of Gilbertese being induced to come to Samoa on false pretenses of various sorts was passed over in silence. The commissioners even concluded that the two women had not died from the effects of their beating, saying that the instrument of punishment, a knotted rope, could not have caused serious bodily harm. On this matter they overlooked significant points: the women were ill beforehand, one of them was kicked by the plantation manager while lying on the ground too weak to move, and neither received medical treatment afterward. As one of the laborers testified, "They were ill and the beating made them worse."

Rather than seeking to isolate the fundamental causes of unrest and dissatisfaction, the commission took the line of least resistance and apportioned blame to certain individuals. The laborers, the commission concluded, were "apt to be corrupted by people who have no correct appreciation of the duties of labourers or the responsibilities of planters." The strike, said the commission, resulted from the baleful influence of two laborers, Holy Joe and Varaveta, a Marshall Islander. The commission painted a picture of previously contented laborers being incited by a couple of subversive men in their midst when, more likely, Holy Joe and Varaveta simply brought massive dissatisfactions into the open. Moreover, the commission's opinions of Holy Joe and Varaveta sat uneasily with those concerning Paul Krause, the plantation manager, whose administration it considered to be "marked by exceeding hardship —often cruelty—and an entire ineptitude or lack of disposition to appreciate the ordinary attributes of humanity." It was not difficult to sustain a case against Krause. In addition to his obvious cruelty, he shocked some of the commission members the evening he struck his wife, "a white woman," in a fit of temper and further disgusted them when it emerged he was having an affair with Varaveta's wife, who was not a white woman. "There has been cruelty practised on the laborers," concluded the commission, "but not in food or clothing or even in the disposition of the owners—but clearly in the maladministration of the superintendent." The very employment of Krause, who had a long-standing reputation for brutality,[29] casts doubt on the commissioners' confidence that Godeffroys wished to treat their plantation laborers with "humanity."

In effect, Godeffroys emerged almost unscathed; the failings and abuses of their system had instead been attributed to a few individuals. The outcome was predictable, and it indicates the company's ability to manipulate a given situation. To an extent the commission itself was stacked, since the German consul's (i.e., the company manager's) representative was a member of the Godeffroy family. Nor was Steinberger, as chairman of the commission, a disinterested party. He had developed close ties with the firm despite its interests being in conflict with those of the Samoans. The previous year, indeed, he had entered into a secret commercial agreement with the firm, guaranteeing title to its dubiously acquired landholdings and ensuring that Godeffroys alone should handle the Samoans' copra tax. Throughout the hearings, therefore, he exercised a subtle influence "to shelter Godeffroys a little" while protesting all along "that justice shall be done."[30] Even the presence of three Samoan delegates was of no help to the Gilbertese. Far from being sympathetic toward fellow Pacific Islanders, the Christianized Samoans regarded the pagan Gilbertese as contemptible heathens to be treated accordingly. Thus by maneuver, political strength, and a measure of luck, Godeffroys survived an inquiry— the only occasion the firm submitted to outside scrutiny—when the merits of the case lay in the other direction. Even so, the local Godeffroy manager, apparently angered that the company had not emerged with a completely clean bill of health, wrote a threatening letter to the commission. It was soon withdrawn, but the fact that it was written indicates the extent to which the firm had become accustomed to freedom of action.[31]

The failure of the 1875 strike impressed upon the majority of Godeffroys' Gilbertese labor force the futility of direct action. The few remaining intransigents were dealt with summarily by Steinberger, in his capacity as premier, who jailed absconders and agitators.[32] With that, the Godeffroy system of plantation employment continued untouched by official regulation until the early 1880s. Between 1882 and 1884, a series of ordinances were drawn up by successive German consuls (by then a post for a career diplomat), but they dealt with the conveyance of laborers rather than their employment.[33] In reality the new legislation was little more than a tactical show of legality: rather than controlling the conduct of the business, it was designed instead to safeguard the company's recruiting rights by giving the impression that German recruiters, as well as British, had to abide by detailed laws. Indeed, the consul discussed the draft ordinance of 1884 with the local company manager, met his wishes on all but a few points, and was careful to do nothing effective. The company continued to treat its laborers in Samoa according to rules virtually of its own making. The firm's interests were regarded as being synonymous with German interests in Samoa, so it could bank on the active support of the local German Consulate on all important matters.

Gilbertese laborers in Samoa fared little better under British employment. This essentially meant working on the plantations of Frank Cornwall, who was formerly a printer of the London Missionary Society but was now deeply involved in land speculation, claiming ownership of 300,000 acres at Magia and Lata. In 1877, Cornwall chartered the *Flirt* for a recruiting voyage to the still-drought-stricken Gilberts, returning with 135 recruits whose physical condition was described as "most wretched." Soon afterward, Cornwall shipped over half the Gilbertese to Lata, on the isolated southwest coast of Savai'i, in the 15-ton cutter *Bertha* and left them with an overseer. Despite their famished condition, the remaining Gilbertese followed a few weeks later, again in the grossly overcrowded *Bertha*, but already there were serious problems at Lata. Within five months, 22 of the 135 laborers had died.[34]

Lata was not the place to send debilitated laborers, especially when no preparations had been made for their arrival. The first party, in fact, had to erect a makeshift dwelling for themselves the day of their arrival. The laborers then had to build their own accommodation in their free time. Their working day was hard, consisting of clearing and planting the rocky and poorly drained terrain. At the same time, their rations were unsuitable and insufficient, and the water supply was inadequate. These inherently unsuitable conditions could have been softened had Harry J. Moors been a less driving and callous overseer, but he made no attempt to cultivate yams, taro, and bananas at Lata and was liberal in his use of the lash. The Gilbertese expressed their distaste for their conditions by absconding, which was an act of desperation given the hazards. One runaway was found by Samoans as he lay dying in a taro patch, having been there for four days. Another runaway was equally unfortunate. Her child died in the bush, and she was returned to the plantation lashed to a pole like a pig, and a severe public flogging was administered by Moors. Even if a runaway was in reasonable health, the chances of successfully escaping were small because Samoans would return them to Lata for a $5 reward.[35]

These happenings at isolated Lata might have continued unchecked but for one of Cornwall's many detractors reporting them to the High Commissioner for the Western Pacific in Fiji.[36] The High Commission officials were the more aghast because Cornwall had been the acting British consul in Samoa at the time of the Gilbertese's arrival in the *Flirt.* Nevertheless, there was little that British officialdom could do given the limitations of their jurisdiction. Both the revised Pacific Islanders' Protection Act of 1875 and the Order-in-Council of 1877, which created the Western Pacific High Commission as the arm of British extraterritorial jurisdiction in the region, dealt inadequately with the labor trade. They were directed only at recruiting and left untouched the questions of conveyance, employment, and repatriation, so the High Commission could not proceed against Cornwall for the overcrowding of the *Bertha* or for labor conditions at Lata. The High Commissioner used these abuses to demonstrate the need to tighten up the legislation bearing on the British labor trade,[37] and an inquiry was held at Lata by Alfred Maudsley, the acting British consul, that substantiated the essential accuracy of the allegations. Important to the inquiry was the ability of the laborers to put their case to Maudsley without Moors, who could speak the Gilbertese language, having to be the interpreter. Many of the laborers had previously worked in Fiji, as had Maudsley, so the two parties conversed in Fijian and shut Moors out of the proceedings being conducted within his hearing.[38] The active displeasure of Her Majesty's Government, however, could only extend as far as refusing to issue Cornwall further licenses to recruit labor unless Moors was dismissed. This duly happened and Cornwall promised in future "to correct any conduct on the part of those under him that may be prejudicial to their rights of his labourers."[39]

Six months later, however, fresh complaints were laid against Cornwall when several of his Gilbertese laborers from Magia plantation—about 20 miles from Apia—presented themselves at the British Consulate saying that they were badly fed and housed. The consul, William Swanston, sent his police constable to investigate, and he confirmed the laborers' complaints, adding that what passed as the plantation hospital was a disgrace. On the credit side, the laborers did not claim to be physically abused. Cornwall was told to remedy the situation, and the only deficiency uncovered by a spot check the following month was a lack of sleeping mats.[40] The matter was settled for the meantime, but only because Magia's proximity to Apia allowed ready monitoring and because Cornwall could be browbeaten into compliance. But Swanston was also aware that the legal situation with regard to laborers on British plantations in Samoa had still to be finalized and what the Gilbertese really needed was an outside protector. As he explained to the acting High Commissioner:

In reference to your remarks that the High Commissioner's Court can, on complaint of the labourers on a plantation deal with the question at issue I submit who is to plead the labourers case; who is to lodge the complaint[?] There is no interpreter here. The labourers cannot lodge a complaint in English and although from fortuitous circumstances, I can listen to what they say [he spoke Fijian as did many of the Gilbertese] I cannot act as counsel, interpreter and judge. Again in reference to the enforcing of a bond against an employer, who is to act on behalf of these people *for they certainly cannot act for themselves* [emphasis added].[41]

Subsequent events would bear out Swanston's words. By 1881 the management of labor at Magia had reverted back to its formerly chaotic state, and Cornwall's financial difficulties were so acute that he left suddenly for Fiji. The British Consulate then assumed the active role of the Gilbertese laborers' protector by setting up a High Commissioner's court and proceeding against Cornwall on behalf of the Gilbertese for back wages and their repatriation costs. The court duly found against Cornwall, who was not represented and awarded the plaintiffs the sum of £900 to be realized from the sale of Magia's cotton crop.[42] A decade later, Gilbertese laborers on German plantations would similarly benefit from having a protector in the local British consul.

Meanwhile, Godeffroys' Samoan branch was going through a period of flux. In 1878, the Godeffroy family in Hamburg responded to impending insolvency by reconstituting its profitable Samoan branch into a new company, the Deutsche Handels- und Plantagen-Gesellschaft, which took over all Godeffroys' possessions in the Western Pacific.[43] It survived the parent company's bankruptcy as the principal German economic interest in the Pacific and embarked on further extension of its plantations in Samoa. Between 1879 and 1882, the DHPG expanded its existing plantations and laid down new ones with the result that the aggregate area under cultivation had increased by nearly 40 percent in those three years, from 4,337 to 6,020 acres.[44] The plantations were still at the cotton stage of cotton-and-copra cultivation because most of the palms were still too young to bear. Thus the question of labor was pressing, because cotton is the more labor-intensive, requiring three times as many laborers than mature coconut palms did per acre and six times as many for new planting.[45] The DHPG managed to keep pace with its labor requirements until 1882 but only by venturing into the Melanesian Islands for recruits (see Table 1).

It was as well for the firm that it began to seek alternative supplies of labor because during the early 1880s the Gilberts all but
dried up as a recruiting area for the DHPG. The company attributed
this to a succession of good seasons and fewer internal wars on
some islands, which removed the "conditions favouring recruiting"; to fiercer competition from Hawaiian, and to a less extent
Tahitian, recruiting vessels; and to a reduction in the firm's trading
activities in the group due to its financial difficulties.[46] It was also
true, however, that the DHPG was so unpopular that the Gilbertese
were wary about working for any planter in Samoa, German or
otherwise.[47]

    This would have mattered less had the DHPG been able to obtain
labor from its new recruiting grounds in Melanesia; but here the
firm was faced with competition from Queensland and Fiji vessels
(see Table 3).[48] Consequently, the DHPG was confronted with an
acute labor shortage at the very time when it needed more laborers
than ever. In 1883, its labor force had dwindled dangerously to
1,177, half of whom were due for repatriation by the end of the
year, and an estimated 820 more recruits were needed.[49] In short,
the 1880s was a time of uncertainty for the DHPG's Samoan plantations: labor was plentiful some years but mostly scarce and always
expensive to recruit. Given the dramatically rising costs of obtaining Melanesian labor, the firm hankered after Gilbertese because
they offered significant economies. Their mortality rate on the
plantations was significantly lower than their Melanesian counterparts.[50] No "beach payments" to chiefs and relatives were needed,

Table 3. Number of Pacific Islanders Recruited for Samoa,
Queensland, and Fiji, 1880–1885

Year	Samoa	Queensland	Fiji
1880	535	1,995	2,361
1881	378	2,643	1,227
1882	264	3,139	2,093
1883	355	5,273	1,550
1884	245	3,289	1,266
1885	512	1,916	295
Total	2,289	18,255	8,892

SOURCES: S. G. Firth, "German Employment and Recruitment of Labourers in the
Western Pacific before the First World War," D.Phil. dissertation, Oxford University, 1973, 33; O. W. Parnaby, *Britain and the Labor Trade in the Southwest Pacific*
(Durham, 1964), 202; Jeff Siegel, "Origins of Pacific Islands Labourers in Fiji," *Journal of Pacific History*, 20:1 (1985), 46.

the Gilbertese came for longer periods of indenture than Melanesians, and the family nature of Gilbertese recruiting reduced the time and costs of the voyages, as well as enabling the company to benefit from the cheap labor of the Gilbertese women and youths.[51] But the good seasons and other circumstances of the early 1880s precluded a sustained recruiting program in the Gilberts, despite a windfall in 1885, when 124 Gilbertese were recruited (see Table 1).

A fresh train of developments occurred in the 1890s, culminating in the DHPG resuming its employment of Gilbertese. On this occasion, however, the Gilbertese were in a position to mount a campaign of resistance to the plantation regime. Briefly, the recurrence of severe drought conditions in the early 1890s resulted in recruiting vessels once again converging on the Gilberts. What success the DHPG had was outstripped by vessels from Fiji and Guatemala, which enlisted almost 1,400 Gilbertese for these destinations.[52] At the same time, the company was experiencing renewed difficulties recruiting in Melanesia.[53] These unwelcome developments coincided with rumors that some Gilbertese chiefs were about to enter into treaties with the United States government, which the DHPG feared would result in the closure of the Gilberts to its recruiting vessels.

Since the Gilberts fell within the British sphere of influence as defined by the Anglo-German Agreement of 1886, Germany asked Britain in 1891 to declare a protectorate over the group to forestall possible American designs. The request was made at the prompting of the DHPG and on the tacit understanding that Germans would be allowed to recruit in the Gilberts as before.[54] Lord Salisbury, the British Prime Minister, was cordial to the idea because he needed German diplomatic support to shore up Britain's position in Egypt against France. The Gilbert Islands Protectorate was duly declared in May 1892, and Sir John Thurston, the High Commissioner for the Western Pacific, was asked to advise on the conditions under which the recruiting of Gilbertese might continue.

In essence, Thurston recommended that Gilbertese laborers, whether or not employed by British subjects, be paid, fed, and treated according to detailed standards but, more important, that their employment be under the control, in the case of Samoa, of the British Consulate in Apia. Laborers were free to complain to the British consul and to be represented by him in any legal proceeding involving their employer; and the consul in turn was free to inspect the places of employment to ensure that the agreement was being observed. The laborers would not be amenable to the German Consular Court or to company discipline as before: their only punishment, apart from cancelling their contracts and repatriating them, was deductions from wages.[55]

Thurston's proposals were accepted and presented to the Germans. Both the DHPG and Biermann, the German consul in Samoa, were fearful of employing Gilbertese under such strict British supervision. Keeping discipline was essential in a work force of "completely savage laborers at the lowest level of human development," Biermann explained, especially as long experience had shown the need for punishment other than fines. The British consul, he predicted, would be concerned only for the laborers' interests, whereas he had to consider the DHPG's interests as well.[56] But the Germans had no valid grounds for complaint, since all prospective employers of Gilbertese, British included, were bound by the same conditions (even if Thurston had framed these conditions with the DHPG in mind). Nor were they in any position to argue. The total number of laborers had dropped from over 1,000 in 1891 to below 700 in 1893, desertions from the company's plantations were of serious proportions, and with some 500 laborers waiting to be repatriated by the end of 1894, the company was facing yet another labor crisis (see Table 4).[57] In January 1894, the DHPG finally repatriated its Gilbertese laborers in the schooner *Aele*, and now needed replacements from that quarter.[58] The following month the firm applied for permission to recruit 120 Gilbertese and was granted a license to engage 49. With the Gilberts still in the grip of a drought, these laborers were readily forthcoming.[59] In April a license for another 49 Gilbertese was granted, and again the *Aele* sailed northwest. When entering Apia harbor on the return voyage, the schooner struck a reef and was stranded until high tide.[60]

It was a dire portent. Now that they had a protector in the British consul, T. B. Cusack-Smith, the Gilbertese took full advantage of the new dispensation of British consular oversight. The first contingent was sent to the Vailele plantation, close by Apia, and within two months Cusack-Smith was investigating complaints that they were on reduced rations. A month later it was clear that disciplinary problems were arising, and Cusack-Smith was called to arbitrate between the management and two Gilbertese charged with desertion and assault. By the end of 1894, the situation at Vailele was deplorable from the Germans' standpoint. The plantation manager stated that three-quarters of his Gilbertese laborers were "useless." They held strikes, often left the plantation, sold their rice rations to neighboring Samoans, and stole coconuts from the trees. The manager refused to have any more Gilbertese, finding that

Table 4. Contract Laborers Employed by the DHPG in Samoa,
1881–1898

Year	Number Employed[a]	Deaths	Runaways
1881	1,426	73	—
1882	1,476	64	—
1883	1,177	85	—
1884	1,249	105	—
1885	1,057	74	—
1886	1,088	123	—
1887	965	194	—
1888	—	—	—
1889	1,074	91	183
1890	1,012	92	164
1891	1,011	172	127
1892	945	130	—
1893	672	66	85
1894	680	58	72
1895	633	50	26
1896	684	12	29
1897	455	30	18
1898	425	31	17

SOURCES: DHPG Directors' Reports, 1881–1898, Senatsaken, Staatsarchiv Hamburg;
Donald Ross, Health Report for Plantations for 1883, 1 January 1884 (copy), encl. in
DHPG to Krauel, 26 January 1885, RKA 2316:25.
[a]There are various census dates. The figures for 1881–1882 and 1884–1887 apply to
1 January; the figures for 1883 apply to 1 September; and from 1889 the numbers
apply to 31 December, except for 1892. Since no data are available for that year, the
figure for 1 January 1893 has been entered instead.

unless he could "terrorize them" he could not make them work.
In fairness to the management (and in contradiction to Biermann's
forecast), Cusack-Smith sanctioned the withdrawal of their biscuit
ration for the next ninety days.[61]

An even worse state of affairs prevailed at Mulifanua, where the
second contingent of Gilbertese had been posted. On Cusack-
Smith's first visit to the plantation, in September 1894, he received
endless complaints all stemming from the failure of the plantation
manager and his overseer to acquaint themselves with the conditions of employment. Throughout, Cusack-Smith interfered as little
as possible, either appealing to the German consul to intercede or
arranging a further visit in the hope that "a better state of things
would exist," but to no avail.[62] Within a month, two Gilbertese
from Mulifanua arrived at the British Consulate saying they had

been assaulted by the same overseer. When Cusack-Smith threatened to lay charges in the supreme court, the DHPG abruptly ended all pretense of cooperation.[63]

There were several dimensions to the problem. The Gilbertese, now that they had a protector in the British consul, were no longer prepared to be ill-treated, but instead fought against the plantation system with panache. The DHPG, on the other hand, was vastly dissatisfied with its Gilbertese laborers, finding "that without the whip, and imprisonment, and irons, they cannot get the same amount of work out of [them] as before."[64] Clearly, the DHPG was unable to adjust to the restraints placed on its traditional methods of disciplining laborers. It is equally apparent that the firm was genuinely surprised at the abrupt change that came over the Gilbertese. It is a nice irony that only the year before, the DHPG described Gilbertese as "extraordinarily useful & strong workers, which we have always preferred to use on our plantations."[65] The firm seemed to forget that in the past their Gilbertese plantation workers had served notice more than once that they were prepared to engage in strikes and other forms of resistance when opportunity offered.

Throughout 1895, the Gilbertese continued to make repeated use of their right to protest about working conditions. In addition to the perennial grievances about their hours of labor being exceeded and their being beaten and whipped by overseers, they were made to work in bad weather, their weekly meat ration was provided only when an ox was sick enough to be slaughtered, and their unmarried women were accosted.[66] Cusack-Smith was on sick leave, however, and their new protector was less assiduous. The acting British consul, C. M. Woodford, was concerned above all with his instructions to avoid friction with his German colleagues. Different in temperament from Cusack-Smith and a proponent of plantation development, he aspired to become the first Resident Commissioner of the British Solomon Islands Protectorate and was not about to prejudice this ambition for the sake of the Gilbertese.[67] Besides, they gave a man of his outlook little cause for solicitous concern. He visited both Vailele and Mulifanua to investigate complaints, most of which he dismissed as being "unfounded," though he had an overseer cautioned for using a whip, and he forced the company to allow the Gilbertese Saturday afternoons off as stipulated in their contracts.[68] Woodford's overall view of the Gilbertese was unfavorable in the extreme. He regarded them as lazy, sulky, and temperamental, and was offended every time he passed Vailele plantation to see a Gilbertese road-making gang sitting down doing nothing. A worse plantation worker could not be found in the Pacific, Woodford concluded: "In fact if I owned a plantation and had the opportunity of engaging a gang of these people free of all expense of introduction I should decline to receive them."[69]

By the end of 1895, the DHPG could see little point in persisting with its Gilbertese work force. The plantation manager at Vailele quite despaired of his Gilbertese laborers ("they may work or not as they like," Woodford reported, "and he will be glad to get rid of them"); at Mulifanua the Gilbertese went back to their houses after the morning roll call or sneaked away from work to spend the day roaming about the plantation.[70] Indeed, the DHPG considered the majority of their Gilbertese laborers a miserable lot who worked with "such demonstrative laziness and lack of enthusiasm" that the whole discipline of the plantation regime was at risk.[71]

Nevertheless, the DHPG made a final effort to solve the problem, but the German government was reluctant to ask Britain to put the Gilbertese under the direct control of the German Consulate in Samoa as the DHPG wanted.[72] When Cusack-Smith returned to Samoa in the latter part of 1895, however, the situation became urgent; Cusack-Smith was more zealous in protecting the Gilbertese than Woodford, and the DHPG was afraid of a further decay of discipline, which could spread to the rest of the work force. From the company's viewpoint, it was vital to gain control over the Gilbertese before recruiting any more of them, or to cease recruiting in the Gilberts altogether. Under the urging of the DHPG's managing director in Hamburg, who was personally acquainted with the Colonial Director in the German Foreign Office, Germany asked Britain to withdraw the condition that the Gilbertese should be placed under British consular control, pleading that the authority of the company was endangered by the existence of two categories of laborers answerable to different consuls.[73] But in Samoa, Cusack-Smith had already rejected a similar request from his German counterpart, and recommended against further recruiting of Gilbertese.[74] Events were bearing out Thurston's earlier statement that "if foreigners only recruit under our regulations and we make it a condition that such laborers at the place of their employment shall be placed under the protection of our Consuls . . . recruiting will cease."[75]

The DHPG's tribulations were not quite over. In January 1896, the entire Gilbertese work force of 49 at Mulifanua trekked the 45 kilometers to the British consulate in Apia to protest against an overseer who had threatened them with a revolver (he said it was in self-defense) and imprisoned one of their countrymen. They were in no mood to be humored and remained on strike for eight days in defiance of Cusack-Smith's orders that they return to Mulifanua forthwith. That was the last straw for the DHPG, who then decided to repatriate their entire Gilbertese work force, glad to be rid of them. Cusack-Smith placed no obstacles in their way, even though the Gilbertese were leaving a year early. They departed on the Swedish barque *Stavanger* in April 1896; and when Cusack-Smith finished his inspection of the vessel, the Gilbertese on board gave him a hearty cheer, which he thought "possibly may be taken to be their recognition that I and Mr. Woodford have done justice by them since they arrived in Samoa."[76] The only other incident involved a handful of the Gilbertese being induced by Harry Moors, by then the leading American merchant in Samoa, to remain behind and work for him instead. The British were annoyed, but the Gilbertese were simply responding to a less restrictive situation and exercising an element of real choice in their decisions.

The Gilbertese plantation workers in Samoa, about 2,500 in all, responded variously to their conditions of employment. The response primarily depended on the level and severity of discipline to which they were subjected. On German plantations from the 1860s through to the 1880s, they generally acquiesced to a reign of near-terror imposed by the Godeffroy/DHPG establishment with the endorsement of an obliging German Consulate. Although this quiescence was occasionally punctuated by short-lived outbursts of resistance, such as the 1875 strike at Mulifanua, company discipline was otherwise pervasive, and the Gilbertese were cowed into submission. The intransigence of contemporary Gilbertese workers in other parts of the Pacific expressed itself on the Godeffroy/DHPG plantations only after 1894, when they were brought within the ambit of a protector in the British consul. Once that happened, the Gilbertese could respond to mistreatment on more even terms. The same applies to the Gilbertese on Cornwall's estates. In their debilitated condition at Lata, and physically isolated from sources of possible assistance, they were again bullied into submission. At Magia, in contrast, they were within reach of Apia and could take their grievances directly to the British Consulate.

A second notable feature of the Gilbertese plantation workers' coping strategies was their ready identification of persons in higher authority to whom they could seek redress. They soon discovered that it was useless to complain to the Godeffroy manager in his then dual capacity of manager/consul, so they appealed to the Samoan government at Mulinu'u instead. Although the tactic backfired, their growing political awareness is evident. This awareness found its fullest expression in the mid-1890s when the British consul became an element in their relationship with their employer. Biermann, the German consul, had feared that his British counterpart would be excessive in representing the Gilbertese's interests, but he was wrong. Instead, the Gilbertese took the initiative: they manipulated the fact of British consular protection in a manner that went well beyond what the British had originally envisaged. They appreciated the essential weakness of their position in the power relationship that defined their conditions of labor on the DHPG's plantations, and they set out to redress the balance using the only weapon at their disposal. As Swanston had remarked fifteen years earlier, "Who is to act on behalf of these people for they certainly cannot act for themselves?" It was only through fortuitous circumstances that the Gilbertese gained a protector, but once this happened, the grudging acquiescence of the past gave way to defiant resistance.

Abbreviations

AA Records	Auswärtiges Amt Records. Potsdam, Deutsches Zentralarchiv.
BCS*	Records of the British Consul, Samoa. Wellington, New Zealand National Archives.
CO 83*	Records of the Colonial Office, Series 83, Fiji. London, Public Record Office.
CO 225*	Records of the Colonial Office, Series 225, Western Pacific. London, Public Record Office.
DHPG Papers	Firmen Archiv, Deutsche Handels-und Plantagen-Gesellschaft der Südsee Inseln zu Hamburg. Stattsarchiv Hamburg.
PMB*	Pacific Manuscripts Bureau, Canberra—manuscript series.
RKA	Records of the Reichkolonialamt. Potsdam, Zentrals Staatsarchiv.
WPHC 4	Records of the Western Pacific High Commission, Series 4, Inwards Correspondence-General. London, Public Record Office.

*Consulted on microfilm at either the National Library of Australia, Canberra, or the Mitchell Library, Sydney.

Notes

A travel grant from the Institute Research Services of the then Darling Downs Institute of Advanced Education enabled us to get together in Sydney to write the first draft of this paper. The earlier version benefited from the comments of Peter Hempenstall, Ralph Shlomowitz, and Owen Parnaby. We are also grateful to John Hancock of Adelaide for providing technical assistance. The final revisions were undertaken at The Flinders University of South Australia, where Doug Munro was awarded a Visiting Research Fellowship in the School of Social Sciences during the first half of 1988.

1. Doug Munro, "The Origins of Labourers in the South Pacific: Commentary and statistics," in Clive Moore, Jacqueline Leckie, and Doug Munro (eds.), *Labour in the South Pacific* (Townsville, 1990), xliv–xlv.

2. Swanston to Gordon, 23 January 1879, WPHC 4, 8/1879.

3. This section is largely drawn from R. P. Gilson, *Samoa, 1830–1900: The politics of a multi-cultural community* (Melbourne, 1970), chs. 6–7.

4. Peter J. Hempenstall, "A Survey of German Commerce in the Pacific, 1857–1914," B.A. thesis, University of Queensland, 1969, chs. 1–3; Stewart Firth, "German Firms in the Western Pacific, 1857–1914," in John A. Moses and Paul M. Kennedy (eds.), *Germany in the Pacific and Far East, 1870–1914* (Brisbane, 1977), 3–25.

5. Kurt Schmack, *J. C. Godeffroy & Sohn. Kaufleute zu Hamburg* (Hamburg, 1938), 145.

6. The cotton boom and land grab is discussed in Malama Meleisea, *The Making of Modern Samoa: Traditional authority and colonial administration in the modern history of Western Samoa* (Suva, 1987), 33–37; Gilson, *Samoa, 1830–1900*, 255–259. Particularly instructive is Godeffroys' purchase of Vaitele, which became one of its major plantations. See Gilson, *Samoa, 1830–1900*, 280, 287–288.

7 Caroline Ralston, *Grass Huts and Warehouses: Pacific beach communities in the nineteenth century* (Canberra, 1977), 121.

8. Auswärtiges Amt (hereinafter AA) to Kusserow, 21 December 1885, RKA 2316:50. Cook Islanders were brought in, but in insufficient numbers, in 1864 and again in 1866.

9. Richard Bedford, Barrie Macdonald, and Doug Munro, "Population Estimates for Kiribati and Tuvalu, 1850–1900: Review and speculation," *Journal of the Polynesian Society*, 89:2 (1980), 236.

10. Roger Lawrence, *Tamana*. Atoll Economy: Social Change in Kiribati and Tuvalu, No. 4 (Canberra, 1983), 10–13.

11. H. E. Maude, Discussion on Roy A. Rappaport, "Aspects of Man's Influence upon Island Ecosystems: Alteration and control," in F. R. Fosberg (ed.), *Man's Place in the Island Ecosystem: A symposium* (Honolulu, 1965), 174.

12. Barrie Macdonald, *Cinderellas of the Empire: Towards a history of Kiribati and Tuvalu* (Canberra, 1982), 20.

13. On Gilbertese laborers to Réunion, see Dorothy Shineberg, "French Labour Recruiting in the Pacific: An early episode," *Journal de la Société des Océanistes*, 78 (1984), 45–50; and to Central America, see David

McCreery and Doug Munro, "The Cargo of the *Montserrat:* Gilbertese labor in Guatemalan coffee, 1890–1908," *The Americas*, 49:3 (1993), 271–295.

14. Munro, "The Origins of Labourers in the South Pacific," in Moore, Leckie and Munro (eds.), *Labour in the South Pacific*, xliv–li; Macdonald, *Cinderellas of the Empire*, 54–57; H. E. Maude, *Slavers in Paradise: The Peruvian slave trade in Polynesia, 1862–1864* (Canberra, 1981), 88–91,188.

15. Jeff Siegel, "Origins of Pacific Islands Labourers in Fiji," *Journal of Pacific History*, 20:1 (1985), 46; Eric Ramsden, "William Stewart and the Introduction of Chinese Labour in Tahiti, 1864–74," *Journal of the Polynesian Society*, 55:3 (1946), 187–214; Colin Newbury, *Tahiti Nui: Change and survival in French Polynesia, 1767–1945* (Honolulu, 1980), 170; G. Delbos, *Nous Mourons de te Voir! (Ti mate ni kan moriko)* (Paris, 1987), 67–73.

16. "Acta betreffend Entwurf eines Gesetzes, betreffend die Beförderung und Beschäftigung eingeborener polynesischer Arbeiter 1875–1876": Bundesrat Session von 1875, no. 86, encl. in Senat to Deputation, Hamburg, 18 October 1875, Staatsarchiv Hamburg, Records of the Deputation für Handel und Schiffahrt, I A 1. 15; *Daily Telegraph* (Sydney), 8 May 1885.

17. S. H. Davies, 1873 Journal, 9–10, Records of the London Missionary Society, South Sea Letters 34/2/D, School of Oriental and African Studies, London.

18. "Acta betr. die Gewährung einer Zinsgarantie aus Reichsmitteln an die Deutsche See-Handels-Gesellschaft in Berlin," AA Records, Handelspolitische Abteilung, vol. 13112:175; John Young, *Adventurous Spirits: Australian migrant society in pre-cession Fiji* (Brisbane, 1984), 285; Macdonald, *Cinderellas of the Empire*, 60–62. The rates of pay were initially $1 (Chilean) per month, often payable in tobacco, plus rations. By the late 1870s, wages had risen to $2 (Chilean) for men and $1 (Chilean) for women and children. See Williams to Clarendon, 9 December 1868, BCS 3/3; Journal of J. C. Williams, 14 December 1869, PMB 37; Swanston to Gordon, 23 January 1879, WPHC 4, 8/1879; "Bilanz pro 1881," DHPG Papers.

19. George Palmer, *Kidnapping in the South Seas* (Edinburgh, 1871), 86; Litton Forbes, *Two Years in Fiji* (London, 1875), 76, 254; Ralph Shlomowitz, "The Fiji Labor Trade in Comparative Perspective, 1864–1914," *Pacific Studies*, 9:3 (1986), 129.

20. Private Journal of J. L. Young, 15 June 1875, PMB 21.

21. Williams to Clarendon, 12 January 1870, BCS 3/3.

22. Journal of J. C. Williams, 14 December 1869, PMB 37.

23. Journal of J. C. Williams, 18 January, 8 March 1870.

24. Cf. Kay Saunders, *Workers in Bondage: The origins and bases of unfree labour in Queensland, 1824–1916* (Brisbane, 1982), 135; Allen F. Isaacman with Barbara Isaacman, *The Tradition of Resistance in Mozambique: Anti-colonial activity in the Zambezi Valley, 1850–1921* (London, 1976), 98–99.

25. James C. Scott, *Weapons of the Weak: Everyday forms of peasant resistance* (New Haven, 1985).

26. Alfred Restieaux, "Holy Joe, a Christianised Native of the Gilberts

(Apaiang [sic]),'' Wellington, Alexander Turnbull Library, Restieaux Manuscripts, Micro ms. 14.

27. Unidentified quotations concerning this episode are drawn from Minutes and Correspondence Relating to Runaway Labourers from Mulifanua plantation, June 1875, Apia, Roman Catholic Mission, A. B. Steinberger Papers (Mitchell Library microfilm FM4 200). Also in BCS 7/3(a).

28. Private Journal of J. L. Young, 18 June 1875, PMB 21.

29. Journal of J. C. Williams, 22 December 1869, PMB 37.

30. Private Journal of J. L. Young, 18 June 1875, PMB 21.

31. Poppe to T. W. Williams, 23 June 1875 (copy), BCS 7/3(a); Private Journal of J. L. Young, 18 June 1875, PMB 21.

32. J. W. Davidson, *Samoa mo Samoa: The emergence of the Independent State of Western Samoa* (Melbourne, 1967), 47; Gilson, *Samoa, 1830–1900*, 323. Steinberger declared to the Commission that "this running off of the labourers if continued could lead to serious results bloodshed &c and it was equal to Rebellion, & showed disrespect to the Government." See note 27, above.

33. Stewart Firth and Doug Munro, "Compagnie et Consulat: Lois germaniques et emploi des travailleurs sur les plantations de Samoa, 1864–1914," *Journal de la Société des Océanistes*, 91 (1990), 119.

34. Some of the relevant correspondence has been printed in *Treatment of Imported Labourers in Plantation of Mr. Cornwall, 1878*, Foreign Office Confidential Print, no. 4022 (London, 1879); "Native Labourers in the Navigator Islands (papers relative to alleged cruelties)," New Zealand, *Appendix to the Journals of the House of Representatives*, A-6, 1879; *New Zealand Herald*, 24 July 1979. On Cornwall's land speculations see Gilson, *Samoa, 1830–1900*, 341, 413–414.

35. Moors later became the leading American merchant in Samoa and has since become best known for his association with Robert Louis Stevenson, whom he systematically swindled. Moors' reminiscences mention Lata but not his activities there. See Harry J. Moors, *Some Recollections of Early Samoa* (Apia, 1986), 24.

36. Hunt to Gordon, 18 February 1878, encl. in Gordon to CO, 2 March 1878, CO 83/16/5790. See also Hunt to Gorrie, 14 February 1879, WPHC 4, 19/1879.

37. Gordon to CO, 2 March 1878, CO 83/16/5790; Deryck Scarr, *Fragments of Empire: A history of the Western Pacific High Commission, 1877–1914* (Canberra, 1967), 177–178.

38. Maudsley to Cornwall, 9 May 1878, BCS 5/2; Maudsley to Gordon, 14 May 1878, BCS 5/23; Alfred P. Maudsley, *Life in the Pacific Fifty Years Ago* (London, 1930), 214–215.

39. Swanston to Gordon, 20 June 1878, BCS 5/23; Swanston to Gorrie, 26 June 1879, copy, encl. in Gorrie to CO, 18 August 1879, CO 225/3/16844.

40. Swanston to Gorrie, 16 January 1879, BCS 3/2; Swanston to Ryan, 16 January 1879, BCS 3/2; Diary of William Swanston, 16–17 January 1879, Suva, Fiji Museum (we thank Fergus Clunie for making available this reference); Ryan to Swanston, 16 January, 17 February 1879, BCS 3/2; Gorrie to Swanston, 4 April 1879, and encls., CO 225/2/10776.

41. Swanston to Gorrie, 4 June 1879, WPHC 4, 53/1879. The role of a

protector in ameliorating plantation conditions has been underlined in Sandra J. Rennie, "Contract Labor under a Protector: The Gilbertese laborers and Hiram Bingham, Jr., in Hawaii, 1878-1903," *Pacific Studies*, 11:1 (1987), 81-106.

42. Graves to Thurston, 25 November 1881, and encl., WPHC 4, 64/1882. For further details see Scarr, *Fragments of Empire*, 79-80.

43. Fritz Stern, *Gold and Iron: Bismarck, Bleichröder and the building of the German Empire* (London, 1977), 396-401; D. K. Fieldhouse, *Economics and Empire, 1830-1914* (London, 1973), 444.

44. "Bilanz pro 1882" and "Bilanz pro 1884," DHPG Papers.

45. Th. Weber, *Ländereien und Plantagen der Deutschen Handels- und Plantagen-Gesellschaft der Südsee-Inseln zu Hamburg in Samoa* (Hamburg, 1888), 12-20, 29.

46. Weber to German Consulate, Apia, 11 May 1883, encl. in Stuebel to Bismarck, 14 May 1883, RKA 2928:50; Stuebel to Bismarck, 27 January 1886, RKA 2316:62; J. A. Bennett, "Immigration, 'Blackbirding,' Labour Recruiting?: The Hawaiian experience, 1877-1887," *Journal of Pacific History*, 11:1 (1976), 3-27.

47. Churchward to Mitchell, 7 October 1885, WPHC 4, 189/1885.

48. Doug Munro and Stewart Firth, "German Labour Policy and the Partition of the Western Pacific: The view from Samoa," *Journal of Pacific History*, 25:1 (1990), 86-91.

49. "Bilanz pro 1882," DHPG Papers; Paul M. Kennedy, *The Samoan Tangle: A study of Anglo-German-American relations, 1878-1900* (Dublin, 1974), 32.

50. Donald Ross, "Health Report for Plantations for 1883," RKA 2316:21-29; encl. AA minute of 23 November 1885, RKA 2316:52; Malet to Hatzfeldt, 27 October 1884, RKA 2793:159-160. The Melanesians' initially high mortality rate probably resulted from their being introduced to a new disease environment against which they had little natural resistance. On this question generally, see Ralph Shlomowitz, "Mortality and the Pacific Labour Trade," *Journal of Pacific History*, 22:1 (1987), 34-55; Ralph Shlomowitz, "Epidemiology and the Pacific Labor Trade," *Journal of Interdisciplinary History*, 19:4 (1989), 585-610; Clive Moore, *Kanaka: A history of Melanesian Mackay* (Boroko, 1985), 244-254.

51. S. G. Firth, "German Recruitment and Employment of Labourers in the Western Pacific before the First World War," D.Phil. dissertation, Oxford University, 1973, p. 64. Planters in Fiji also found that Gilbertese were cheaper to recruit than Melanesians. See Shlomowitz, "The Fiji Labor Trade in Comparative Perspective," 122-123.

52. Commander of S. M. *Sperber* to German Consulate in Apia, 25 May 1891, encl. in Schmidt to Caprivi, 2 December 1891, RKA 2547:85-88; McCreery and Munro, "The Cargo of the *Montserrat*," 289; Siegel, "Origins of Pacific Islands Labourers in Fiji," 46.

53. Firth, "German Recruitment and Employment," 46-48.

54. The most detailed account of the background to the declaration of the Gilbert Islands Protectorate is W. P. Morrell, *Britain in the Pacific Islands* (Oxford, 1960), 272-275.

55. Thurston to Ripon, 5 September 1893, and encls., CO 225/42/19620.

A copy of the conditions of employment of Gilbertese plantation workers in Samoa is in CO 225/51/5257.

56. Biermann to Caprivi, 8 March 1894, RKA 3217:24–26; DHPG to Prussian Embassy, 12 January 1894, encl. in Bülow to Caprivi, 16 January 1894, RKA 2317:10–14.

57. Biermann to Caprivi, 15 December 1893, RKA 2317:10–14; Stewart Firth, "German New Guinea: The archival perspective," *Journal of Pacific History*, 20:1 (1985), 99.

58. DHPG Directorate's report, 1893:13, DHPG Papers.

59. Cusack-Smith to Thurston, 15 February 1894, WPHC 4, 73/1894; *Western Samoa Herald*, 17 February 1894; 21 April 1894. Copies of the licences are in RKA 3217:92, and in CO 225/45/10693.

60. Cusack-Smith to Thurston, 2 June 1894, WPHC 4, 161/1894; *Samoa Times*, 28 April 1894; *Western Samoa Herald*, 23 June 1894.

61. Cusack-Smith to Biermann, 23 June 1894, copy, encl. in Cusack-Smith to Thurston, 3 August 1894, WPHC 4, 223/1894; Cusack-Smith to Thurston, 18 August 1894, WPHC 4, 227/1894; Cusack-Smith to Thurston, 14 December 1894, WPHC 4, 10/1895. (The Consul's despatches to the High Commissioner are also in BCS 5/24).

62. Cusack-Smith to Beckmann, 16 October 1894, encl. in Cusack-Smith to Thurston, 17 October 1894, WPHC 4, 277/1894.

63. Cusack-Smith to Beckmann, 22 October 1894, encl. in Cusack-Smith to Thurston, 17 October 1894, WPHC 4, 280/1894; Cusack-Smith to Thurston, 14 December 1894, WPHC 4, 10/1895.

64. Cusack-Smith to Thurston, 17 October 1894, WPHC 4, 277/1894.

65. DHPG to Kolonialabteilung des AA [Colonial Division of the Foreign Office], 6 January 1893, RKA 2336:25.

66. Woodford to German Consulate in Apia, 14 April 1895, copy, encl. in Schmidt-Leda to Hohenlohe, 17 May 1895, RKA 3217:69.

67. Anderson to Woodford, 8 February 1895, copy, encl. in FO to CO, 8 February 1895, CO 225/48/2541; Woodford to Kimberley, 25 March 1895, BCS 3/7. On Woodford generally, see Ian Heath "Charles Morris Woodford: Adventurer, naturalist, administrator," in Deryck Scarr (ed.), *More Pacific Islands Portraits* (Canberra, 1978), 193–209, 271–275.

68. Woodford to Berkeley, 11 July 1895, WPHC 4, 235/1895; Woodford to Salisbury, 8 August 1895, confidential, encl. in FO to CO, 27 September 1895, CO 225/49/1 7053.

69. Woodford to Berkeley, 20 August 1895, WPHC 4, 319/1895.

70. Woodford to Berkeley, 20 August 1985, WPHC 4, 319/1895; DHPG Directorate's report 1895, DHPG Papers.

71. DHPG Directorate's report 1895, DHPG Papers.

72. Meyer-Delius to KA, 3 November 1894, extract; and DHPG to German consulate at Apia, 31 May 1895, copy, encl. in Schmidt-Leda to Hohenlohe, 16 June 1895; and DHPG to KA, 30 July 1895, RKA 2317:47, 73, 75–76.

73. Meyer-Delius to Kayser, 2 November and 16 December 1895; AA to Hatzfeldt, 6 January 1896, RKA 2317:78, 84, 86–88.

74. Cusack-Smith to Schmidt-Leda, 3 December 1895, copy, encl. in Schmidt-Leda to Hohenlohe, 3 December 1895 (also in CO 225/51/1960); Salis-

bury to Hatzfeldt, 14 March 1896, RKA 2317:90–91, 100–101; Cusack-Smith to FO, 3 December 1895, BCS 3/7; FO to CO, 13 February 1896, CO 225/51/3405.
75. Thurston to Meade, 2 October 1895, CO 225/49/18609.
76. Cusack-Smith to Salisbury, 28 January 1896, encl. in Cusack-Smith to Thurston, 20 February 1896, WPHC 4, 60/1896; Cusack-Smith to Thurston, 5 March 1896, WPHC 4, 94/1896; *Samoa Times,* 11 April 1896.

5

"We Do Not Come Here to Be Beaten": Resistance and the Plantation System in the Solomon Islands to World War II

Judith A. Bennett

The development of plantations in much of Melanesia was late in comparison to the large islands of tropical Polynesia. Between 1870 and 1911, over 30,000 contracts of indenture were entered into by Solomon Islanders for plantation work away from their home archipelagoes.[1] During the 1890s, however, changing political circumstances laid the foundation for this turbulent recruiting frontier to become a more settled plantation economy and the gradual transition whereby Solomon Islanders' labor power was reserved for plantation work within the group. The first step occurred in 1893, when Great Britain declared a protectorate over those Solomon Islands not already claimed by Germany, and then proceeded to ignore the new addition to empire. It was only in 1896 that Britain reluctantly approved the appointment of the first Resident Deputy Commissioner of the Solomons, C. M. Woodford, who was responsible to the High Commissioner for the Western Pacific based in Fiji. Even this limited arrangement was only struck on the understanding that the new protectorate had to be self-supporting and that, once federated, the Australian colonies would take over control, as they were expected to do with British New Guinea. With Australian Federation almost a reality and the British Colonial Office breathing down his neck, Woodford set out to make the Solomons a viable economic proposition.[2]

Woodford saw only two options: to foster trade by Europeans with local producers or to follow the example of other territories in encouraging commercial plantations. He chose the latter because the former would have been too slow in developing the fiscal scale necessary to sustain a government infrastructure. To achieve this, Woodford had to bring together the capital, the land, and the labor in the Solomons. The Pacific Islands Company, which had among its principals Lord Stanmore, former governor of Fiji, soon showed interest in the Islands. The Colonial Office, supported by Woodford, granted the company a concession to occupy "wasteland" in the Solomons. Delays concerning the nature and extent of the title to be conveyed resulted in the Company selling its concession to England's William Lever in 1906. Lever wanted to invest in copra to guarantee a supply of oil for his expanding soap-making ventures. Lever had also purchased freehold in the Solomons the year before. Other companies, including the subsidiaries of the Queensland-based Burns, Philp and Company, were formed with mainly Australian capital, as the world boom in tropical products such as copra and rubber was inspiring confidence among investors. Like Lever, Burns Philp obtained some land under a Certificate of Title on a lease of 999 years. These concessions covered land that Woodford deemed to be virtually unoccupied and therefore unowned and "waste," so the government had title to lease. Most investing companies wanted a more secure title. Levers, Burns Philp, and others purchased both undeveloped and partly developed land from established traders who had used their knowledge of local communities to buy up cheaply large tracts of coastal land once the Protectorate had been established. By 1914, about 5 percent of all Solomon Islands land had been alienated—a significant amount in terms of the accessible coastal areas suitable for coconut growing. At least £1,000,000 flowed into the Solomons.[3]

Woodford and the first investors in Solomons' plantations believed labor would be abundant, especially since Queensland had ceased to import Islanders in 1904 when reorganization of the sugarcane industry, plus pressure from Australian unions, made it no longer profitable. The demand soon began to outstrip the supply, so Woodford lobbied to have the Solomons closed to Fijian recruiters. He was to succeed in 1911. Except for a virtually nominal right to recruit in the Solomons by Germans for Samoa, the labor force seemed to be now the preserve of the Solomons planters.[4] A compliant government had facilitated the mixing of the elements of land, capital, and labor and anticipated a secure economic future for the Islands.

The Plantation System and the Planters

The present discussion will focus primarily on the plantation as a social system within the Solomons as a whole, but particularly in relation to the small communities that supplied labor. However, dominating as it did the Islands' economy, the plantation made heavy demands on government. Government policy influenced, and was influenced by, the plantation as a social system. The internal economic dimension in turn interacted with, or, more exactly, reacted to, economic forces from the world beyond the Solomons. In a sense then, the plantation system represented the cutting edge of world capitalism on the domestic economies of the Solomons.

In the Solomons, the government played a key role in the establishment of the plantation system. At no time did planters ever complain of not having enough land, although they regarded some types of title as more secure than others. Capital, at least in the honeymoon years from ca. 1905 to 1913, had not been difficult to attract. However, the government and the planters found that their plans for expansion of the plantation system were retarded by two internal factors: first, the government's inability to guarantee the safety of its subjects as plantations extended into the frontier and second, the reality of a limited labor supply.

Woodford's establishment in 1896 had been scant: eight Fijian policemen, a whaleboat, and occasional assistance from a visiting British warship, of little use in pursuing people into the interior.[5] In these early years, the punitive expedition became the main means by which the government extended its control. This is not the place to detail every expedition, but the pattern was one of overreaction whenever a white man was killed and otherwise virtual neglect of native societies, at least until the early 1920s. Punitive expeditions, such as those mounted to apprehend the killers of Porret and Burns, were generally clumsy affairs headed by a solitary government officer and involving a militia made up of "friendly tribes" and avenging traders, who often killed or captured innocent parties. Ineptitude on such a scale left a legacy of distrust toward the government.[6]

Woodford had received some help in 1898 when the High Commissioner for the Western Pacific appointed A. W. Mahaffy to Gizo. With a force of twenty-five Solomon Islander police in a confiscated war canoe, Mahaffy subdued the ferocious headhunters of Gizo, Munda, and Roviana.[7] The government opened stations in the Shortlands in 1907, at Aoke on Malaita in 1909, and a temporary one at Masi in the disturbed Marovo lagoon area in 1910. Thus, until 1914, when a station was established at Aola, Guadalcanal, three government officers plus about fifty-five local police, along with the Resident Commissioner and his establishment of about ten officers at the capital, Tulagi on Nggela, were the sum total of the administration. With this, Woodford was expected to control a scattered and diverse population of an estimated 150,000–200,000 indigenous people and between 200 and 400 Europeans.[8]

Plantation managers perforce became their own militia and

police. The government tacitly expected this of them. For example, Vella Lavella was in a disturbed state following two punitive expeditions in 1901 and 1908.[9] In this maelstrom of distrust, the government approved the establishment of a plantation because "it is a distinct advantage to encourage settlement on the coast lands of Vella la vella as by this means the more troublesome bush tribes can be better brought under control."[10] Even on more populous Malaita, notorious for its uncaptured killers of white traders and recruiters and for its incessant "payback" killings and local wars, the government in 1909 had approved the sale of 10,000 acres of land to the Malayta Company before the establishment of the government base on the island. The government hoped that company's principals, A. H. and E. Young, whose sister Florence founded the Queensland Kanaka Mission, would continue their work of achieving "wonders in taming such natives that come under their influence" on Malaita.[11]

Plantations often became armed camps. In 1908, Levers' manager at Uki had to arm his labor for fear of raids by Malaitans.[12] The Malayta Company did the same at Su'u and Baunani on west Malaita throughout 1911 to 1917, with little assistance from the government officer at Aoke. In 1910, it was not unusual when planters left laborers unsupervised on new plantations, to supply them with arms "for purposes of protection."[13]

For white men, there was little to offset the physical insecurity of an isolated plantation as a place of employment. Few planters had their wives with them. Salaries were not appealing—managers were paid from £100 to £500 and overseers, £60 to £200 a year, depending on the size of the plantation and the number of laborers.[14] Their counterparts in the East received double that.[15] Housing was sometimes primitive and leave paid only if the man returned to the job.[16] Consequently, when hiring in Sydney, Melbourne, Brisbane, or even in the Solomons, plantation companies could not afford to pry into the history of applicants, sometimes to their subsequent regret.[17] Managers and their overseers, moreover, were isolated from other Europeans. Loneliness plagued them, especially the young, inexperienced men from the southern cities. Some sought solace in alcohol;[18] others in liaisons with indigenous women or, worse still, with indigenous men. Socially unacceptable to their superiors and many other Europeans, such practices also jeopardized their authority with the laborers and occasionally led to open conflict, a jail sentence, deportation, or even death.[19] The scourge of the European, especially the pioneer opening new land, was malaria and complications such as blackwater fever, pneumonia, anemia, and influenza. Within six months, almost 50 percent of once fit young newcomers had to leave the islands or died as a result of the ravages of malaria.[20] Quinine was both a prophylactic and a treatment for malaria, but there were differing opinions as to safe dosages. Planters often forgot to dose themselves and soon succumbed.[21] Among the most common effects of malarial episodes were depression, a feeling of hopelessness, and even suicide.[22] Forms of paranoia were another outcome.[23] Isolated, lonely, often inexperienced and vulnerable, their judgment frequently clouded by malaria or sometimes by drink, the white men, be they planter-trader, manager, or overseer, did not have a secure life.

In time, the planters had another worry that better government would do little to relieve. Mahaffy's report on the Protectorate in 1908 submitted to the High Commissioner in Fiji touched on two realities underpinning the labor supply: first, that the population was declining and second, that the ill-treatment of laborers would only create "great difficulty in inducing native labor to recruit for their plantations."[24] There is evidence to suggest that the population, at least in the western islands and San Cristobal, had been declining, but was expected to stabilize by the 1920s and early 1930s.[25] Certainly, calculations were too optimistic. The first census, in 1931, counted 94,000, compared to earlier estimates of 150,000 to 200,000.[26] Pioneer planters soon found they could not obtain enough local labor to clear and develop the land, even after Queensland and then Fiji closed.[27] Levers especially was concerned, as it had failed to meet development clauses regarding its 200,000-acre concession in the western Solomons, concentrating its energies on its freehold land.[28]

As early as 1909, Woodford suggested the introduction of indentured Indian coolies, as had been done in Fiji in 1879.[29] Levers and Burns Philp intermittently lobbied the Colonial Office for this from 1911 to 1915.[30] Although not unsympathetic to the importation of free labor, the Colonial Office heeded the Secretary of State for India who, in the face of a developing nationalist movement, refused to permit labor to go under indenture, as the companies wanted in order to guarantee a stable, tractable work force.[31] Newly federated Australia also opposed the move, urging that, as in Papua under Hubert Murray, the Melanesians work with the planters where possible, for the gradual development of the country.[32] The Colonial Office maintained this stance with slight variations until World War II, when it became a nonissue because of war damage in the Solomons and India's independence.[33]

For the foreseeable future in the Solomons, the government and the planters had to utilize the existing labor force as best they could. The government was in a position to view this more globally. Given the exigencies of Colonial Office policy at the time, the Solomons government had to continue to support itself. To do that, the plantation economy had to be safeguarded. At the interface of management and labor, planters were shortsighted, aiming for the solution of the problem then and there instead of looking to the long term. However, until it was convinced that overseas labor was a virtual impossibility, the government of the Solomons, embodied in Woodford, was slow to introduce effective laws to protect and foster the labor force.

The government had made regulations relating to labor as early as 1897, but these were premised on the indenture system. Evolved from the English Masters and Servants Acts, these and subsequent labor regulations before World War II provided a scale of wage deductions if a laborer "fails to show ordinary diligence." An amendment in 1915 provided for a heavy fine or three months in jail if a laborer were convicted three times of "unlawfully absenting himself from the service of his employer." Strikes, organized protests, and trade unions and the like were forbidden. In addition to a minimum wage, the employer had to provide food, shelter, clothing, a mosquito net, and other customary rations such as tobacco and soap, as well as transportation to and from the plantation.[34] The 1897 regulation set contracts at two years, authorized government inspection of plantations, and required the repatriation of laborers, but it did not establish a procedure for ratification of contracts and inspection of conditions.[35] Persuaded by planters, who found new recruits inefficient in their first two years, Woodford permitted three-year contracts in 1908, an extension repudiated by the Colonial Office in 1911.[36] In practice, these regulations were of limited value. Any disputes between management and labor needed settling quickly. Without regular visits by a government officer, any appeal to the government for arbitration would necessitate a long and costly trip from the plantation to Tulagi or the district station.[37]

Government intervention in the contractual arrangements between management and labor came largely as a result of the need to maintain the limited labor supply in the light of what was happening to the laborers on plantations and how they were dealing with this. Through their resistance, laborers, despite many constraints, were to have a considerable input into the evolution of the plantation system in the Solomons.

Resistance: Rewards and Restrictions

Solomon Islanders were in a seemingly powerless position on the new plantations. Hedged by the colonial government, supervised by exacting and often violent masters, cut off from their own societies, and isolated from counterparts on other plantations without any institutions to act collectively, laborers appeared ill-equipped to fight the excessive appropriation of their labor, their rewards, and their dignity by the planters. But they did, by what James C. Scott calls "everyday forms of resistance": "Footdragging, dissimulation, false compliance, pilfering, feigned ignorance, slander, arson, sabotage, and so forth."[38] These forms of struggle"require little or no coordination or planning; they often represent a form of individual self-help; and they typically avoid any direct symbolic confrontation with authority or éite norms."[39]

Given the parameters of indigenous social constraints and the imposed colonial institutions of repression (which will be elaborated below) this kind of resistance was ideally suited to the Solomon Islander laborer.

From the Solomon Islanders' perspective, one of the earliest acts of oppression was the appropriation of their labor for an unfair wage. In the 1900s, the planters discovered that, in general, recruiters were only able to hire young men who were "practically savages . . . unaccustomed to field work."[40] Emotive language aside, there was truth in this statement. Experienced hands who had returned from overseas, particularly Queensland, did not want to work as common laborers on the new plantations because of the comparatively low wages. In the days before Queensland closed (1904), they could demand of planters in the Solomons double the minimum Queensland wage of 10 shillings a month. Fiji absorbed some of these until it too closed in 1911. Although the minimum wages in Queensland and Fiji were £6 and £3 per annum, respectively, for new recruits, "old hands" were paid far higher rates.[41] The Solomons' rate settled at 10 shillings a month mainly on the calculation of Levers' accountants. The closing of Queensland, Fiji, and, by 1914, Samoa removed all the options but one for Solomon Islands labor and undercut entirely the only bases for wage negotiation remaining in the hands of young men offering for work. Those returners who went to local plantations received a shock: less than 1 percent of the labor force was paid 10 to 15 shillings a week, and the majority received only the minimum wage of 2 shillings and sixpence a week.[42]

Plantation Society on the Frontier (ca. 1897–1917)

For Solomon Islanders, the degree of adjustment necessary for plantation life, be it overseas or in the Solomons, was considerable. The young relied on their elders and "passage masters" to negotiate with each other and with the recruiters who plied the coastlines of Malaita, Guadalcanal, and San Cristobal for the preferred plantations and for the best "beach payment," which was a solatium, a compensation to the recruit's elders for the absence of their relative.

The passage masters were usually older men with several years of experience on overseas plantations. They knew their own societies and the white man's world. Fluent in pidgin and often in basic English, they played a crucial role in the process of recruiting, especially on Malaita, where it was not safe for recruiters to go ashore at many "passages"—places where recruiters anchored offshore and which gave the easiest access to the mountainous inland areas. Such men were compradors, an essential link in the articulation of the capitalist and domestic modes of production. Rewarded by the recruiters, they also had close economic ties with clan elders and "big men" of the related societies in the hinterland.

New recruits had to quickly pick up pidgin—the lingua franca of the plantation—in order to know what the Solomon Islander boss-boy and the overseer were saying to them.[43] Laborers sometimes found themselves working alongside old enemies. Where they existed, neighboring villages, though intriguing to the curious, were also the abode of potentially dangerous strangers. Such enemies and strangers could offend an individual's particular socio-religious values just as the *masta* could. Women were few and usually married to other laborers. Family life, except for this handful, was nonexistent.[44]

Laborers found that the clock dominated the routine of the white man's plantation day, so different from the ebb and flow of seasonal activities in their own society. The workday usually started before sunup and ended at dusk—one monotonous round of eating, sleeping, and working, punctuated by a half day off on Saturday and a rest day on Sunday and virtually no other variation except a day off and a feast at Christmas and perhaps Easter.[45]

The pioneering work was constant and difficult: hardened Australian bushmen contracted with Levers to clear virgin land in 1913, but they gave up and went home after four weeks, as the work was too onerous.[46] Clearing land was particularly dangerous to health, because the dense undergrowth favored the malarial mosquito, *Anopheles*. As they worked among the fallen trees and scrub, the laborers cut their legs and feet and ended up with septic ulcers, so difficult to heal in a pre-antibiotic era.

Once the coconut palms were bearing, copra making too was hard work—the nuts had to be collected from the ground, usually in carts, and then husked. After husking, the nuts were split and the "meat" extracted to be stacked in the drier. The fires had to be watched to keep them at a constant pitch and to prevent the copra and the copra house from catching fire. The dried copra was put into bags, "rammed" tightly, sewn and weighed, and then stacked in a shed until the copra boat arrived. Weeding underneath the palms, digging and clearing drains, catching beetles, which were beginning to be pests, and collecting and cutting firewood were other regular plantation tasks usually allocated as "light work" to very young or physically slight recruits. Married men sometimes worked with their wives as cooks for the laborers. A likely youth might be singled out either because of his intelligence, cleanliness, or gentleness to work as a domestic in the "master's" house.

Another difficulty was spiritual vulnerability for this predominantly pagan group, who were animists and venerators of ancestral spirits. The strange place, strange people, and distance from clan priests and shrines kept newcomers uneasy. The illnesses that befell them were believed to have, in Western terms, a supernatural origin activated by sorcerers or vengeful ancestral spirits, and were thus all the more frightening. The plantation store with its high-priced goods found out another kind of vulnerability. As in the plantation gambling sessions, laborers were tempted to risk indebtedness to relieve the monotony of plantation life, but could end up going home virtually empty-handed and shamefaced to their families.[47]

In the early days of plantation development, the most difficult element for laborers to deal with was the "master." This was particularly so on larger company plantations, where salaried managers and overseers had no vested interest in the plantation worker. These men maintained a strict demarcation line between themselves and their labor, partly from fear and partly because they regarded them as inferior beings and savages. Owner-planters on smaller concerns, though not always paragons of enlightened management, took a more personal interest in keeping their labor well and content, if for no other reason than the money they had invested in their recruitment. Often these small operators could afford to hire not more than ten to twenty laborers, so they got to know them as individuals, whereas the company overseer might have one hundred men under his direct control.[48]

Gross abuse of labor was at its worst before 1916 and almost without exception was on the larger holdings of Levers, Burns Philp's subsidiaries (Solomon Islands Development Company and Shortland Islands Plantations Ltd.), and of Billy Pope, who sold out to the Malayta Company in 1910. Beating and flogging of laborers were common; shooting or killing, all too frequent. In 1976, old Solomon Islanders recalled more cases than the government documents record, probably reflecting the ability of some managers on isolated plantations to disguise a homicide and the reluctance for further inquiry by a government lacking a judicial commissioner with the requisite authority to try murder cases in Tulagi until 1914.[49]

Solomon Islanders retaliated in kind, well deserved "by the rough standards of native justice" of the frontier.[50] Physical violence as the ultimate sanction and form of resistance was so common in this period it soon became part of the mores of plantation society. When seventeen-year-old Charles Bignell took up his position of overseer in charge of seventy laborers at Ghojoruru, Santa Isabel, in 1908, he had to cope with a fight with them or perhaps suffer the fate of his predecessor George O'Neal, whom they had killed.[51] This ritual of newcomers being "tried out" by laborers was to persist beyond the 1900s and 1910s. A man had to prove he had a right, by might, to be an authority over laborers. This was particularly so with Malaitans, who, on average, were physically stronger than say, men from San Cristobal.[52] Malaitan societies placed high value on physical violence and killing as sanctions. New overseers would be provoked by a deliberately disobedient individual, and then a group of his friends and kinsmen would throw themselves at the man. If the overseer acquitted himself well, laborers accepted the odd kick in the backside or hit over the ear he doled out to them —given there was justification. If subsequently, however, an overseer was consistently unjust, the group would wait for a vulnerable moment to beat him up. In general, both management and laborers knew this was simply part of plantation life.[53] Visiting European commentators sometimes thought such attacks on lone overseers cowardly,[54] but failed to understand that the overseer often forced his will on the labor with the help of a revolver or a whip, far from any government officer; nor could they see that, to most Solomon Islanders, helping a kinsman was a sacred duty.[55]

Word of such conflicts spread among potential recruits. Returners, plus local crewmen on recruiting and trading ships, soon acquainted their descent group and passage masters with the most notorious plantations. Avoidance became the best resistance.[56] Mahaffy had noted this trend as early as 1908.[57] Passage masters and kinsmen of the young recruits advised them to sign on for safer plantations. However, there was a way around this. Recruiters were paid by planters to recruit, so they "recruited for plantations with a good reputation and sent the recruits to plantations with bad reputation," leaving the destination off the list submitted to the government official in Tulagi.[58]

Another way laborers resisted a harsh master was to desert and try to get back to their homes. Many laborers simply ran away, and only their master and their friends knew of it, either because they had been illegally recruited or because records of their contracts were poorly kept.[59] Runaways often disappeared and were presumed dead, either eaten by crocodiles[60] or killed by men from the surrounding villages. Such were the trials that faced deserters. The planters knew this, just as they knew that men whom they recruited locally could melt into the bush when they found plantation life distasteful. In contracting men from distant islands, planters capitalized on the Islanders' fear of the unknown to bind them to the plantation.[61] By the late 1910s, desertions declined as government control of the Solomons consolidated; they remained at less than .1 percent yearly throughout the 1920s. By then, runaways usually ended up in court or back on their employer's plantation.[62]

Government Regulation of the Plantation System

The killings, beatings, recruiting of underage boys, and desertions had been graphically described in Mahaffy's report of 1908. He well understood that Solomon Islanders would not continue to tolerate such treatment. Emphasizing that the treatment of labor was "grave" and counterproductive, Mahaffy urged the High Commission to introduce new labor regulations and a mobile inspectorate.[63]

These came into effect in 1910. Among other things, they set the minimum age of recruiting at fourteen (amended in 1911 to sixteen), provided for the establishment of plantation hospitals where more than fifty men were employed, set maximum hours of work at nine per day, five on Saturday, with Sunday as a rest day, and no more than a total of fifty hours a week to be worked without the worker's consent and overtime payment.[64] Resident Commissioner Woodford's support remained more lip-service than practice. The first Inspector of Labour, appointed in 1908, had been too busy with supervising the repatriation of Queensland returnees to concentrate on local plantations.[65] When William Bell became inspector in 1911, he was somehow to do his job without a ship. Moreover, few planters could obtain copies of the 1910 regulations, as Woodford had failed to have a sufficient number printed.[66] Bell did not long remain silent and complained to the Secretary of State for Colonies. Bell realized that punishment for refractory laborers as set out in the 1912 amending legislation was all but useless: it provided a penalty of two hours' pay for every one hour of work lost by disobedience or carelessness on the worker's part, as decided by a magistrate. Bell recommended that if a laborer were imprisoned because of some breach of the regulations, his indenture be extended to compensate for the time lost to the employer or a heavier fine inflicted at the magistrate's discretion; otherwise lack of inspection and the isolation of plantations would mean that planters would exact their own punishment as in the past.[67] Bell did manage to make recruiters clearly indicate the destination of their recruits at the "signing" of the indenture at Tulagi, thus sewing up the last loophole for the disreputable planter to obtain recruits. This act alone, Bell maintained, was a major factor in the noticeable improvement of plantation conditions by 1914.[68]

Colonial Office attention, more and more, focused on labor matters in the Solomons. Reports of the more violent of Levers' managers and of the Protectorate government's failure to fully enforce the new regulations filtered back to Great Britain just when William Lever was trying to convince the Colonial Office of his company's suitability as a potential employer of immigrant coolie indentured labor. A terrible dysentery epidemic raged through plantations in 1913–1914, killing, on average, 4.8 percent of the laborers, but on some plantations, mainly Levers', as many 10 percent. Reported desertions quadrupled. Investigations revealed the almost total lack of sanitation and medical care on plantations, on recruiting vessels, and at the transit depots of Levers and Burns Philp near Tulagi. The Colonial Office began asking a lot of questions of the High Commissioner and Lever.[69] Woodford retired in 1914, and his successor, Barnett, actively supported the labor inspectorate. By 1915, Bell had his ship and was able to inspect 36 percent of all plantations. All were inspected in the following year.[70]

The wise planter now took the advice of the inspector. From Boroni, the manager explained to his directors in Australia, "I have had to erect all new labour houses as the old ones did not satisfy our friend the Labour Inspector."[71] Prosecutions in open court began to follow blatant abuse of the regulations.[72] Levers, wishing to find favor with the Colonial Office, sacked managers and overseers who ended up in court and replaced them with more humane and better-educated men.[73] By 1917, over half the laborers were supervised by married men with families, who were less susceptible to the loneliness that in the early years had sometimes distorted judgment and occasionally embroiled them in compromising sexual affairs with Solomon Islanders.[74]

Strategies of Management and Labor:
Adaptation and Accommodation

Management began to modify the plantation routine to suit the working patterns of Solomon Islanders. Solomon Islanders found the 6 A.M. to 11 P.M., midday to 5 P.M. (or nine working hours a day after the 1912 Amendment), working to the clock to be alien; working to complete a specific task was not. Levers introduced task work on their plantations in 1919, mainly for copra cutting. The most common daily task was to cut three bags of green copra weighing 150 lbs. each (450 lbs. when dried yielded about 236.5 lbs.). Here and there, planters tried to set a more onerous task, but word of this tended to spread, so employers would find it hard to get good recruits.[75] Some employers accepted three partly filled bags while others did not, so when laborers went to a more exacting employer, resistance was likely. Laborers were adept at avoiding the full task. Some would "birdcage" the green copra in the bags, giving it the appearance of being well filled, but in fact it was half-full. Others rammed only the bottom and top layers of copra into the bag, with the middle section loosely packed. Others somehow got hold of smaller copra sacks and so filled their quota more quickly. To attempt to standardize the task, the government defined it, albeit vaguely, in 1921: "In the case of a male that extent of task work which can be performed by an ordinary able-bodied adult male in six hours, working steadily at such work."[76]

The task system of copra cutting was common by the mid-1920s. The larger companies extended the concept to other tasks such as road construction and drain digging. Sometimes called "contract" work, the task system differed from less quantifiable work, such as catching beetles, which remained under the old nine hours daily routine. Laborers under this routine, as signaled by the plantation gong or bell, were "bello" workers. The contract or task system suited the rhythm of the laborer and the pocket of the planter. It meant less need for the supervision and constant checking of the old "day work" system, which in turn reduced conflict between laborer and overseer and thus reduced costs.[77] Once output was being weighed, an added bonus, from the employers' viewpoint, was that the copra cutters "invariably try to cut a little over the task so as to avoid any rejection of their task at the Drier when it is brought in on account of being underweight."[78] Some employers exploited this by not openly weighing the copra, until in 1922 the government intervened.

Laborers' complaints assisted the inspector in his duties. One offense commonly reported in the late 1910s and the early 1920s was any shortness of the food and tobacco ration.[79] Although a scale of rations was set in 1910 by the government, abuses before the introduction of inspections must have been common, but received little government notice. On the San Cristobal estates of Fairley, Rigby and Company, the manager, George Mumford, trying to manage a small, struggling operation, told his employers of his methods in 1914: "Now as regards plantation expenses I can assure you that I am cutting things. . . . [specific calculations], this is less than ½ the allowance and we cannot safely give less as we never know when a labour inspector will come prying around looking for faults."[80]

Four years and some inspections later, Mumford had increased the bulk of the food, but the protein component remained half the legal ration, as long as the inspector kept away.[81] The improvement was due in part to the inspection, and in part to the reaction of the laborers. With time and inspectors, they were learning what was theirs by right, as Mumford knew when he admitted that "he had to feed his 'wild Malayta bushmen' or there would be trouble."[82]

Laborers were not averse to self-help when it came to varying the regulation diet. Especially after the introduction of the task system, men often spent their spare time fishing, hunting, or growing local vegetables. They also bartered rations like soap and tobacco to neighboring villagers for fresh food, as well as betel nut. Although forbidden by their employers, laborers on Levers' plantations stole and ate fresh coconuts when they had the chance, and occasionally killed plantation cattle for food.[83]

The regular inspection of plantations opened a way of redress for Islanders.[84] Even before inspections became routine, disgruntled laborers began to realize that with the new regulations, there was a possibility of some respite if they could get a hearing by a government officer.[85] In 1918, new government stations opened at Kira Kira (San Cristobal) and Tanabulli (Santa Isabel). District officers assisted in the inspection of plantations and heard cases. In 1921, the government appointed a second labor inspector. Thus by the early 1920s, laborers had the means to report their problems. They soon became accustomed to doing so to government officers, who were seen by the majority as being reasonably fair in their investigations. Much to the chagrin of planters, laborers had learned the elements of the crime of assault from kinsmen in the native police. They often managed to see the inspector before the master could, laying a charge against their employer.[86] They were also aware of strategies of resistance that would provoke a master to hit first: "The methods usually adopted are loud 'asides' commentary on orders issued by the white man; a swaggering, insolent bearing when addressing him; 'back slack' when reproved; a debate or discussion, tantamount to disobedience, bearing on his ability or inability, to do work allotted to him—and so on insufferably."[87]

Of course, the majority of planters by the early 1920s understood their labor better and had acquired some fundamental management skills. Although they commonly complained about "trouble-mongers" or the occasional "bush-lawyer" among the laborers, experienced planters and managers routinely relied on "tact," "a sense of humour," and "get[ting] amongst them [laborers] and show[ing] them their work" as strategies of effective labor control. The everyday running of plantations by these methods was rarely a matter for record, whereas the deviations from the norm were concerns for note by inspectors.[88]

Some planting companies even took the initiative to improve the health of their labor. From 1917, Levers employed their own medical superintendent. By 1922, regular medical bulletins written by the superintendent were being distributed to managers.[89] When the world pandemic of Spanish influenza struck the Solomons in 1919, both Levers and Burns Philp had purchased thousands of shots of a vaccine and inoculated their laborers. Despite government attempts to stop the spread of the disease by quarantine and by suspending recruiting, the influenza virtually affected every district, spread by interisland vessels after its introduction on the Burns Philp's steamer *Marsina*. As the government provided no health services to the bulk of the people, no accurate Protectorate-wide statistics were collected, but the death rate was thought to be 4 percent of those infected in villages and was probably higher on Malaita. With more specific treatment on plantations, as advised by the government, the death rate among laborers infected was only 1 percent. The government further assisted planters in the care of themselves and their employees in 1921 when it published *A Brief*

Medical Guide for the Layman on the Recognition, Treatment, and Prevention of the Common Diseases of the Solomon Islands.[90] That same year Levers and Burns Philp provided financial support for a survey of hookworm infestation among their laborers because, in the words of the government medical officer,

> I have seen the physique of some natives considerably improved after being treated for hookworm. By increasing the physique of the native his working efficiency is also increased. Consequently if the working efficiency of the native is improved by only five percent through treating him for hookworm, much good will be accomplished and the native labour force will become more effective.[91]

When sickness struck plantation workers, they tended to blame it on the breaking of a taboo by themselves or their relatives, which offended ancestral spirits, or on sorcery. This perception was so palpable that even the labor inspectors were forced officially to attribute two deaths on plantations in 1921 to "native superstitions."[92] In the late 1920s at Levers' Faiami plantation, Russell Islands, four laborers were so ill they were taken to the hospital at Tulagi, where they died, probably from beriberi. A kinsman of one and a fellow laborer believed that his relative died because he had failed to make recompense for an offense against custom before he left home.[93] Most experienced planters knew how pervasive such beliefs were and tried to allay the fears of laborers and to move them temporarily from an area where deaths had occurred.

Planters were less sympathetic to other forms of interplay between the material and nonmaterial world among their laborers. Now and again a disturbed Malaitan would go "cranky" and run amok, attacking people and lashing out at things.[94] Europeans, uninterested in the finer points of indigenous cosmology, attributed this to the "devil," a pidgin term that encompassed a multitude of states and complaints. Carden Seton, a planter in the Shortlands, dealt with sufferers by getting two of their strongest countrymen to hold the heads of the possessed in a bucket of water until the "devil" left them.[95]

To protect themselves from adverse spiritual forces, pagan Malaitans carried with them from home some relic or a few pieces of shell valuables dedicated to their *akalo*, or ancestor spirit. They kept these on the plantation and offered them in supplication when they sought their *akalo*'s protection. Sometimes, perhaps because of offenses on the plantation or the death of a relative at home, a laborer would buy a pig to sacrifice to the ancestors. Given the constraints of plantation life, the requisite religious forms could not always be followed. Laborers created substitutes. When no pigs were obtainable at Lofung in the Shortlands, a laborer helped himself to one of the planter's calves and killed it for the ancestors.[96] Other societies' sanctions were not so readily understood or placated. The Malaitans feared the Guadalcanal sorcerer's powerful magic *(piro)* and the deadly *vele* man; the Guadalcanal people believed that Malaitans could make sorcery against them; and in turn, Malaitans also feared the "devil devil" of San Cristobal.[97]

An increasing number of Solomon Islanders were Christians in the 1920s. By 1930, even on Malaita, almost half the population followed the new religion.[98] On most plantations Christianity was as much a means of control of overseers and managers as of laborers.[99] Missionaries, both brown and white, encouraged adherents to take their problems with management to the government officer. Although violence continued as a standard way of resisting an over-demanding master, the Christian laborer did have new moral issues with which to contend. At Tetepare (SIDC), Batini Bosuri wrote a letter to a friend outlining his dilemma: "We sign long [in front of] government—we do not come here to be beaten. We want to beat him [the overseer]—but our teacher Napthali says we don't beat because we belong God now. Tell Turner [Officer in Command of Constabulary]. We report to government in labour office and I said I want to beat if master beats us—he said no—we must tell government.[100]

Christianity was sometimes the basis for new friendships between people from different regions. Men who had previously only called kinsmen "friend" began to apply this term to former strangers. They sometimes returned with their new friends to the latter's villages, stayed on, found a wife, and married.

Pagan laborers had their own solidarity, but ancient animosities based on linguistic, regional, and "tribal" differences were close to the surface because laborers were not all that far removed from their home in either time or space. However, ideas were exchanged between groups. The use of pidgin on plantations and recruiting vessels enabled Malaitans, for example, to discover that as inhabitants of the one island, they shared many religious concepts.[101] Nonetheless, disputes between different groups were common. As Malaitans constituted two-thirds of the labor force, they were usually involved. A minor falling-out, resulting, for example, from the slowness of the cook boys to bring lunchtime rice out to the copra cutters, could lead to an exchange of swearing in pidgin or in a vernacular. Malaitans were especially sensitive to any swearing that insulted their ancestors and would demand compensation in cash, as laborers were far from sources of traditional valuables. If this was not forthcoming, a brawl could result. Unexpected differences sometimes arose. At Aruliho, Guadalcanal, a planter shot hordes of semiwild cats that had become a nuisance. The Malaitans collected and ate them, much to the open disgust and derision of the Guadalcanal men.

Pagans often found themselves at odds with the Christians. In many Melanesian societies men feared the procreative power of women. On Malaita particularly, people believed that this capacity in women, unless hedged with safeguards and confined to limited areas, could pollute and debilitate the males as well as offend the ancestors. Christians had abandoned such beliefs, so conflicts arose when Christian couples lived among pagan Malaitans. If a woman drew water from a tank used by Malaitan pagans, they considered this polluted them and would demand compensation, usually £1 to £5. If a woman were menstruating while cooking the laborers' meals, the pagans would object. Likewise, a woman about to give birth was forced to do it out in a bush in a place that was often as alien and frightening to her as to the Malaitan men. A planter on Bagga (Mbava) refused to reemploy one of his best bossboys because the latter forced a woman to do this, the baby dying as a consequence. Of course, it was not only the Christians with whom the pagan Malaitans quarreled regarding their value system. Several Guadalcanal and San Cristobal societies did not have such rigid traditional beliefs, but as Malaitans were usually in the majority, these kinds of arguments were common.[102]

To avoid such disputes, both planters and men preferred to have distinct groups of laborers accommodated in houses with eight to twenty of their fellows. This was less a capitalists' ploy to prevent the growth of a workers' class consciousness, and more a practical measure to prevent men being hurt or even killed, albeit in the interests of efficiency. That it in some way hindered the exchange of ideas was incidental. As more Christian laborers offered for work, Malaitan spokesmen made it clear that they wanted pagans to be housed apart, especially from married Christians.[103]

The recruiting of single, unaccompanied women was anathema. This was as much a reflection of the rigid view of the position of women in Solomons Islands societies as it was of the European perception that all women "belonged" to some male—a father, an uncle, a brother, or a husband.[104] In the Solomons as in other agricultural-based subsistence societies, young women represented the means for the physical and social reproduction of the clan. Male clan elders controlled the "flow" of women by the provision of bride payments to young men. This was the basis of the power of gerontocracy, particularly in the eastern Solomons. To preserve the value of such women before marriage, ideologies flourished that condemned adultery and, on Malaita, even premarital sexual intercourse, with sanctions invoking the wrath of the ancestors on the reproductive and productive resources of the living. The same ideologies also kept the women confined to the village home and hearth and, in doing so, guaranteed their continuity, as women played a central role in socialization of the young and transmitting cultural information.[105] These ideologies also constituted barriers to large-scale permanent migration of families into possible plantation settlements.[106]

The government early forbade recruiting of single, adult women in 1909 and had been quick to enforce this when a Levers' recruiter on the *Ruby* accepted two single women from north Malaita. Woodford perceived native women, rather than the men who sought them out, to be so much a "fruitful cause of trouble" on plantations and elsewhere that he refused to have them, even with their husbands at government headquarters at Tulagi. More important, village-bound females guaranteed the reproduction of the labor force at minimum cost to government and planters.[107]

Although the occasional official considered the idea of institutionalizing prostitution on plantations, Colonial Office policy precluded this, as did undoubted mission opposition and the difficulty of establishing a system that would not offend, in some way, the mores of the various Solomon Islands societies, including the conservative and proprietary Malaitans.[108]

Plantation laborers sometimes formed both licit and illicit associations with women in nearby villages, especially on San Cristobal and Guadalcanal, but in general the villagers disapproved, and so liaisons were few and difficult to arrange.[109] In the Russells and Kolombangara, neighboring villages were sparse, so there, where the concentration of unmarried males was greatest, opportunity for contact with women was least.[110] Explicit labor-line songs of sexual fantasy only intensified the men's needs. Homosexuality was common on the larger plantations. Government officers deplored these "unnatural offences" but could see no solution.[111] Management did not take much notice, as "the majority adopt the view that they are not concerned with the morals of their labor but only their work and production of figures required by the Principals."[112] Relationships tended to be between young recruits of about sixteen and older men in their thirties. Although such relationships were pecu-

liar to the plantation (and the labor lines in Tulagi), older men, usually Malaitans, were very possessive, often having given the young
man's kinsmen gifts of calico and tobacco to seal the arrangement.
Jealous outbursts often resulted in fights, but this rarely interfered
with the daily routine of the plantation. The mores of most Solomon Islands societies, especially on Malaita, opposed any form of
male homosexuality. In the light of the prevailing sexual ideology,
it was an admission of male weakness as much as a source of shame.
This episodic homosexuality was also an immediate way of resisting
the dehumanizing features of plantation life, and perhaps too of an
overdemanding older generation at home.[113]

Unlike overseas plantations, there were no "bright lights" in the
Solomons to relieve the monotony of the laborers' existence. Tulagi
was the only center that could be described as a town and was
designed primarily to serve the purposes of a colonial administration. By about 1914, Levers and Burns Philp had stores attached to
their labor transit depots on islets adjacent to Tulagi. The Chinese,
who had come to the Solomons as cooks, carpenters, and tailors,
moved into trading and by the 1920s were well established in stores
along the foreshore of Tulagi. There were few women at Tulagi, all
of them European or Chinese, and virtually all married. The most
that laborers in transit could hope for at Tulagi was a shopping
spree at the company stores or the preferred Chinese shops, plus a
meal or two in Chinatown. Former Queenslanders and laborers who
as boats' crew or domestic servants had developed a taste for alcohol could get it "under the counter" from certain Chinese, though
doing so was illegal and risky. Any roistering after dark on Tulagi
was done in secret because of a government curfew on native
people.

Prior to 1920 the major plantation companies, Levers and Burns
Philp, tried to confine laborers and their money to their stores at
Ghavutu and Makambo, respectively. Under the cover of night,
Chinese traders or local Nggela people would smuggle men across to
Chinatown by canoe to shop. The companies ceased their opposition by the early 1920s, as laborers had to go to Tulagi to sign off at
the Labour Department and because W. R. Carpenter had opened a
store there and objected to any unfair cornering of the "pay-off"
business. Generally, it was the Chinese who filled the laborers'
trade boxes, receiving in return most of the laborer's wages collected at the completion of his contract. Especially after a wage
increase in 1924, the Tulagi area traders did very well as the cash
advance (beach payment) given by recruiters to the recruit of up to
one-quarter of the two years' wages did not go entirely to the

recruit's elders. The balance, after the advance and spending at the plantation store, was a minimum of half the wage. At the completion of his contract this was paid to the laborers in front of a government official, usually at Tulagi or, if the east-west Tulagi axis was not traversed, another government center such as Gizo. Planters naturally wanted a greater proportion of the wage paid on the plantation, where their store goods sold at 100 percent markup. The government refused, well aware that not only the laborer needed protection but also his community, which relied on him for goods and cash on his return.[114]

Spending opportunities, sexual contact, fellow workers, housing, rations, and masters were all aspects of plantation life that varied with time and place. Laborers on the smaller plantations suffered less than those on the larger ones, while local casual laborers employed on a monthly basis had the best of two worlds.[115] However, there were changes in the government's administrative policies that affected everyone, no matter where or for how long they worked. In the early 1920s, two of those changes were to be instrumental in creating resentment toward the government that frequently found expression in the social system of the plantation.

Solomon Societies and the Plantation System

The small-scale societies of the eastern Solomons—on Malaita, Guadalcanal, and San Cristobal—subsidized the plantation system. They did this by bearing the costs of the reproduction of the labor supply, supporting and maintaining the laborer even when he was not a productive member of society—during infancy, old age, and illness.[116] With plentiful land, the communities could provide sustenance, although they were occasionally stressed because of disasters like cyclones. These societies were labor reserves that "produced" the laborer who, despite low wages, went off to work on plantations because of the power his elders had over him, especially in relation to obtaining a future wife. For the planter, although he could not get enough labor (at least until the 1930s), here was a continuous supply of men in their most productive years who cost him nothing to sustain except while on the plantation. It rankled the planter that he could not readily persuade many sound, healthy men to extend their contracts. Condemning employers who tried devious means to impoverish laborers in transit and thus induce them to "sign back," Bell, then Malaitan District Officer, summed up planter attitudes and the relationship between the capitalist and the domestic economy:

Many employers of native labour in the Solomon Islands appear to think that they should receive all their money back from the labourer forthwith, and also the labourer back, unless he has become useless and then he can be dumped on his people. They appear to think that the natives should be treated as cattle, and they do not for one moment consider the people who have suffered pain, and years of worry and toil, in order to bring the natives, who become labourers for the white man, into the world and to provide for them until they reach a state of usefulness. The natives would [sic] not continue to propagate their own species under cattle conditions and solely for the benefit of the white man.[117]

Just as distance and lack of communication acted to prevent any concerted industrial action by employees, so too did employers find they could not organize themselves to resist the increasing beach payment. Despite the formation of the Solomon Islands Planters' Association in 1914, someone was always opting in or out of any such informal arrangement, depending on the demand for labor. So the planters called for the government to abolish the beach payment or to limit its value. In 1922, K. J. Allardyce, a special commissioner from the High Commission in Fiji, investigated labor matters in the Protectorate with the aim of improving the entire system. Among his recommendations was the abolition of the beach payment, a change supported by both the High Commissioner and the Colonial Office.[118] The government had perceived the old beach payment system as one in which the "recruit is sold by the Chief or people to the recruiter who is the highest bidder and the purchase money is paid in trade."[119] The new system, the perceptive Allardyce believed, would place more of the planters' outlay into the hands of the individual laborer, rather than into the hands of the passage masters, village elders, or even the recruiters. However, the trade-off was higher than some planters had anticipated. The minimum wage was to increase by 100 percent (i.e., £6 to £12 per annum), with a beach pay of £6 in cash being the maximum advance allowed on the sum of the year's wages. Recruiting vessels were no longer permitted to carry trade goods either to "give away" or sell.[120]

Overall cost to the planter of around £90 total outlay for a recruit on a two-year contract remained about the same after April 1924, when the new system came into force, as it had been in 1922.[121] Although some planters complained they had gained nothing, the change was to their advantage. First, they could operate on a more businesslike footing, as they would now "know exactly what their costs will be."[122] Second, and more significant, the set beach payment meant that the real wage was fixed, and market forces were unlikely to push it up without government intervention. By the time the recruit was at the plantation, the contract had been ratified, so bargaining for above the minimum wage was impossible. Planters encouraged the belief that the minimum wage was fixed. The recruiter had no brief to offer higher wages and acted as a barrier between employee and employer just before the point of actual agreement for indenture in front of the government official. The new regulations came into force as the price of copra began to recover after the slump of 1921–1923, but renewed demands for labor saw no increase of the planters' costs that the limited supply had effected formerly by means of the beach payment.[123]

The change disadvantaged Solomon Islanders. The old passage masters and their descent groups gradually declined in status as they could do nothing to bargain for an increased beach payment—though until the 1930s some passage masters managed to get a "present" of 10 shillings or £1 from the recruiter for themselves because they convinced him that they still had a vital function. (By then their fate was sealed with the ubiquity of the introduced law, the pidgin language, and the use of well-known Malaitan crewmen as agents on recruiting vessels.) At first, laborers' communities saw the new system as a boon, as they believed the wage was £24 plus the £6 beach payment instead of £18 plus the £6. It was not until 1926 that they understood, when the first returners under the new system arrived home, that the £6 was merely an advance. Recruiters had not bothered to explain this in case of a boycott. The villagers found that cash was of no immediate use except to pay the tax, as for some years there were few stores and traders on Malaita and southern Guadalcanal. The result of the loss of the negotiable beach payment and the imposition of the tax, which took about one-tenth of the entire new wage bill, was a decline in the real disposable earnings for the laborers' community.[124]

In return for the tax, the government set aside funds to establish a "Native Administration" in 1922. Each major island or district was divided into subdistricts under the supervision of a district officer: by 1926 Malaita had 22, Eastern district (San Cristobal, Santa Ana, Santa Catalina, Ulawa, and Ugi) had 6, Guadalcanal 25, Gizo district 9, and Shortlands 2. Each subdistrict had a paid headman, and within the subdistrict each major central village had a headman and a village constable. This cost less than 2 percent of the Protectorate's annual expenditure and served to bring villagers under more direct government scrutiny. As well as enforcing regulations relating to sanitation, communal services, and sanctions against adultery, swearing, disorderly conduct, and the like, the headman was required to notify the district officer of the movement of people in and out of the district. Thus tax defaulters and runaways from plantations could be more easily detected.[125]

The imposition of the tax meant cash was a necessity, as the government refused taxes in kind. This had far-reaching economic and political effects on the societies of the eastern Solomons. In time, the clan elders became indebted to their juniors, who returned from the plantations with money for the tax. The goods they brought home could be readily absorbed into the linkages in the exchange system. But because an agent, external to the existing economic structures, required a cash payment as tax with no concomitant reciprocal obligation for goods or services in the customary mode, the elders increasingly found themselves the clients of young men.

Government control of the indigenous population tightened as an outcome of the tax and the Native Administration ordinances. Resentment at their enforcement was the proximate cause of the killings of government officials in inland Guadalcanal in February 1927 and on Malaita in the Kwaio district in October. In retaliation for the killing of District Officer William Bell, cadet K. C. Lillies, and local policemen in Kwaio, government reprisals involving a north Malaita militia were swift but inhumane. Investigations by the Colonial Office produced a critical review of administration. The more isolated bush areas of Guadalcanal and Malaita came rapidly under the government's edicts. By 1930, except for Christian and sparsely populated Choiseul, long neglected by government, the frontier was no more.[126]

Solomon Islanders grudgingly or gratefully accepted the Pax Britannica, but government interference with various customs, especially ways of settling disputes involving adultery, rankled the elders because it weakened their control over the young and undermined their authority. The introduced law seemed too easy on adulterers, who were fined or put in prison for a few months, growing fat on government rations.[127] Under the old system, death or a very heavy fine was the remedy.

For major crimes like murder, the law seemed still capricious and racially biased, just as it had appeared in the pioneering plantation era. Events on plantations in 1922, coinciding with the introduction of the hated tax and the "Native Administration," produced near-insurrection among sections of the population of southern Malaita. At Levers' Pipisala estate a man from Small Malaita (Maramasike) called Otainau refused to obey a command of the overseer, Firth, because the job was not part of the laborer's normal tasks. Firth assaulted him. Assisted by two tribesmen, Otainau, in turn, assaulted Firth. Not content to leave it at that, Firth and two of his friends, Dawson and McGregor, beat Otainau so severely that he died.[128] In the same year on Santa Isabel, C. Maxwell had been charged with the killing of another Malaitan, Oreta. The court acquitted and deported all the Europeans involved in these cases.[129] In revenge, on Small Malaita the people sought three "white heads." Some attacked Hillien, the master of the *Hawk*, one of Levers' recruiting vessels, severely wounding him. Another recruiter, Thomson, reported "a meeting of some 700 men who each gave from 1–10 strings [of shell valuables] each—a boxful of which is to be the reward of the successful murderers."[130] District Officer Bell instructed recruiters to make sure they used a second or covering boat while recruiting there and at Takataka in the 'Are'are district, which was restive too. However, a virulent illness "similar to bubonic plague" on Maramasike left the people "dying like sheep," putting any plans for wholesale resistance on hold, but anger toward the government, as well as Levers, persisted.[131] Five years later, in August 1927, an inexperienced overseer at Manning Strait (SIDC), J. Cameron, was killed by a man called Gousie (or Tomasi), who was from Uru in the Sinalagu region of Malaita. After arguments spread over several days about the way Gousie worked and his "go slow" tactics, Cameron finally hit him from behind with a *loia* cane and swore at him. Gousie felled Cameron, then hit him in the head with an axe, killing him. This case, which might have really tested the impartiality of the white man's law, did not get to court. Terrified, Gousie had run away and hanged himself, his body then being mauled by crocodiles.[132] Six weeks later, enough time for news by letter or word of mouth to reach Malaita, Bell and Lillies and their policemen were killed by Basiana and his Sinalagu people on Malaita.[133] In 1925 and 1928, two overseers at Levers' Tenaru and Lungga estates faced groups of aggrieved laborers who had ganged up on them. In each fracas a laborer died as a result of injuries, and the overseers, C. W. E. Smith and H. M. Rikard Bell, respectively, faced manslaughter charges. They appear to have been discharged and left the Solomons, confirming to the bulk of the population the arbitrariness of the white man's law.[134]

These were extreme cases, which demonstrated that brutal and sometimes fatal violence, although sporadic, continued as part of the plantation social system and had repercussions on the wider society. Low-level violence remained a ubiquitous and bilateral sanction of plantation society, especially on the large, company-owned plantations. Changes to the labor regulation penal provision in 1921 were designed to compensate the employer for loss of time by labor by an extension of the indenture period. Its administration was too cumbersome. Consequently, many planters continued to keep disputes out of court and did not apply for an extension of anything less than a month. When conflict arose, physical force was still a common resort. Only when someone was seriously hurt or considered the outcome grossly unfair did the magistrate's court hear of it.

Within this context of an acceptable level of physical force punctuated by the extreme of unlawful killing, the 1920s were otherwise fairly quiet years for the average laborer. The round of recruiting, working, payment, and repatriation had become routine. The government regularly checked plantation conditions and legislated to double the cash wage, creating an illusion of greater reward for the worker. In terms of the individual, the laborer was perhaps better off than before 1924, but his community was poorer, and the economic power of his elders weakened with the loss of the old beach payment and the annual demand for tax money.[135]

From the planters' viewpoint, plantation development, for much of the period since World War I to 1929, had reached a plateau. The planters regarded the market dislocations of World War I as transitory, but a slump in world prices for copra during 1921–1923 highlighted a consistent trend. In contrast to the years before World War I, world production of edible oils was exceeding demand. Planting of new areas all but ceased in the Solomons, with a mere 3,000 acres being added to the 57,000 under cultivation between 1924 and 1929. The emphasis was on maintenance, with a gradual increase in production being the result of more acres of palms reaching their full capacity.

The Great Depression and Plantation Society

The even tenor of life on plantations in the 1920s came to an abrupt close in 1929–1930 as wider economic imperatives impinged on the Solomons' economy. The Great Depression saw a dramatic fall in the world price for vegetable oil. By 1931 the price of copra, most of it low-quality smoke-dried, in the Solomons was down to £9 to £10 a ton as opposed to £4 to £9 a ton production cost, depending on factors such as size, location, age, and fertility of the plantation.

Of these production costs the main outlay was for laborers, each one costing a total of £35 to £40 per year.[136]

Planters made various economies. Although smaller concerns tried a variety of schemes involving local labor on short-term agreements or time-expired men "profit sharing" and growing their own food on plantation land, the continuing decline in prices eventually made these unworkable.

Under the straitened economic conditions, the surviving planters, especially Levers and Burns Philp, put far greater demands on indentured labor in the 1930s. Although all laborers could subsist on their land, they were not free to remain at home. As the numbers recruited fell, the laborers' communities also increased demands on those recruited because the tax money still had to be found for the stay-at-homes and the wherewithal to obtain trade goods. In a sense, the laborer was cornered, and the planters knew it. Even before the Depression, some planters tried to obtain a greater "task" of their labor than was conventional. The Labour Department had acted in 1928, establishing set payments for copra cut in excess of the "task."[137] By 1931 this overtime working or bonus system was common, with threepence being paid for every 28 lbs. in excess of the task of 450 lbs. on Levers' plantations. This meant £1 was paid for every ton of copra produced. The average copra cutter produced 600 lbs. a day and some earned 15 to 25 shillings in bonuses a week.[138] It meant he also worked longer. On some of Levers' estates in the Russells, this may have indirectly increased the deaths due to beriberi. As the ration scale of 1921 was deficient in vitamin B_1, longer working hours meant less opportunity to obtain dietary supplements from gardening and fishing. On Levers' plantations overseers strictly enforced the ban on laborers eating the company's coconuts, a source of vitamin B_1. When there were six deaths due to beriberi on one of Levers' estates in 1933, the Colonial Office intervened. A revised ration that included "green" coconut, mealies, and a yeast supplement rectified the situation until a new ration scale became law in the late 1930s.[139]

Planters preferred the "bonus" system to the task system because it could save them money. They could not easily reduce the cost of recruiting, rations, the tax, the fees for repatriation, for ratification of the contract, and for medical examination, amounting to about £40 a laborer each year. However, instead of four laborers (in return for wages plus general costs) each producing 450 lbs. daily under the old system, the planters could employ only three laborers each producing 600 lbs. (an additional 150 lbs. each) and

still obtain the same production—1,800 lbs. And all the planter had
to pay was the extra bonus payment. There were no general costs
for the redundant fourth worker. This system was effective, especially on the larger plantations.[140] In part it explains how the level
of copra production from the mid to late 1920s was maintained
throughout the 1930s with only half of the labor force of the earlier
period (see Table 1).

Plantation management feared similar retrenchment. By 1933,
wages for overseers had been reduced from an average of £20 a
month to £16, practically the basic or minimum wage in Australia.
Both Levers and Burns Philp consolidated neighboring estates and
closed down their more distant and less profitable plantations dur-

Table 1. Labor Employed in Solomon Islands

Year	Labor Employed at Beginning of Year	Recruits for Year	Total Labor Employed
1915	1,403	2,855	4,255
1916	—	—	—
1917	2,303	1,967	4,270
1918	2,898	1,888	4,786
1919	3,462	2,028	5,490
1920	—	—	—
1921	4,182	2,668	6,796
1922	3,743	2,400	6,143
1923	3,965	2,188	6,152
1924	3,704	2,062	5,766
1925	3,509	2,232	5,741
1926	3,703	2,665	6,368
1927	3,755	2,360	6,115
1928	3,840	2,176	6,016
1929	3,166	2,005	5,171
1930	3,454	1,909	5,363
1931	3,189	1,112	4,301
1932	2,187	1,726	3,913
1933	2,430	1,103	3,583
1934	3,410	1,163	3,578
1935	1,871	1,122	3,093
1937	2,059	1,264	3,607
1938	2,560	1,129	3,993
1939	2,478	1,015	3,796
1940	2,278	1,023	3,459

SOURCE: Murray Chapman and Peter Pirie (eds.), *Tasi Mauri: A report on population and resources of Guadalcanal weather coast* (Honolulu, 1974), Table 2:3.

ing the periods of lowest copra prices in the 1930s. Some managers and overseers lost their jobs and faced life in their homelands, where unemployment and privation were rife.[141] So management tried to extract greater production from labor, but not simply by relatively benign reorganization of the payment system. An incident in 1931 illustrated the return to methods typical of the years up to about 1917. When Frank Soro, a Nggela man supposedly doing "light work" as fitted his slight physique, could no longer keep up with the rest of a road gang after five hours of hauling stones on Levers' Pipisala estate, the overseer, Flegg, hit him several times. Flegg then had Soro tied up under the midday sun. Weakened and ill with dysentery, Soro died the next day, 18 October 1931. A superficial inquiry saw Flegg fined £25. Levers dismissed him. By the time "native sources" had drawn the attention of the Labour Department and ultimately the Resident Commissioner and the Chief Judicial Commissioner to the full story, Flegg had left the Protectorate. He was charged with manslaughter but never brought to trial. The Colonial Office censured the officials, with good reason. The subsequent inquiry had revealed that besides this crime, several of Levers' managers on their Russell Island plantations set their own punishments for "lazy" and wayward laborers. If "contract" laborers failed to complete a task in good time as defined by management, they had to work late or on Saturdays and Sundays. An overseer on neighboring West Bay estate punished a laborer for taking goods without permission by withholding his tobacco ration. He had also thrashed a laborer who came at him with a knife. Levers' management also interpreted the definition of "light work" to their advantage. "Light labourers" (including women) received half the regular wage—quite a saving if they were put on heavy tasks, such as, in the Soro incident, road construction and the carrying of stones. For workers on "bello" work (nine hours a day) the time taken to walk to and from where the work was located on the plantation was not counted as part of the nine hours. These practices were not confined to Levers. On Carpenter's estates on Santa Isabel, a new manager increased the task from 145 lbs. a day to 155 lbs. with no extra recompense for the laborers.

These were standard methods of "disciplining" laborers when management did not want to spend time and effort on resorting to the magistrate's court, and of maximizing output. Levers had been in trouble with the labor inspector in 1928 for breach of the "light work" clause. Withholding the tobacco was a common ploy, as no record of this was available for the inspector to see. Sometimes planters tried to reduce the food ration, but this was likely to cut the laborers' output, as companies like Levers realized. In 1937, at Carpenter's Estrella Bay estate (Santa Isabel), reduced rations met with a virtual strike of thirteen men led by Sifomamata of Narabu, Malaita. As the manager explained to the inspector, "Occasionally they work, but often they just give me cheek. They argue and shape up as if to fight." The inspector found in favor of the men. The employer was in breach of contract, so the men were not obliged legally to obey orders. Throughout the 1930s, laborers did not hesitate to report any consistent denial of the full ration.[142]

The strained nature of the relationship between company, management, and labor during the 1930s was obvious to experienced government officers:

> It is a well known fact that Plantation overseers in the Protectorate are held responsible by their employers for obtaining the maximum output of copra of which an estate is capable and they are aware that failure to produce the required amount may lead to summary dismissal. They are also encouraged to take the law into their own hands in dealing with lazy or refractory labour, rather than prosecute under the provision of the Labour Regulations, the loss of time initially involved in such proceedings being regarded as a further reduction in the productive capacity of the unit concerned.[143]

Laborers resisted the increasing exploitation. In the early 1930s there was "a curious epidemic of copra store and drier fires."[144] It was virtually impossible for the planters to discover the cause. From the laborers' viewpoint, fires were thus a successful form of resistance to planters. Except for the small uninsured operator, they were nothing more than an inconvenience to planters.[145] There were other "incidents." Despite the reduced labor force, convictions of laborers for misdemeanors and crimes and for offenses against the Labour Regulations multiplied more than threefold for the period 1930–1939 in comparison to the years 1920–1929.[146]

Although management maintained production in the early 1930s, the price of copra continued to plummet. The Solomons' ports price per ton was £4.61 in 1932 and £1.53 in 1934. To reduce overheads, the planters attempted to have the tobacco ration deleted, but the government refused because of the potential loss of excise duty.[147] Some planters wanted a wage cut for labor as early as 1930.[148] By 1933 all demanded it. In December 1934, the government set the minimum wage at the old pre-1923 level of 10 shillings a month. This saved the planters about 8 to 15 shillings a ton production cost, but it reduced the laborers' take-home wage by half. The bonus was also halved.[149] Instead of simply stating the industry could no longer pay the old wage and return a profit to planters, Resident Commissioner Ashley maintained that wages were only "pocket money" for the Solomon Islanders to subsidize the living that laborers made by subsistence agriculture: an admission if ever there was one that village community bore the reproduction cost of labor. Ashley claimed few knew the "value of money," and so the reduction was of no great consequence to them.[150] That did not reflect the feeling of the Solomon Islanders concerned. The immediate response was a boycott of recruiters on Malaita, Guadalcanal, and even far-flung Santa Cruz, without any apparent concerted organization beyond that of localized descent networks and the links created by crews on ships and returners. Recruiters predicted it would not last much more than a year, probably because of pressure for tax money. Its impact was lessened by the fact that there was a decreased demand for labor.[151] Nonetheless, for almost two years on Malaita, the "good type laborers," men proud of their own worth, would not accept work at the reduced wages. Their places were taken by physically inferior youths, those who normal times were least likely to find the wherewithal to obtain wives, in societies where there was a surplus of males to females.[152]

There were other ways of resisting. As late as December 1936, at Takataka, Malaita, men accepted the cash beach payment in advance from the recruiter but gave him false names and home villages and then disappeared. This happened so much that the District Officer warned the people of the region that they were getting a bad reputation.[153] There was a more sophisticated variant of this tactic for the young men: to agree verbally to sign on at 10 shillings a month, take the beach payment of £3, distribute it among their relatives, then once they were at Tulagi refuse to ratify the contracts, claiming that the recruiters had told them that the old rate of £1 a month prevailed. The recruiter lost heavily, as he was legally bound to return the recruit to his passage and stood little chance of reclaiming the scattered £3 beach payment. This practice had sprung up in the late 1920s, but it intensified after 1934. It persisted, so much so that recruiters met with the Resident Commissioner in mid-1937 at Aoke in the hope of finding legal redress. By December 1939, the Colonial Office advised that, rather than apply penal sanctions, a civil action follow any such repudiated contracts.[154] Strikes were forbidden and punishable under the penal clauses of the indenture. They did not occur on plantations on any significant scale, but in 1935 on one of W. R. Carpenter's vessels, the crew, who had learned of strikes elsewhere in the South Pacific, attempted one in order to obtain a better wage. As there were always more men wanting to work on ships than there were vacancies, this isolated attempt failed without any organization to unify the men.[155]

The wage cut of 1934 reduced the laborers' tolerance threshold to the tighter conditions on plantations in the Depression. Recorded desertions increased sixfold, despite the fewer numbers under indenture, and remained high for three years, by which time all the first intake on the lower wage had completed their contracts.[156] With the outbreak of war in Europe in 1939, copra prices went up, but the northern shipping lanes were blocked by the enemy, so Solomons' production fell from 21,666 tons in 1939 to 12,299 tons in 1940, of which 75 percent came from Levers' plantations.[157] Many planters were now totally ruined. Overseers' wages were reduced to £12 a month. On the larger plantations, further redundancies of European staff, as well as indigenous laborers, seemed inevitable. Because Levers' output was absorbed by their soap factory at Balmain, Sydney, they sustained production and profit as long as the war kept out of the Pacific, since the Australian market could not be supplied by European competitors.[158] On their plantations in the Russells, in 1940 a new feature on resistance seems to have developed. In that year and early 1942, assaults by laborers on employers markedly increased. This was attributable, in part, to "slackness on the part of the manager, frequently the previous one" who had allowed discipline to deteriorate. The laborers involved were Malaitans from Uru and Sinalagu—communities demoralized and disaffected by undisciplined punitive actions perpetrated by the government militia that rounded up the Kwaio people implicated in the Bell killings of 1927. On Levers' Mbanika and Yandina plantations (Russell Islands), there was for the first time "a certain amount of collusion between labourers on these estates" against management. Two "professional-troublemakers" among the laborers organized a group of twenty-five men, nine on Mbanika and the rest on Yandina, to violently resist their employers.[159] It seems likely that Levers' consolidation of management units in 1936 and the use of mobile work gangs across plantations may have made it easier for the men to spread the word of resistance from one plantation to another.[160] Certainly, the seeds of organized resistance were present for the first time across plantations, but resistance was still based almost entirely on kinship and "tribal" society lines.

Outside the structure of the plantation, Solomon Islanders were capable of discussing their differences as well as matters of common concern across language, "tribal," and even island lines. This was demonstrated in 1939. Responding to concerns voiced among his co-religionists, Richard Fallowes, Melanesian Missionary (Anglican) priest, formerly on Santa Isabel, attended a series of meetings with people there, on Savo, and on Nggela. Representatives from Malaita, Guadalcanal, San Cristobal, and the Russells also took part. Fallowes assisted local leaders in drawing up petitions to the High Commissioner, Harry Luke. One of the petitions from a Nggela headman, John Pidoke, asked the High Commissioner to give the Malaitans special help by means of increased wages for plantation workers and boat crew members—evidence that Islanders not only could transcend parochial interests but also felt the prevailing plantation wage inadequate. As well, in return for their tax, the petitioners wanted technical education, basic medical care, and relatively minor adjustments to assist their social and economic betterment. Other matters relating to crucial customary concerns—the punishment of adultery, marriage, compensation for social offenses —were considered, and the right to appeal to a higher court of law and for a "trained man" to assist in such appeals were requested. There was also a request for some kind of "Native Assembly" to represent Islander concerns. Despite these written petitions and the High Commissioner's discussion with Fallowes and with various headmen at San Cristobal and Aola (Guadalcanal), and Nggela, the demands of the "Chair and Rule" movement were largely ignored by the government. Although it deported Fallowes and told Solomon Islanders that they could not understand the needs of "the natives," the common concerns of many Solomon Islanders remained.[161] The development of native courts to rule on customary law was already under way in places such as Guadalcanal, but the Chair and Rule movement may well have spurred the government's initiatives on this and the establishment of councils on Malaita, Nggela, and Gizo in 1939–1941.

On north Malaita, by 1941 there were councils of elders at Maluu and Fo'ondo (To'ambaita), Lau lagoon in the Baelelea area (northeast Bali), in the Baegu area (east Bali), and the Langa Langa lagoon. There was also a council functioning on Small Malaita with others beginning in Kwaio and 'Are'are, along with a single court to arbitrate on customary law for Kwara'ae. The government, seeing increasing social instability in village society, was trying to prop up the old system of control by validating the authority of the elders in local matters and in the adjudication and punishment according to customary laws. All this involved much discussion of what "custom" and the elders' "ancient rights" were by the district officers and more particularly by the Islanders.[162] Despite their dislike of the government, the elders supported the councils and courts and requested legislative support to give their authority more clout.[163]

The Pacific War halted this process. Japanese forces invaded the Solomons in May 1942. Practically all expatriate civilians fled, and most plantations were soon battlegrounds. The war, in destroying the plantation infrastructure and removing the politically influential planters,[164] provided the High Commission with an "opportune time to review the whole question of policy in respect of economic development and to decide whether the old system is to continue or if we [the High Commission] should endeavour to replace it with something better and more suitable to local circumstance and the well-being of the native population." H. Vaskess, the Secretary of the High Commission, who had firsthand knowledge of the Chair and Rule movement, summed up the old order in 1943: "The old system was in fact if not in theory, forced labour at very low wages —forced by the imposition of a poll tax, by the desire for trade goods and the total lack of other means of earning money, and organised on a system of two-year indentures based on severe penal sanctions."

Vaskess recommended radical measures to give Solomon Islanders a greater role in the cash economy, including the transfer of plantations to them because the plantation system would be uneconomic without some form of "forced labour."[165] Many of his ideas were implemented, but some not as quickly or as thoroughly as he had suggested.[166] One thing was clear—the indenture system with all its attendant abuses was to be dismantled. After the war, civil contracts were limited to one year despite objections by Levers. In 1946, steps were taken to legalize trade unions.[167] In the absence of large numbers of plantation owners, the government introduced various subleasing and partnership schemes with neighboring villages to get plantations producing. When planters did return in significant numbers in 1947–1948, many were happy to continue with similar arrangements and the profit-sharing schemes that had evolved as a survival mechanism by the small planter during the Depression.[168]

The war had a tremendous impact on Solomon Islanders. During the military occupation, the government established the Solomon Islands Labour Corps (SILC) in 1942. It functioned until early 1946. Islanders, still from the major reserves of Malaita, south Guadalcanal, and San Cristobal, joined the corps as laborers. The corps was stationed at every military establishment but concentrated at

American bases on north Guadalcanal and, to a lesser extent, the Russells and Tulagi. Thousands of young Solomon Islander men passed through the SILC, which had a maximum strength of about 2,500. Interaction among different groups of Islanders and between them and the mainly American troops was intense, unlike the isolating conditions on scattered plantations prewar. The men received £1 a month in official wages, but earned a lot more from generous Americans for souvenirs and personal services. They discussed political matters with the Americans. Among the SILC men, the incipient class formation, evolving under the plantation system, suddenly flourished, with the SILC men developing a new consciousness and identity. With some obvious antecedents in the Chair and Rule movement, a new movement, called Maasina Rule, evolved a broadly based ideology, redolent with millennial expectations, preaching brotherhood and self-determination for the Malaitans, including a central council—a rejection of the government-sponsored district councils of the elders. Five of its nine leaders were members of the fundamentalist, but island-based, SSEM. These men, except for Timothy George, who was more a venerable figurehead, were young, mainly in their twenties, like most men in the SILC. Maasina Rule placed great emphasis on the codification of custom, even as to how it related to Christianity. Much time was to be spent on census taking and on constructing clear political and administrative hierarchies within each district and village.[169]

This was a creative time for Malaitans, and it needed to be. Their island and others were in a bad way. The sudden migration of villagers out of potential war zones and away from their gardens had caused malnourishment. The SILC had overrecruited, in some areas leaving few healthy men to make gardens. A failure of the sweet potato crop on Malaita had exacerbated the problem in 1943. The entire Solomons population declined, in some places, such as Guadalcanal, by as much as 11.8 percent.[170] In 1945, Maasina Rule leaders revived an idea of thirty years earlier, after seeing the apparent success of the Americans' intensive gardening of 2,000 acres on north Guadalcanal. They reasoned that their labor should be concentrated on the rehabilitation and development of their own lands, not on those of the planters. They planned new, large settlements with big commercial farms on the coast. Malaita was declared *tapu* to recruiters.[171]

This had no significance to the planters, because there were none in the Solomons at this time. In fact, in early 1945 there were no European civilians except for the odd missionary. The first half-dozen planters had returned on the *Southern Cross* in March 1946, much against the government's wishes. These people faced enormous difficulties in even getting themselves and their scarce supplies transported to plantations, let alone going to other islands to seek recruits. Any laborers they employed in the following months were from villages adjacent to plantations.[172]

However, the government now wanted labor in the new capital, Honiara, on north Guadalcanal, for a variety of new projects. By August 1946, the Solomon Islanders' cash reserves from the wartime "dollar invasion" of about £100,000 had dropped to £40,000. The Chinese, left behind on San Cristobal in the wake of the European civilian evacuation, had provided retail outlets in the immediate postwar period.[173] Moreover, prewar inflation of traditional valuables and subsequent intensive buying of these by souvenir-hungry Americans had greatly reduced their availability, as had diversion of labor from exploiting the sources of the shell as well as from its manufacture. It is not surprising then that Maasina Rule leaders relented and were prepared to let some of their men go. They needed to supplement funds being collected for a new "tax" for the movement. This money was spent on both development and "custom"—traditional religious ceremonies of the pagans. Malaitan leaders reasoned that with a higher wage fewer men need leave Malaita. In December 1946, the leaders demanded a monthly wage of £12—the average prewar wage of plantation overseers and the amount requested by the Chair and Rule petitions. However, with the Solomons' cash economy virtually nonexistent, both government, soon to be a major employer of labor, and planters refused to accept this, the going wage being £2 a month.[174]

The Maasina Rule leaders consequently called for a boycott of plantations and other employment away from Malaita in 1947. The Malaitans held huge meetings to discuss this and other matters. Messengers traveled around Malaita, Nggela, Santa Isabel, Guadalcanal, and San Cristobal spreading the news. Because of this and other more significant opposition to the administration, the government arrested, tried, and imprisoned the leaders for sedition in 1947–1948. Despite this, the boycott was a clear signal to the government that any return to the old system, be it on plantations or the government's expanding labor lines, was anathema. And it proved Islanders could organize themselves to resist on a large scale, given the appropriate conditions.[175]

Resistance continued on Malaita. The government imprisoned hundreds, until the jails could not cope. Labor recruiting was at a standstill. The large coastal communities created their own frictions. Sexual intrigues, conflict over taboos, hardship for the women and children, and the easy spread of infectious diseases took their toll of the "brotherhood." The millennial hope of "cargo" and help from the beneficent Americans failed to appear, and many followers of Maasina Rule became disillusioned. In order to break the stalemate, the government released the Maasina Rule leaders in 1950 when they promised to cooperate. Plantation labor was not easy to recruit because the men preferred working in centers like Honiara where government projects were concentrated. Nonetheless, laborers gradually came forward for plantation work, and production of copra commenced. However, with the failure to reestablish the old plantation system, more and more it was "native" copra, grown and produced by Islanders, as Vaskess had envisaged.[176]

Conclusion

Throughout much of the period 1900–1942, resistance by laborers to the inequities of the plantation system was neither large-scale nor coordinated. Groups of laborers on any one plantation had little or no contact with men on other plantations because of isolation. There were a few exceptions, such as Levers' estates in the Russells, but it was on just these plantations that discipline by the master was most rigid and penal clauses most often invoked. The two-year contract period meant an annual turnover of about half the employees. Men who consistently "signed back" were either boss-boys, obliged by their position to side more with management than laborers, or men who were frightened to return home because of a broken taboo or because their fecklessness had left them with nothing to take to their elders. Such men were unlikely to win followers among the laborers to resist management.

The endemic, small, and fragmentary socio-political units characteristic of most of the labor reserves also slowed the development of a workers' consciousness. And laborers were still close enough to home to be captives of their respective cultures. Although Christianity was the basis of new relationships among old enemies, not all laborers, particularly the bush Malaitans, shared this potentially uniting ideology. Ancient fears of old enemies were strong, and the suspicion of sorcery handicapped some attempts by laborers to learn about each other, just as much as did structural aspects of the organization of plantation life. With little or no formal education, laborers could talk to each other and the government in pidgin or broken English. As Europeans held pidgin somewhat in contempt, ordinary laborers did not often have the chance to articulate their demands for a fairer deal on plantations; nor were they able to find legal redress other than the assistance of the labor inspector or, in some cases, a sympathetic missionary.

In no way does this minimize the forms of resistance offered by Solomon Islanders. In terms of the constraints their particular cultural origins imposed on them and those enforced by the plantation system, Solomon Island laborers resisted successfully because they, as a group, refused to be cowed into submitting to the unregulated appropriation of both their labor and sometimes their lives on the plantation, and their taxes at home. To look for some overarching organization or organizer of resistance is to impose the bureaucratic structures of a different society, that of the colonizer.[177] What Scott says of peasant society is true of the migrant laborers:

> The intrinsic nature and, in one sense, the "beauty" of much peasant resistance is that it often confers immediate and concrete advantages, while at the same time denying resources to the appropriating class, and that it requires little or no manifest organization. The stubbornness and force of such resistance flow directly from the fact that it is so firmly rooted in the shared material struggle experienced by a class.[178]

It was in resistance of an everyday sort to common, shared impositions that groups of Solomon Islanders from different societies began the process of discovering themselves as a class.[179]

In spite of these social and imposed limitations, then, resistance to oppressive aspects of the plantation system and of laws that fostered it in the wider society was ubiquitous. Though sporadic, it ebbed and flowed as part of the larger tide of the plantation-based economy of the Solomons. During the early years, the establishment of plantations in hostile territory put workers under tremendous pressure, as did the maintenance of production by reduced numbers during the Depression. Periods of steady, if spontaneous, resistance were punctuated by more intense and active opposition in the wake of other crises for the Solomon Islanders, such as the imposition of the male head tax in 1921–1923 and the reduction of the plantation wage in 1934.

The most common form of resistance was physical violence against the master. A distinction must be made between "try-out," which set boundaries of compliance and resistance, itself a product of violent frontier conditions, which a European newcomer and, for that matter, a group of laborers had to go through to establish mutual, if macho, respect and that violence toward a master as retaliation for blatant injustice. Both processes—the "try-out" and physical resistance—could be set in motion by the laborers' use of "go slow" tactics, insolent bearing, "back talk," and the like. These antagonized overseers, especially the young and inexperienced, and often involved them in fights from which they emerged second best. Often they let the matter drop so as not to waste time and money with a hearing at a distant court. Sometimes, too, violence between management and labor arose from personality clashes or weakness rather than oppression. Individuals, brown and white, had their own particular fears engendered by their belief system, illness, loneliness, sexual frustration, and strange surroundings. Under these stressful conditions, minor misunderstandings could flare up into heated confrontation.

Such incidentals aside, resistance in the form of physical violence was probably at its apex before 1917, as was the violence planters perpetrated on their laborers. Laborers fought back against maltreatment of kinfolk, the violation of socio-religious beliefs, being verbally abused, being forced to work overtime, short rations, and excessive tasks as well as neglect, beatings, blows, woundings, and even deaths meted out by overseers and planters. In the pioneering phase of plantation development, there were no inspections, so laborers had little redress except self-help. If statistics of violent crimes against the person as well as misdemeanors are any indication, the 1930s also saw more of this kind of resistance than the 1920s, as employers tried to extract more work from fewer workers. Desertion was another form of resistance and appears to have been more frequent before 1917 than in the 1920s, even in the face of great risks. There was another spate of desertions following the wage cut of 1934, despite the network of district officers and headmen who were poised to catch runaways. The wage cut also saw a two-year boycott of plantations by the better-class potential laborers. The boycott had antecedents in the earliest form of passive resistance—the avoidance of the notorious plantations by men in the 1900s and 1910s. Yet these protests against the system remained disconnected and largely uncoordinated except for the circulation of ideas through informal linkages typical of small-scale societies.

During the 1930s other forms of resistance appeared. The burning of copra driers was popular for a while, but more sophisticated methods evolved. Apparent recruits accepted the cash beach payment, then disappeared into the bush or repudiated contracts at Tulagi. This retaliation for the reduced wages ultimately meant losses for the planter.

On the plantations, employers and laborers tried to outwit each other with new strategies and ruses. To obviate conflict, to reduce supervision and costs, the planters, by the 1920s, had accepted the task orientation of their workers rather than working to the clock. Dishonest overseers increased the conventional task, exploiting its lack of precise definition by the government. Laborers, in turn, resisted these demands by devising ruses to make their copra bags appear full. Planters then took to weighing the copra. In the 1930s, with increased stringencies, some overseers again tried to increase the weight of the task. Laborers refused to work or adopted a "go slow" approach. The planters called for the labor inspector or magistrate to enforce the contract. The labor inspector supported the laborers. Long before the 1930s, the laborers had learned that the labor inspector was of use to them. By the 1920s, appeal to the inspector was common, and Solomon Islanders' understanding of the new law was often very exact. The inspectorate was an outlet, often a very successful one, for the laborers' feelings of resentment toward the planter. Because of this it was also useful for the planters and the government in that it probably prevented the development of large-scale concerted resistance by laborers.

As exemplified in the inspectorate, the Protectorate government's role was to mediate between management and labor. Doubtless, there was a strong element of justice and humanity in the motivations of the formulators of policies, both at the Protectorate and Colonial Office level. What also cannot be doubted was the pragmatism of the government's institutional arrangements to preserve and protect the labor supply for the plantation economy. Without that economy there would be no revenue for government. The British Treasury had made it clear that the Solomons had to be self-supporting. The various revisions of the labor regulations aimed at preventing overexploitation of labor by shortsighted planters, just as the Native Administration and Tax ordinances in the early 1920s aimed at keeping the labor reserves policed and the numbers of workers available for plantations at the maximum possible.

Improvements in the laborers' lot as embodied in successive labor regulations came from administrators, yet they took their cue from the representations and the reactions of the laborers via the labor inspectors, rarely from the planters. Had the workers had a corporate organization, the lag between abuse or problem and solution so commonly evident on the plantation would have been less— if such a potentially threatening organization had been allowed to survive by the planters. The planters, too, despite their own association, could have achieved their goals more rapidly if they had been more unified and less in competition with one another. What they failed to achieve—the control of a limited supply's power to increase the price paid to it when demand was high—they persuaded the government to do by the abolition of the beach price. Thus, government intervention in employer-employee relations was not all one-way. In the eyes of the Solomon Islanders, the fact that the government set a wage meant it was a fixed, not merely a minimum, wage, a misconception planters fostered. One problem that the government failed to solve before World War II to the dissatisfaction of all concerned was the cumbersome mechanism associated with the penal aspects of breaches of indenture. Much of the violence on plantations was directly traceable to the nature of indentured contracts. The solution was found after the war in the law of civil contract. The war provided the impetus to change the labor system of the Solomons. The government displayed its pragmatism again. Rather than expropriate plantations for the local people and pay the planters for them, the government simply marked time until sheer economics overtook the system. Without the prewar indenture system, with its penal clause and low wage structure, expatriate-owned plantations, at best, were only marginally economic. Many planters and plantation companies soon abandoned their deserted lands or sold relatively cheaply to the local people, the government, or highly capitalized companies.

Thus, the government's role was an interventionist one in the plantation system. Regulations and practices relating to and emanating from that system permeated the wider society, as well as the plantation.

Likewise, resistance was not confined geographically to the plantation. Southern Malaitans attacked a recruiter and were poised for payback killings in 1922 when they heard the outcome of the trials of killers of Malaitan laborers on plantations in the Russell Islands and Santa Isabel. Even the killings of Bell and his men occurred within the context of numerous abuses to east Kwaio laborers over the years.

Conversely, events affecting the wider society touched potential laborers. The Native Administration regulations circumscribed the old prerogatives of the elders, and that in turn threatened to destroy their authority over the young. The tax forced men to go to work on plantations because of the need of the community to find cash. This created more demands on village society, but it was also a new way by which the young could resist the complicity of their elders in the oppression of the plantation system.

Before World War II, the tax in the eastern islands was not quite so high that the continued existence of the subsistence economy was threatened—for the government to do that would have thrown the reproduction cost of the worker onto the plantation system and made it uneconomic. Yet the tax was the death knell for the elders' authority within their own societies. From the days of the overseas labor trade into the plantation era, the elders dealt with the comprador passage masters to send their young away to get the goods of the white men to bolster their own authority and power in the community. These elders in turn dealt with other elders to get young women as wives for returners from plantations. This was sanctioned by indigenous ideologies. But by the late 1920s, the government and the missions, including returners from the overseas labor trade, either by force or persuasion, introduced alternatives that allowed these ideologies to be questioned more than ever before.

The passage masters and elders were linkages between two modes of production—the domestic and the capitalist. When the need for tax arose, only the young men could earn it. Unlike reciprocal traditional exchanges, the tax money flowed outward from the village, beyond the subsistence economy, failing to create indebtedness in the government and failing to provide elders with the means to make effective economic and social demands on that government. The abolition of the beach payment on goods, aided by the spread of pidgin, saw the passage master's role shrink in the late 1920s and atrophy entirely by the early 1930s. Elders became indebted to their juniors for the inflow of money for the tax, and all the more rapidly after 1934 when the lower wage prevailed. The demands of the elders to respect the old ways could be increasingly ignored as their economic power waned and alternatives to the old beliefs flourished. Unless there was a virtual revolution or a new age, the sole hope for the failing gerontocracy was in the only type of reciprocity offered by the government—the elevation of the elders' status in the new district councils and courts. By 1942, it seemed that the elders might be able to use central government initiatives to regain their former position.

It was no wonder that the young leaders of Maasina Rule wanted no truck with these councils and courts, but aimed for a central one in which, presumably, they would have the power. Any local councils needed to deal with minor concerns would have their appointees on them. In Maasina Rule's brotherhood, the young men saw a real probability for recognition of their importance in the economy of a changed Solomons, both in the local and wider society. And it is no wonder that the elders too supported Maasina Rule, glorifying as it did in its ideology *kastomu* or "custom"—the old traditions and ways (often not all that old, and not all that traditional). The elders would have a major role, probably as custom chiefs (or *araha*), busying themselves in the lengthy codification of these matters. Since the composition of the central Malaitan council was never clearly defined, perhaps they could still play the old power games on an even larger stage. The fact that fewer young men would be beyond their control on Malaita must also have appealed, especially since some of their earnings were set aside for "custom." The elders could see in all this seemingly new order a possibility of reviving and shoring up their hegemony, which had flourished probably as never before from the early years of the overseas and internal labor migrations until the 1920s. And all this could be done, it now seemed, without the patronage of an interfering and alien government and their often puppet headmen. In its millennial features, Maasina Rule provided a new order where the old patterns of exploitation and status degradation, within both the wider and the particular societies, were overturned. Practically all Malaitans both young and old saw the movement as a way of registering opposition to the exactions of the plantation system and its treatment of labor. With apparently something for everyone, the movement had wide support.

The plantation system was central to the capitalist economy of the Solomons. Because that economy demanded labor, it inevitably impinged on the subsistence economy, finding a niche in existing socioeconomic exchange patterns, where it could command collaborators. Resistance to the demands of the plantation system was most violent on the plantation interface of management and labor, but from the earliest, it extended beyond those boundaries into broader relationships between, on one hand, the government and the capitalists and, on the other, eastern Solomon Islands societies. By the 1930s, the plantation economy and its attendant administrative props had made such demands that one generation was resisting the other at the very heart of village society. Such resistance then called forth political movements to try to deal as much with the deep, divisive social conflict as with the impositions of an alien government.

Abbreviations

AA	Australian Archives (Canberra).
AC	Advisory Council.
ADC	Acting District Commissioner.
ADO	Acting District Officer.

AHC	Acting High Commissioner.
ANU	Australian National University.
AR	Annual Report.
ARC	Acting Resident Commissioner.
BP	Burns Philp.
BPA	Burns Philp Archives, Rare Book Room, Fisher Library, University of Sydney.
BSI	British Solomon Islands.
BSIP	British Solomon Islands Protectorate.
BSIP-AR	Annual Report of BSIP.
BSIP-CS	BSIP Confidential Series.
CAO	Commonwealth Archives Office.
CO	Colonial Office (United Kingdom).
CO 225	Records of the Colonial Office, Series 225, Western Pacific, PRO.
Col. Sec.	Colonial Secretary.
CRSI	Correspondence relating to the Solomon Islands.
DC	District Commissioner.
DO	District Officer.
FRCP	Fairley Rigby company papers, University of Melbourne Archives.
Gov.	Government.
HC	High Commissioner.
HMSO	Her Majesty's Stationery Office.
KR	King's Regulation.
LD	Labour Department.
LTR	Land Titles Record.
MMP	Methodist Mission Papers, Mitchell Library, Sydney.
PIM	Pacific Islands Monthly.
PM	Prime Minister.
PRO	Public Record Office, London.
QSA	Queensland State Archives, Brisbane.
QR	Queen's Regulation.
RC	Resident Commissioner.
Reg.	Regulation.
Sec.	Secretary.
Sec. S	Secretary of State.
SI	Solomon Islands.
SIDC	Solomon Islands Development Company.

SILC	Solomon Islands Labour Corps.
SS	South Seas.
SSEM	South Seas Evangelical Mission.
UA	Unilever Archives.
VKTCA	Vanikoro Kauri Timber Company Archives, University of Melbourne Archives.
WP	Woodford Papers, Department of Pacific and Southeast Asian History, ANU (microfilm version).
WPA	Western Pacific Archives, Suva.
WPHC	Western Pacific High Commission.
WPHC 4	Records of the Western Pacific High Commission, Series 4, Inwards Correspondence-General, PRO.
WRCP	W. R. Carpenter and Company Papers. In writer's care.

Notes

1. Doug Munro, "The Origins of Labourers in the South Pacific: Commentary and statistics," in Clive Moore, Jacqueline Leckie, and Doug Munro (eds.), *Labour in the South Pacific* (Townsville, 1990), xxxix–li. For a discussion of labor reserves in Melanesia, see Colin Newbury, "The Melanesian Labor Reserve: Some reflections on Pacific labor markets in the nineteenth century," *Pacific Studies*, 4:1 (1980), 1–25.

2. Judith A. Bennett, *Wealth of the Solomons: A history of a Pacific archipelago, 1800–1978* (Honolulu, 1987), 104–105

3. Bennett, *Wealth of the Solomons*, 106, 126–146.

4. Deryck Scarr, *Fragments of Empire: A history of the Western Pacific High Commission, 1877–1914* (Canberra, 1967), 283–284.

5. BSIP-AR 1896, WPHC 4, 477/1896; BSIP-AR 1898, WPHC 4, 286/1898.

6. Bennett, *Wealth of the Solomons*, 106–112.

7. Woodford to O'Brien, 17 April 1898, WPHC 4, 9/1898; Woodford to O'Brien, 11 September 1899, and encls., WPHC 4, 285/1898; Mahaffy to Woodford 1 August 1898, and encls.

8. Bennett, *Wealth of the Solomons*, 397–404; Mahaffy to HC, 8 April 1910, CO 225/91; im Thurn to Crewe, 4 August 1910, CO 225/92; Woodford to May, 28 February 1912, and encls., WPHC 4, 831/1908; Woodford to Major, 6 May 1912, WPHC 4, 1149/1912; BSIP-AR, 1914–1915, 1917–1918 (printed); Mahaffy, Memo on duties of District Magistrates 16 December 1911, WPHC 4, 1196/1911; Memo: The Labour Question in the Pacific Islands 1909, CO 225/87.

9. Woodford to O'Brien, 28 December 1901, and encls., WPHC 4, 41/1902; Nicholson to Danks, 4 October 1909, MMP; Nicholson to Danks, 7 December 1909, MMP; Nicholson to Danks, 27 January 1910, MMP; Goldie to Danks, 7 February 1910, MMP; Nicholson to Danks, 1 March 1910, MMP; Goldie to Danks and encl., 11 March 1910, MMP; Nicholson to Danks, 10 April 1910, MMP; Goldie to Danks, 29 May 1910, MMP; Nicholson to Danks, 15 June 1910, MMP; Frank Burnett, *Through Polynesia and*

Papua: Wanderings with a camera through southern seas (London, 1911), 145–157; Woodford to HC, 23 September 1909, and encls., WPHC 4, 1121/1909; Clarence T. J. Luxton, *Isles of Solomon: A tale of missionary adventure* (Auckland, 1955), 96.

10. Barnett to Major, 23 June 1909, WPHC 4, 782/1909.

11. Mahaffy to Major, 21 December 1908, WPHC 4, 281/1911.

12. Meek to RC, 30 March 1909, and encls., WPHC 4, 206/1909.

13. Edge-Partington to Woodford, 7 July 1911, WPHC Malaita Series (Mala), 66/1911; Edge-Partington to Woodford, 6 December 1911, WPHC (Mala), 97/1911; Walsh to Woodford, 24 April 1911, WPHC (Mala), 22/1912; Walsh to Woodford, 5 September 1912, WPHC (Mala), 55/1912; Edge-Partington to Woodford, 16 June 1913, WPHC (Mala), 64/1913; Campbell to RC, 29 January 1914, WPHC (Mala), 73/1913; Bell to Barnett, 17 November 1915, WPHC (Mala), 38/1915; Bell to Barnett, 11 June 1916, WPHC (Mala), 4/1916; Bell to Barnett, 20 November 1917, WPHC (Mala), 29/1917; Barnett to HC, 27 January 1917, and encls., WPHC (Mala), 489/1917; Mahaffy to im Thurn, 8 April 1910, CO 225/91. See also Mumford to Fairleys, 21 May 1914, FRCP; Mumford to Fairleys, 25 December 1914, FRCP; Hermes to Sec. S, 12 January 1909, CO 225/89; Tillotson to Under Sec. S, 2 September 1910, CO 225/94; Meek to May, January 1912, WPHC 4, 61/1905.

14. Lucas, Reports on SI Plantation Properties, 1 October 1910, BPA; Mumford to Fairleys, 1 January 1919, FRCP; LD-AR 1913, WPHC 4, 801/1914; May to Harcourt, 20 June 1911, encl. in CO 225/96; Peter Plowman, interview (see list of informants in Bennett, *Wealth of the Solomons*, 501–503).

15. J. S. Phillips, *Coconut Quest: The story of the search for the Solomon Islands and the East Indies* (London, 1940), 54.

16. Mumford to Fairleys, passim, FRCP; Phillips, *Coconut Quest*, 54.

17. M. Hirst, Memoirs, typescript in BPA; Peter Plowman, interview; Woodford to HC, 5 February 1914, WPHC 4, 801/1914; LD-AR 1913; Woodford to Major, 27 January 1909, and encls., WPHC 4, 206/1909; Tillotson to Under Sec. S, 2 September 1910, CO 225/94; Major to Earl of Crewe, 13 March 1909, and encls., CO 225/85; Woodford to im Thurn, 14 October 1908, and encls., WPHC 4, 693/1908; Woodford to [HC?], May 1904, encl. in WPHC 4, 136/1904; FRCP: Mumford to Fairleys, 22 August [1915?], FRCP.

18. [?] to Fairley 10 July 1912, FRCP; Lucas to Burns, 28 July 1908, BPA; Lucas to Burns 1 April 1910, BPA; Mahaffy to HC, 21 December 1908, WPHC 4, 281/1911.

19. [H. M. Bell], *Man No Good: An autobiography of the South Seas* (London, 1936), 26–44; Re Bell, or Rikard-Bell as he was also known, see LD-AR 1928, WPHC 4, 1426/1929; WPHC 4, 770/1928; Hermann Norden, *Byways of the Tropic Seas: Wanderings among the Solomons and in the Malay Archipelago* (London, 1926), 30; Barley to HC, 23 July 1917, WPHC 4, 2540/1917; Charge against F. E. Gilbert, 31 July 1917, WPHC 4, 2843/1917; Woodford to May, 22 January 1912, and encls., WPHC 4, 515/1912; Tommy Elkington, interview; C. M. Woodford, *Protectorate of the British Solomon Islands: Statistics to 31 March 1909* (Sydney, 1909), 12; im Thurn to Earl of

Crewe, July 1908, and encls., CO 225/76. See also Woodford to im Thurn, 25 October 1906, CO 225/91; Mahaffy to im Thurn, 8 April 1910, CO 115/91.

20. Jack McLaren, *My Odyssey* (London, 1923), 245; Eric Muspratt, *My South Sea Island* (London, 1931), 112–113, 121, 223–225, 233–234, 255; Caroline Mytinger, *Head-hunting in the Solomon Islands around the Coral Sea* (New York, 1943), 89–90, 100; Mumford to Fairley, 20 September 1920, FRCP; Mumford to Fairleys, 1 January 1919, FRCP; Lucas, Reports on SI Plantation Properties, 1 October 1910, BPA; Peter Plowman, interview.

21. "A few hints on Malaria and its treatment," n.d., VKTCA; Hirst, Memoirs, 13; Muspratt, *My South Sea Island*, 110–111; Mumford to Fairleys, 10 January 1918, FRCP. See also Lucas, Reports on SI Plantation Properties, 1 October 1910, BPA.

22. Peter Ryan, *The Hot Land* (Melbourne, 1969), 34. Hirst, Memoirs, 12; Muspratt, *My South Sea Island*, 107; Mytinger, *Head-hunting in the Solomon Islands*, 100; im Thurn to Sec. S, 27[?] April 1906, CO 225/76; Mahaffy to im Thurn, 8 April 1910, CO 225/91. Re suicides see WPHC 4, 698/1908.

23. Muspratt, *My South Sea Island*, 133–137.

24. Mahaffy to Major, 21 December 1908, WPHC 4, 281/1911.

25. Bennett, *Wealth of the Solomons*, 113, 193.

26. Census of BSIP for 1931, 23 December 1931, WPHC 4, 284/1932.

27. Scarr, *Fragments of Empire*, 283; im Thurn to Lyttelton, 2 November 1905, CO 225/69.

28. Mahaffy to Major, 21 December 1908, WPHC 281/1911; Vernon, Notes on SI, 7 December 1912, WPHC 4, 63/1913.

29. Woodford to im Thurn, 26 December 1909, and encls., WPHC 4, 774/1909.

30. Vernon, Notes on SI, 7 December 1912, WPHC 4, 63/1913; WPHC 1154/1913; RC to HC, 16 May 1913, and encls., WPHC 4, 63/1913; Woodford to HC, 26 June 1914, and encls., WPHC 4, 1784/1914.

31. Stubbs to Under Sec. S, 23 October 1911, and encls., WPHC 4, 1505/1913.

32. Woodford to im Thurn, 26 December 1909, and encls., WPHC 4, 774/1909.

33. Bennett, *Wealth of the Solomons*, 236–238.

34. KR III of 1910, CRSI, A1108, vol. 57, AA; KR VIII of 1912, WPHC 4, 1605/1912.

35. SI (Labour) Reg. 1897, encl. in Papers relating to Labour Trade, Premier's Department, vol. 85, QSA.

36. Woodford to im Thurn, 13 February 1908, and encls., WPHC 4, 281/1911; Vernon, Notes on SI, 7 December 1912, WPHC 4, 63/1913. See also Colonial Sec. to im Thurn, 31 July 1808, CO 225/81.

37. LD-AR 1913, WPHC 4, 801/1914.

38. James C. Scott, *Weapons of the Weak: Everyday forms of peasant resistance* (New Haven, 1985), 36. Scott provides an illuminating discussion of the validity of this position on pp. 289–303.

39. Scott, *Weapons of the Weak*, 29.

40. Scott, *Weapons of the Weak*, 29.

41. Ralph Shlomowitz, "Markets for Indentured and Time-expired

Melanesian Labour in Queensland, 1863-1906: An economic analysis," *Journal of Pacific History*, 16:2 (1981), 77-79; Ralph Shlomowitz, "The Fiji Labor Trade in Comparative Perspective, 1864-1914," *Pacific Studies*, 9:3 (1986), 118-120.

42. Bennett, *Wealth of the Solomons*, 118; Rigby to Fairley 30 November 1912, FRCP; Mumford to Fairleys, 23 May 1915, 4 December 1912, FRCP; Richard A. Herr and E. A. Rood (eds.), *A Solomons Sojourn: J. E. Philp's log of the "Makira," 1912-1913* (Hobart, 1978), 87-88; LD-AR 1913, WPHC 4, 801/1914; LD-AR 1914, WPHC 4, 698/1915; LD-AR 1915, WPHC 4, 875/ 1916. See also Vernon, Notes on SI, 7 December 1912, WPHC 4, 63/1913.

43. Judith A. Bennett, "Some English-based Pidgins in the Southwest Pacific: Solomon Islands pidgin," in Stephan A. Wurm (ed.), *New Guinea and Neighbouring Areas: A socio-linguistic laboratory* (The Hague, 1979), 64-72; Bennett, *Wealth of the Solomons*, 164, 187, 190.

44. Mahaffy to HC, 8 April 1910, CO 225/91; Bennett, *Wealth of the Solomons*, 172-175; Woodford to im Thurn, 15 February 1908, CO 225/83.

45. Bennett, *Wealth of the Solomons*, 171, 182.

46. Vernon, Notes on SI, 7 December 1912, WPHC 4, 63/1913.

47. Woodford to im Thurn, 15 February 1908, CO 225/83; Hector Mac-Quarrie, *Vouza and the Solomon Islands* (Sydney, 1946), 16-17; *People*, 18 March 1958, 43; Bennett, *Wealth of the Solomons*, 175-177, 185.

48. Bennett, *Wealth of the Solomons*, 178-182; Butchart, Statement of 24 April 1908, CO 225/81.

49. LTR: 177-003-1, nos. 226-229/70, Lots 4, 5, 6, and 7 LR 50, Yandina, Banika. Bennett, *Wealth of the Solomons*, 154; Woodford to Edge-Partington, 24 December 1911, WPHC 4, 2161/1911; James A. Boutilier, "Killing the Government: Imperial policy and the pacification of Malaita," in Margaret Rodman and Mathew Cooper (eds.), *The Pacification of Melanesia* (Latham, Md., 1983), 61. See also MacQuarrie, *Vouza and the Solomon Islands*, 22-24; Rigby to Fairleys, 12 May 1913, FRCP. For examples of recorded homicides and assaults, see Papers relating to murder of Patrick Brown, 3 April 1908, WPHC 4, 190/1908; im Thurn to Earl of Crewe, 7 July 1908, and encls., CO 225/81; Mahaffy to May, 21 December 1908, and encls., WPHC 4, 281/1911; Woodford to im Thurn, 24 August 1908, and encls., CO 225/83; Major to Earl of Crewe, 9 June 1909, CO 225/85 (cf. Hermes to Sec. S, 12 January 1909, CO 225/89); Norris to Barnett, 15 December 1915, WPHC 4, 258/1916; Barnett to HC, 31 August 1915, and encls., WPHC 4, 2735/1915; *Sydney Truth*, 12 May 1917, and encls., WPHC 4, 1437/1918; Barnett to HC, 30 August 1915, and encls., WPHC 4, 2735/ 1915; Bennett, *Wealth of the Solomons*, 154; im Thurn to Earl of Crewe, 9 June 1909, CO 225/85; LD-AR 1913, WPHC 4, 801/1914; Woodford to Escott, 7 February 1914, WPHC 4, 647/1914.

50. Mahaffy to HC, 21 December 1908, WPHC 4, 281/1911. See also Woodford to im Thurn, 24 August 1908, CO 225/83.

51. Margaret Clarence, *Yield Not to the Wind* (Sydney, 1982), 12-14.

52. Clarence, *Yield Not to the Wind*, 84; Walter G. Ivens, *The Island Builders of the Pacific* (London, 1930), 308; Mahaffy to im Thurn, 8 April 1910, CO 225/91; Report of Labour Commission in BSIP, 1929, WPHC 4, 827/1930. See also Roger M. Keesing and Peter Corris, *Lightning Meets the*

West Wind: The Malaita massacre (Melbourne, 1980), passim. Roger Keesing sees such generalizations about the capabilities of Malaitans as a form
of stereotyping. If this were so, then virtually all planters shared this
"stereotype." See Roger Keesing, "Plantation Networks, Plantation Culture: The hidden side of colonial Melanesia," *Journal de la Société des
Océanistes,* 82–83 (1986), 163–170.

53. Barley to Sec. of Gov., 13 September 1927, encl. in WPHC 4, 3423/
1927; Gizo district–AR, 1929, WPHC 4, 1422/1929; Report of Labour
Commission in BSIP, 1929, WPHC 4, 827/1930. J. Svensen, "Early Pioneering of the Solomon Islands," undated typescript, Department of Pacific
and Southeast Asian History, ANU; Hirst, Memoirs, and interview; Plowman, interview; Tommy Elkington, typescript of interview with official of
Department of Labour, Honiara, 1968; S. G. C. Knibbs, *The Savage Solomons as They Were and Are* (London, 1929), 73; Mytinger, *Head-hunting
in the Solomons,* 88–90; James A. Brook, *Jim of the Seven Seas: A true
story of personal adventure* (London, 1940), 193; Bell, *Man No Good,* 7–36;
Romesio Ngura, Aliesio Tavoruka, Sove Kimbo, interviews.

54. Mytinger, *Head-hunting in the Solomons,* 88.

55. Roger M. Keesing (ed.), *'Elota's Story: The life and times of a Solomon Islands big man* (Brisbane, 1978), 64; Bishop of Melanesia, Memo on
Native Code, 14 June 1922, WPHC 4, 1448/1922.

56. Mumford to Fairleys, June 1915, FRCP; The Labour Question in the
Pacific Islands, 1909, CO 225/87; For the effectiveness of this, see Ashley
to Shortland Islands Plantation Ltd., May 1935, WPHC BSIP Series, F34/11/
6; interviews with various informants.

57. Mahaffy to May, 21 December 1908, WPHC 4, 281/1911.

58. LD-AR 1913, WPHC 4, 801/1914. See also Rigby to Fairleys, 15 October 1913, FRCP; Rigby to Fairleys, n.d., Dec. [?] 1913, FRCP.

59. Bell to Sec. S, 10 April 1913, WPHC 4, 937/1913; Bell to Barnett,
28 August 1913, and encls., WPHC 4, 1995/1913; Sec. of Earl of Crewe to
Lever, 5 May 1909, CO 225/85; Woodford to HC, 5 February 1914, LD-AR
1913, WPHC 4, 801/1914. For examples of recorded desertions see Lucas to
Burns, 28 July 1908, BPA; Lucas to Burns, 1 April 1910, BPA; Woodford to
im Thurn, 22 August 1908, WPHC 4, 442/1908; Levers to im Thurn, 30
August 1908, CO 225/82; Edge-Partington to Woodford, 1 November 1910,
WPHC (Mala), 67/1910; Woodford to Edge-Partington, 18 November 1910,
WPHC (Mala), 28/1910.

60. Alike Ghandokiki, interview.

61. Rigby to Fairleys, 14 September 1912, 21 November 1912, 16 May
1913, FRCP.

62. LD-AR 1913, WPHC 4, 801/1914; LD-AR 1914, WPHC 4, 698/1915;
LD-AR 1915, WPHC 4, 875/1916; LD-AR 1916, WPHC 4, 701/1917; LD-AR
1917, WPHC 4, 1284/1918; LD-AR 1919, WPHC 4, 535/1920; LD-AR 1921,
WPHC 4, 350/1923; LD-AR 1922, 4, WPHC 1554/1923; LD-AR 1923, WPHC
4, 1121/1924; LD-AR 1924, WPHC 4, 1197/25; LD-AR 1925, WPHC 4, 1170/
1926; LD-AR 1926, WPHC 4, 1510/1927; LD-AR 1927, 4, WPHC 4, 1835/
1928; LD-AR 1928, WPHC 4, 1426/1929; LD-AR 1929, WPHC 809/1930.

63. Mahaffy to May, 21 December 1908, and encls., WPHC 4, 281/1911.
See also Woodford to im Thurn, 24 August 1908, CO 225/83.

64. Correspondence and papers relating to BSI, 1901–1915, CRSI A1108, vol. 57, AA: PM Dept.; KR III of 1910 as amended in KR VIII in 1912, encl. in WPHC 4, 1605/1912; Major to Sec. S, 11 February 1909, encl. in WPHC 4, 281/1911.

65. Woodford to im Thurn, 13 February 1908, encl. in WPHC 4, 281/1911.

66. Woodford to im Thurn, 13 February 1908, encl. in WPHC 4, 281/1911; Notes on SI, December 1912, WPHC 4, 63/1913; Bell to Sec. S, 10 April 1913, and encls., WPHC 4, 937/1913; Bell to HC, 28 August 1913, and encls., WPHC 4, 1995/1913; LD-AR 1913, WPHC 4, 801/1914; LD-AR 1914, WPHC 4, 698/1915. See also Rigby to Fairley, 1 January 1913, FRCP.

67. Mahaffy to HC, 21 December 1908, and encls., WPHC 4, 281/1911; Notes on SI, 7 December 1912, WPHC 4, 63/1913; Bell to Sec. S, 10 April 1913, and encls., WPHC 4, 1995/1913; Footaboory, Ambuover, and other to Woodford, 17 October 1912, WP; Bell to RC, 15 October 1913, WPHC 4, 2299/1913; KR XI of 1911 and KR VII of 1912, PM's Dept., correspondence and papers relating to BSI, 1910–1915, CRSI A1108, vol 57, AA. See also LD-AR 1917, WPHC 4, 1284/1918.

68. LD-AR 1913, WPHC 4, 801/1914.

69. LD-AR 1913, WPHC 4, 801/1914; Sec. S to Levers, 6 May 1906, CO 225/85; Major to Earl of Crewe, 13 March 1909, and encls., CO 225/85; Memo: The Labour Question in the Pacific Islands, 1909, CO 225/87; Barnett to Eyre Hutson, 31 August 1915, and encls., WPHC 4, 2738/1915; Barnett to Eyre Hutson, 15 September 1915, and encls., WPHC 4, 2747/1915; Barnett to Eyre Hutson, 16 September 1915, WPHC 4, 2748/1915; Bonar Law to Sweet-Escott, 9 June 1915, WPHC 4, 1987/1915; Bonar Law to HC, 11 January 1916, WPHC 4, 661/1916; LD-AR 1914, WPHC 4, 698/1915; LD-AR 1915, WPHC 4, 875/1916; LD-AR 1916, WPHC 4, 701/1917; Barnett to HC, 28 August 1914, WPHC 4, 2316/1914; Barnett to Bell, 26 October 1914, WPHC 4, 2811/1914; Bell to Barnett, 11 November 1914, WPHC 4, 3033/1914.

70. LD-AR 1915, WPHC 4, 875/1916; LD-AR 1916, WPHC 4, 701/1917; LD-AR 1917, WPHC 4, 1284/1918.

71. Mumford to Fairleys, 25 December 1914, FRCP.

72. LD-AR 1917, WPHC 4, 1284/1918; Jackson to HC, 16 August 1915, WPHC 4, 2505/1915.

73. Bonar Law to HC, 11 January 1916, WPHC 4, 661/1916; Woodford to Escott, 15 June 1914, WPHC 4, 1779/1914; LD-AR 1917, WPHC 4, 7284/1918.

74. LD-AR 1917, WPHC 4, 1284/1918; Woodford to Escott, 16 October 1913, WPHC 4, 2298/1913 . See note 19, above.

75. Welchman in *Southern Cross Log* (Sydney), 6 March 1906; Keesing (ed.), *'Elota's Story*, passim, LD-AR 1919, WPHC 4, 535/1920; LD-AR 1921, WPHC 4, 350/1923; LD-AR 1922, WPHC 4, 1554/1923; LD-AR 1923, WPHC 1121/1924; LD-AR 1924, WPHC 4, 1197/1925; LD-AR 1925, WPHC 4, 1170/1926; LD-AR 1926, WPHC 4, 1510/1926; LD-AR 1927, WPHC 4, 1835/1928; LD-AR 1928, WPHC 4, 1426/1929; Isabel District–AR 1931, WPHC 4, 1214/1932; Vera Clift, interview; *People,* 19 March 1958, 42, 44; *Planter's Gazette,* December 1922, pp. 19–20. See also Ashley to Fletcher, 19 May 1930, and encls., WPHC 4, 1758/1930. This system *(uku pau)* was in vogue

in Hawaii at the turn of the century and appears to have developed earlier in the West Indies when slavery was abolished. See Edward D. Beechert, *Working in Hawaii: A labor history* (Honolulu, 1985), 24. Whether Levers learned of it from elsewhere or adopted it as a result of the immediate Solomons situation is unclear.

76. Memo regarding Labour on Estates, 9 January 1924, UA.

77. LD-AR 1919, WPHC 4, 535/1920; LD-AR 1921, WPHC 4, 350/1923; LD-AR 1922, WPHC 4, 1554/1923; LD-AR 1923, WPHC 4, 1121/1924; LD-AR 1924, WPHC 4, 1197/1925; LD-AR 1925, WPHC 4, 1170/1926; LD-AR 1926, WPHC 4, 1510/1926; LD-AR 1927, WPHC 4, 1835/1928; LD-AR 1928, WPHC 4, 1426/1929; Allardyce, Report on Labour Conditions 17 April 1922, WPHC 4, 1094/1922; Ashley, Report 8 August 1931, and encls., WPHC 4, 2907/1931.

78. Memo regarding labor on estates 9 January 1924, UA.

79. LD-AR 1913, WPHC 4, 801/1914; LD-AR 1914, WPHC 4, 698/1915; LD-AR 1915, WPHC 4, 875/1916; LD-AR 1916, WPHC 4, 701/1917; LD-AR 1917, WPHC 4, 1284/1915; LD-AR 1919, WPHC 4, 535/1920; LD-AR 1921, WPHC 4, 350/1923; LD-AR, WPHC 4, 1554/1923; LD-AR 1923, WPHC 4, 1121/1924; LD-AR 1924, WPHC 4, 1199/1925; LD-AR 1925, WPHC 4, 1170/1926; LD-AR 1926, WPHC 4, 1510/1927; LD-AR 1927, WPHC 4, 1835/1928; LD-AR 1928, WPHC 4, 1426/1929; LD-AR 1929, WPHC 4, 809/1930; Pinching to RC, 17 September 1920, and encls., WPHC 4, 2608/1920; Thompson to RC, 7 April 1922, WPHC 4, 1035/1922; Bennett, *Wealth of the Solomons,* 171–172.

80. Mumford to Fairleys, 25 December 1914, FRCP.

81. Mumford to Fairleys, 20 March 1916, FRCP.

82. Mumford to [Fairleys ?], September 1918, FRCP.

83. LD-AR 1919, WPHC 4, 535/1920; LD-AR 1921, WPHC 4, 350/1923; LD-AR 1922, WPHC 4, 1554/1923; LD-AR 1923, WPHC 4, 1121/1924; LD-AR 1924, WPHC 4, 1197/1925; LD-AR 1925, WPHC 4, 1170/1926; LD-AR 1926, WPHC 4, 1510/1926; LD-AR 1927, WPHC 4, 1835/1928; LD-AR 1928, WPHC 4, 1426/1929; Bennett, *Wealth of the Solomons,* 171, 177; *People,* 19 March 1958, 44.

84. See for example LD-AR 1921, WPHC 4, 350/1923; LD-AR 1923, WPHC 4, 1121/1924.

85. Samuel Bau, Alvelti Tongareva, interviews; Woodford to Lucas, 11 December 1912, BPA.

86. Bennett, *Wealth of the Solomons,* 112, 399, 403; LD-AR 1921, WPHC 4, 350/1923; LD-AR 1922, WPHC 4, 1554/1923; Solomon Islander informants, 1976; *Planters' Gazette,* August 1922, 19–20; May 1922, 7.

87. *Planters' Gazette,* May 1922, 7.

88. *Planters' Gazette,* August 1922, 19–20; December 1922, 11–12. See also *Planters' Gazette,* December 1922, 18. For an example of a typical "trouble-monger" at home and abroad, see Levers' manager to RC, 20 April 1938, and encls., WPHC BSIP Series 34/1/7.

89. LD-AR 1917, WPHC 4, 1284/1918; *Planters' Gazette,* May 1922, 17–18.

90. Bates to HC, 14 April 1919, WPHC 4, 1050/1919; Crichlow to RC, 16 July 1920, WPHC 4, 2491/1920; LD-AR 1919, WPHC 4, 835/1920; LD-AR 1921, WPHC 4, 350/1923. Part of the lower death rate could be attributed

to a regular and often improved diet and superior living conditions on plantations in comparison to many home villages, especially in the bush. This would be particularly so in areas where periodic food shortage was common. See Woodford to im Thurn, 13 February 1908, WPHC 4, 830/1908; Kidson to HC, 8 February 1929, WPHC 4, 634/1929.

91. Fulton to Greene, 15 April 1921, WPHC 4, 1152/1921; Crichlow, Report on Hookworm, 1921, WPHC 4, 1416/1922.

92. LD-AR 1921, WPHC 4, 350/1923. See also H. Ian Hogbin, *Experiments in Civilization: The effects of European culture on a native community in the Solomon Islands* (London, 1939), 86–87.

93. Jotam Finau, interview.

94. Hogbin, *Experiments in Civilization*, 86; Mumford to Fairleys, August 1914, 22 April 1917, FRCP; Grassick to Middenway, 16 November 1932, and encl., WPHC 4, 1090/1933. See also Report of Labour Commission on BSI, 1929, WPHC 4, 827/1930.

95. Georgina Seton, interview.

96. Jo Ariana, Jack Mainagwa, Joseph Afe'ou and elders of Igwa village; Georgina Seton, interviews. For a similar use of shell valuables in nineteenth-century Queensland, see Clive Moore, *Kanaka: A history of Melanesian Mackay* (Boroko/Port Moresby, 1985), 69.

97. Judith A. Bennett, "Cross-Cultural Influences on Village Relocation on the Weather Coast of Guadalcanal, Solomon Islands, c.1870–1953," M.A. thesis, University of Hawaii (Honolulu, 1974), 10–12; Mumford to Fairleys, August 1914, FRCP.

98. David Hilliard, *God's Gentlemen: A history of the Melanesian Mission, 1849–1942* (Brisbane, 1978), 274–275; Hogbin, *Experiments in Civilization*, 173.

99. For an example of the limited involvement of missions in the control of plantations, see David Hilliard, "The South Sea Evangelical Mission," *Journal of Pacific History* 4 (1969), 41–54; Bennett, *Wealth of the Solomons*, 144–145, 196–197.

100. Letter, translation, 19 July 1934, encl. in WPHC BSIP Series F34/3/1.

101. Hogbin, *Experiments in Civilization*, 49; Thomas, Beni Kai, Josepa Odofia, Samuel Ramuhuni, Nelson Anuanuiabu, Bese Ghaura, Daisi Sauha'abu, Matthew Mairotaha, Thomas Ta'ae'ke'kerei, Kaspa Arubae'awa, John Bana, interviews. See also H. Ian Hogbin, *A Guadalcanal Society: The Kaoka speakers* (New York, 1964), 5–6. In the labor trade days, men from diverse societies often formed a common, if temporary, identity when separated from home in places like Queensland. See Peter Corris, *Passage, Port and Plantation: A history of Solomon Islands labour migration, 1870–1914* (Melbourne, 1973), 88–89.

102. Vera Clift, interview; Bennett, *Wealth of the Solomons*, 173; Report of Labour Commission in BSI, 1929, WPHC 4, 827/1930. Since working in collaboration with Shelly Schreiner, Roger Keesing has revised his views in the polluting aspects of women in Kwaio society, stating that they too have a *tapu* (Kwaio *abu*) quality about them in various circumstances. Be that as it may, they were still confined to predictable roles in Kwaio society and remained largely confined spatially to areas limited by male interdict. Only pacification (and in some areas Christianization) has reduced the brute

physical force men used, often in an arbitrary manner, against women. See Roger M. Keesing, "Ta'a Geni: Women's perspectives on Kwaio society," in Marilyn Strathern (ed.), *Dealing with Inequality* (Cambridge, 1987), 59–106.

103. Bennett, *Wealth of the Solomons*, 172–173; Report of Labour Commission in BSI, 1929, WPHC 4, 827/1930.

104. Report of Labour Commission in BSI, 1929, WPHC 4, 827/1930; Brownlees to Sec. to Gov., 18 August 1939, WPHC BSIP Series 34/4; Woodford to im Thurn, 8 July 1909, CO 225/89; LD-AR 1913, WPHC 4, 801/1914; LD-AR 1914, WPHC 4, 698/1915; LD-AR 1915, WPHC 4, 875/1916; LD-AR 1916, WPHC 4, 701/1915; LD-AR 1917, WPHC 4, 1284/1918; LD-AR 1919, WPHC 4, 535/1920; LD-AR 1921, WPHC 4, 350/1923; LD-AR 1922, WPHC 4, 1554/1923; LD-AR 1923, WPHC 4, 1121/1924; LD-AR 1924, WPHC 4, 1197/1925; LD-AR 1925, WPHC 4, 1170/1926; LD-AR 1926, WPHC 4, 1510/ 1927; LD-AR 1927, WPHC 4, 1835/1928; LD-AR 1928, WPHC 4, 1426/29; LD-AR 1929, WPHC 4, 809/1930.

105. Bennett, *Wealth of the Solomons*, 13, 29–30, 114, 121–122, 153, 183, 189; R. M. Keesing, *Kwaio Religion: The living and the dead in a Solomon Island Society* (New York, 1982), 218–228. For how Kwaio men kept their culture intact through the isolation of their women from the outside world, see R. M. Keesing, "Kwaio Women Speak: The micropolitics of autobiography in a Solomon Island Society," *American Anthropologist* 87:1 (1985), 27–39. For similarities with African situation, see Claude Meillassoux, *Maidens, Meal and Money: Capitalism and the domestic community* (Cambridge, 1981), 61–82.

106. Statement of Charlie Farutarava, in Report of Labour Commission in BSI, 1929, WPHC 4, 827/1930.

107. Woodford to im Thurn, 20 August 1908, and encls., CO 225/83; Woodford to Under Sec. S, 8 July 1909, CO 225/89; Barnett to Major, 23 April 1909, and encls., WPHC 4, 517/1909; Report of Labour Commission in BSI, 1929, WPHC 4, 827/1930. See also Workman to Sec. of Planters' Association, 24 June 1920, encl. in WPHC 4, 1912/1920.

108. Bennett, *Wealth of the Solomons*, 173–174; Memo on amendments to SI Labour regulations of 1897, and encls., CO 225/86; Woodford to im Thurn, 24 August 1908, and encls., CO 225/83; Statements of Malaitans, Report of Labour Commission in BSI, 1929, WPHC 4, 827/1930; Hogbin, *Experiments in Civilization*, 99, 113, 115, 209–211; Keesing (ed.), *'Elota's Story*, passim.

109. Bennett, *Wealth of the Solomons*, 173–174; Keesing (ed.), *'Elota's Story*, 45; Hogbin, *Experiments in Civilization*, 163–164.

110. In 1931 in the Russells, there were 735 adult male (aged sixteen and over) compared to 45 adult female villagers of a total village population of 199; on Kolombangana, there were also 735 adult male laborers compared to 70 adult female villagers of a total village population of 258. See LD-AR 1931, WPHC 4, 1228/1932.

111. Bennett, *Wealth of the Solomons*, 174; Waddell to Sec. to Gov. [?], August 1939, WPHC BSIP F34/4.

112. Keegan to Sec. to Gov., 14 August 1939, encl. WPHC BSIP F34/4 Part II.

113. Bennett, *Wealth of the Solomons*, 174. Sometimes older kinsmen

disapproved of these liaisons on the plantations, even to the extent of laying a curse on their younger relative (Georgina Seton, interview).

114. Bennett, *Wealth of the Solomons*, 205–208; Report of Labour Commission in BSI, 1929, WPHC 4, 827/1930.

115. Bennett, *Wealth of the Solomons*, 178–179, 182–184. In 1924 the government ruled that all casual labor was on a monthly basis unless specified otherwise in a contract. Payment of wages was monthly as opposed to daily, when casual laborers had worked and were paid by the day. See LD-AR 1924, WPHC 4, 1197/1925.

116. This discussion is based on the approach taken in Claude Meillassoux, "From Reproduction to Production," *Economy and Society*, 1:1 (1972), 93–105, and Meillassoux, *Maidens, Meal and Money*, passim.

117. Bell to RC, 16 November 1922, WPHC 4, 661/1923.

118. AC Minutes, 10 November 1921, encl. in WPHC 4, 25/1922; AC Minutes 8 December 1922, encl. in WPHC 4, 185/1923; Thomson, Comments on Allardyce's Report, 11 July 1922, and encls., WPHC 4, 1714/1922; Workman to HC, 8 July 1920, and encls., WPHC 4, 1912/1920; Workman to HC, 19 October 1920, and encls., WPHC 4, 2902/1920; Allardyce, Report on Labour Conditions, 17 April 1927, WPHC 4, 1094/1922.

119. Greene, Memorandum, 17 March 1924, WPHC 4, 243/1924.

120. Allardyce, Report on Labour Conditions, 17 April 1922, WPHC 4, 1094/1922; Greene, Memorandum, 17 March 1924, WPHC 4, 243/1924; LD-AR 1924, WPHC 4, 1197/1925. See also *Planters' Gazette*, April 1921, 2–3, May 1923, 6–8.

121. Allardyce, Report on Labour Conditions, 17 April 1922, WPHC 4, 1094/1922; AC Minutes, November 1931, WPHC 4, 3698/1922. Another source in 1922 reckoned the cost at £83 before abolition and at £85 after (*Planters' Gazette*, May 1923, p. 7). For a comparison with the Queensland situation at the turn of the century, see Moore, *Kanaka*, 173–174. Inflation, especially after World War I, must also be considered.

122. WPHC 1094/1922: Allardyce, Report on Labour Conditions, 17 April 1922. See also WPHC 3343/26: Barley to Gov. Sec., 23 August 1926.

123. Kane to the HC, 23 September 1926, and encls., WPHC 4, 3343/1926; *Planters' Gazette*, May 1923, 8, 17.

124. Bennett, *Wealth of the Solomons*, 163–164.

125. KR XVII of 1922, encl. in WPHC 4, 2768/1922; List of appointments for 1926–1927, WPHC 4, 2558/1926; BSIP-AR 1927; Kane to AHC, 6 October 1921, and encls., WPHC 4, 2912/1921; Kane to HC, 6 May 1922, WPHC 4, 1448/1922; Rodwell to RC, 24 August 1922, and encls., WPHC 4, 2290/1922.

126. Bennett, *Wealth of the Solomons*, 123, 211–212, 214–216, 242. For a study of the Malaita killings, see Keesing and Corris, *Lightning Meets the West Wind*. Note that the imposition of the tax also created challenges elsewhere in the Solomons. Western Solomon Islanders successfully organized mass resistance. This was facilitated by the Roviana lingua franca and the network of Methodist missions. See Bennett, *Wealth of the Solomons*, 214, 246–252.

127. Bennett, *Wealth of the Solomons*, 261, 277–278, 281.

128. Greene to Col. Sec., 10 April 1922, WPHC 4, 561/1922; Preliminary hearing of charges against Dawson, Firth, and McGregor, 27 February 1922, WPHC 4, 1000/1922.

129. Turner to RC, 18 May 1922; WPHC 1202/1922: 4 May 1922, and encls.; WPHC 3059/1921.

130. Kane to HC, 28 September 1922, WPHC 4, 2989/1922.

131. *Planters' Gazette*, December 1922, 13; Hugh Laracy, *Pacific Protest: The Maasina Rule movement, Solomon Islands, 1944–1952* (Suva, 1983), 12; Kane to HC, 19 April 1927, and encls., WPHC 4, 1497/1927; Keesing and Corris, *Lightning Meets the West Wind*, passim.

132. Campbell to Officer commanding Native Constabulary, 5 September 1927, WPHC 4, 3423/1927; Kane to HC, 2 November 1927, encl. in WPHC 4, 3250/1927.

133. By this time, some Malaitans were writing letters from plantations to relations at home. See Kane to HC, 19 April 1927, WPHC 4, 1497/1927.

134. Kane to HC, 25 April 1925, and encls., WPHC 4, 1181/1925; Acting RC to AHC, 10 March 1928, WPHC 4, 770/1928. See also LD-AR 1928, WPHC 4, 1426/1929. Rikard-Bell subsequently wrote a book under the pseudonym of Clarke Jameson called *Man-No Good: An Autobiography of the South Seas* (see note 19, above).

135. Bell to RC 15 October 1913, WPHC 4, 2299/1913; Hill to HC, 10 January 1921, and encls., WPHC 4, 806/1921; Report of Labour Commission in BSI, 1929, WPHC 4, 827/1930; Native laborers extension of contracts re KR XV of 1921, WPHC BSIP F34/1.

136. Bennett, *Wealth of the Solomons,* 219–220; Report of Labour Commission in BSI, 1929, WPHC 4, 827/1930.

137. Bennett, *Wealth of the Solomons,* 221–222; *Planters' Gazette*, April 1921, p. 5; LD-AR 1928, WPHC 4, 1426/1929.

138. LD-AR 1931, WPHC 4, 1228/1932; Bennett, *Wealth of the Solomons*, 226; William Albert Robinson, *Deep Water and Shoal* (London, 1937), 203.

139. Hetherington to RC, 8 August 1933, and encls., WPHC 4, 2869/1933; Meeting of AC, October 1936, encl. in WPHC 4, 2467/1936. Allardyce in his 1922 report on labor urged the inclusion of "ripe coconut" in the laborers' ration, but the government did not legislate for this. See Allardyce, Amendments to regulations, 17 April 1922, WPHC 4, 1092/1922.

140. Bennett, *Wealth of the Solomons*, 226–227. Cf. K. Buckley and K. Klugman, *"The Australian Presence in the Pacific": Burns Philp, 1914–1946* (Sydney, 1983), 236–237. There is strong evidence that recruiters, to compete, were forced to reduce actual recruiting fees by about a third during the Depression. See Clift and others to ARC, 21 December 1923, WPHC 4, 243/1924; Ashley, Report, 19 August 1931, WPHC 4, 2907/1931; Elkington, typescript; Notes on front of the file Native Labourers—Repatriation, WPHC BSIP F34/10.

141. Bennett, *Wealth of the Solomons*, 221, 226–227, 236; Clift to HC, 6 November 1933, WPHC 4, 3461/1933; Repatriation of Mr. Marks to Australia, [n.d.], WPHC 4, 2931/1931. See also Barley to HC, 6 April 1933, and encls., WPHC 4, 3571/1932.

142. Keegan, Inquiry at Mendaña estate, Isabel, 22 September 1937 and encls., WPHC BSIP F14/1[?]; Ashley, Report 8, August 1931, and encls., WPHC 4, 2907/1931; LD-AR 1924, WPHC 4, 1197/1925; Barley to HC, 12 December 1932, WPHC 4, 3842/1932. W. R. Carpenter acquired the Isabel estates through foreclosure, but their owners, Clift and Clift Pty. Ltd., reg-

istered a further mortgage with Burn Philp. See LTR: 089–002–5, Kahige; LTR: 108–003–1, Fera Island.

143. Barley to HC, 12 December 1932, and encls., WPHC 4, 3842/1932.

144. Vaskess, Notes, Burning of Copra Driers, 12 August 1934, WPHC 4, 1073/1934.

145. Mytinger, *Head-hunting in the Solomons,* 101–112; Vaskess, Notes, Burning of Copra Driers, 12 August 1934, and encls., WPHC 4, 1073/1934; See also *PIM,* 24 August 1934.

146. Annual Reports of Labour Dept., 1919–1939, encl. in WPHC 4, 535/1920; WPHC 4, 350/1923; WPHC 4, 1554/1923; WPHC 4, 1121/1924; WPHC 4, 1197/1925; WPHC 4, 1170/1926; WPHC 4, 1510/1927; WPHC 4, 1835/1928; WPHC 4, 1426/1929; WPHC 4, 809/1930; WPHC 4, 755/1931; WPHC 4, 1228/1932; WPHC 4, 506/1933; WPHC 4, 920/1934; WPHC 4, 1612/1935; WPHC 4, 1598/1936; WPHC 4, 2744/1937 (also numbered 4, 2774/1936); WPHC 4, 1628/1938; WPHC 4, 2469/1940; WPHC 4, 2399/1941; Elkington, typescript, 1965.

147. Meeting of AC, October 1934, encl. in WPHC 4, 2722/1934; BSIP-AR 1931; Widdy to HC, 6 March 1941, WPHC BSIP F34/11/6.

148. Meeting of AC, October 1930, WPHC 4, 3269/1930; Meeting of AC, November 1931, WPHC 4, 3698/[1936]; Meeting of AC, October 1933, WPHC 4, 4028/1933.

149. Minutes of Proceedings, 8 March 1934, and encls., WPHC 4, 3461/1933; Meetings of AC, October 1934, WPHC 4, 2722/1934; Ashley, Report on the Position of the Copra Industry in SI, 14[?] May 1934, WPHC 4, 1850/1934; Ashley to HC, 6 December 1935, and encls., WPHC BSIP 48/4.

150. Ashley to HC, 28 October 1936, encl. in WPHC 4, 2307/1934.

151. Report of Manager, Makambo Branch, 1935 MR, BPA; District Diary Santa Cruz, October 1935, encl. in WPHC 4, 1587/1935; District Diary Malaita, December 1934, encl. in WPHC 4, 1587/1935; Malaita AR, 1936, encl. in WPHC 4, 1052/1936; District Diary Santa Cruz, October 1936, WPHC 4, 1585/1935; Guadalcanal AR, 1935, WPHC 4, 2593/1936; Guadalcanal AR, 1938, WPHC 4, 1779/1939.

152. District Diary, Malaita, October 1936, WPHC 4, 1585/1935; Bennett, "Cross-Cultural Influences on Village Relocation," 54–55.

153. District Diary, Malaita, December 1936, WPHC 4, 1585/1935.

154. Ashley to HC, 21 June 1935, encl. in WPHC 4, 2307/1934; *PIM,* 23 July 1935, p. 63.

155. Manager to R. B. Carpenter, 2 May 1929, WRCP; see also Manager to W. R. Carpenter 4 May 1929, WRCP; *PIM,* 21 February 1935, p. 22.

156. Annual Reports of Labour Dept. 1928–1939, see note 146, above.

157. Agricultural Report, 1 April 1940–31 March 1941, WPHC 4, 3054/1941. Most of Levers' production was "hot-air" dried and of good quality. See Bennett, *Wealth of the Solomons,* 228–229.

158. Hewitt to HC, 10 May 1937, encl. in WPHC 4, 1605/1937; Meeting of AC, October 1930, encl. in WPHC 4, 3296/1930; Meeting of AC, November 1938 encl. in WPHC 4, 3215/1938; WPHC BSIP CF 33/6, vol. I and vol. II encls.

159. Wright to Sec. Gov., 5 May 1941, WPHC BSIP 35/1/4[?]. See also Santa Isabel AR, 1940, WPHC 4, 1486/1941.

160. Levers' Pacific Plantations Limited, "Brief History of Levers' Plantations in the Pacific," typescript copy in Department of Pacific and Southeast Asian History, ANU, n.d.; Keegan to RC, 7 June 1937 encl. in WPHC BSIP F14/1[?].

161. Bennett, *Wealth of the Solomons*, 259–263.

162. Bengough, Quarterly Report, Malaita District, 5 July 1941, WPHC BSIP 26/RC/41. See also Bennett, *Wealth of the Solomons*, 280–282. Respecting the establishment of local government, cf. Frederick Osifelo, *Kanaka Boy* (Suva, 1985), 22–23.

163. Bengough, Quarterly Report, Malaita District, 31 March 1941, WPHC BSIP 13/RC/41.

164. Bennett, *Wealth of the Solomons*, 153, 165–166; Minutes of Advisory Council, October 1953.

165. H. H. Vaskess, *Post War Policy, Reconstruction and Reorganization of Administration* (Suva, 1943), passim.

166. Bennett, *Wealth of the Solomons*, 301–302.

167. Noel, Memorandum on the problem of Labour Supply and Recruitment in the BSIP, 3 April 1945, WPHC BSIP F15/18 (vol. I); *Minutes of Advisory Council*, October 1953 and February 1954; *PIM*, June 1946, 27; March 1947, 25; January 1948, 69.

168. Bennett, *Wealth of the Solomons*, 303, 306, 308; *PIM*, January 1946, 7; March 1946, 8; March 1947, 11.

169. Distribution of Labour, 31 December 1943, WPHC Papers Relating to Labour Corps; *PIM*, October 1943, 16; August 1944, 17–18; Bennett, *Wealth of the Solomons*, 289, 290, 292–293; Laracy (ed.), *Pacific Protest*, 20–21, 98–109, 112, 122, 136–149; Ian Frazer, "Maasina Rule and Solomon Islands Labour History," in Moore, Leckie, and Munro (eds.), *Labour in the South Pacific*, 191–203.

170. Hogbin to Trench, October 1943, WPHC Papers relating to Labour Corps; *PIM*, October 1943, 16; Bennett, *Wealth of the Solomons*, 289; Paul Wright, "Population: The project census," in Murray Chapman and Peter Pirie (eds.), *Tasi Mauri: A report on population and resources of the Guadalcanal weather coast* (Honolulu, 1974), 3–37.

171. Laracy (ed.), *Pacific Protest*, 88–89; *PIM*, December 1946, p. 26; Bennett, *Wealth of the Solomons*, pp. 298, 409.

172. *PIM*, February 1945, 10, 35; November 1945, 43; March 1946, 8; August 1946, 11; October 1946, 6–7, 38–39.

173. *PIM*, November 1945, 65; August 1946, 11; Laracy (ed.), *Pacific Protest*, 25; Bennett, *Wealth of the Solomons*, 288, 306, 308.

174. *PIM*, November 1945, 65; April 1946, 30; October 1946, 6–7; December 1946, 26; Laracy (ed.), *Pacific Protest*, 21, 25–26, 108–109.

175. Laracy (ed.), *Pacific Protest*, 21–22, 25–26, 165–167; Bennett, *Wealth of the Solomons*, 294–295.

176. Bennett, *Wealth of the Solomons*, 295–296, 306; *PIM*, January 1948, p. 34; Keesing (ed.), *'Elota's Story*, 161–164.

177. Scott, *Weapons of the Weak*, 297–300.

178. Scott, *Weapons of the Weak*, 296.

179. Scott, *Weapons of the Weak*, 296–297.

6

"Nonresistance" on Fiji Plantations: The Fiji Indian Experience, 1879–1920

Brij V. Lal

Fiji acquired an unenviable reputation as the employer of Indian indentured labor. Throughout the entire period of indenture, between 1879 and 1920, it invited odious comparisons with other Indian labor-importing colonies, especially in the West Indies. The Government of India concluded in 1895 that "the relations between the employers and the immigrants [in Fiji] are not as satisfactory as they are in the West Indies," while the Colonial Office in London believed that Fiji was "no doubt rather worse than Mauritius and British Guiana."[1] C. F. Andrews, the Indian nationalist sympathizer, writing at the end of indentured emigration to Fiji in 1916, concurred: "What we have seen with our own eyes in Fiji is far worse than anything we had seen before. The moral evil in Fiji appears to have gone deeper."[2] The tasks were heavier, prosecution of laborers was more frequent, and the suicide rate in the indentured community was among the highest in the world. In short, it was in Fiji that the indenture system existed "at its worst."[3]

Yet, during the forty years of indenture, the Indian indentured laborers mounted few organized protests against the plantation authorities and the oppressive conditions under which they lived and worked. The sum total of such activity amounted to two short-lived strikes in 1886 and 1907 and a few other minor demonstrations of dissatisfaction. In this respect, the contrast with the West Indies is striking. In British Guiana between 1886 and 1890 alone, there were over one hundred strikes. Trinidad was quieter, but as Kusha Haraksingh suggests, "not very noticeably so."[4] The two Fiji strikes were spontaneous movements, local in scope and lacking in firm leadership and strategy. Being essentially defensive actions to protest the withholding of some privileges to the workers or the abuse of the labor ordinances by the employers, they were quickly and effectively put down by the authorities. Hugh Tinker observes

aptly that the "coolies hardly ever took action in support of
demands for new gains: almost always, they protested because the
management tried to take away some existing portion of their
agreed conditions."[5]

The paucity and the ultimate failure of collective resistance does
not mean that the Indian laborers were indifferent to their plight,
docile, or submissive. On the contrary, this chapter argues, the
absence of strikes reflected the indentured laborers' acute appreciation of the realities of the plantation system and the structure of
power in the larger colonial society. The planters enjoyed almost
unfettered authority over their laborers, and the ease and effectiveness with which they were able to punish them, year after year,
for even minor breaches of the labor ordinances, starkly reinforced
the futility of overt action. Not surprisingly, then, the laborers
chose accommodation and individual acts of passive resistance to
cope with an almost impossible situation. Eugene Genovese's observation about slave resistance in America applies with equal force to
the Fiji situation: "If a people, over a protracted period, finds the
odds against insurrection not merely long but virtually uncertain,
then, it will choose not to try. To some extent this reaction represents decreasing self confidence and increasing fear, but it also
represents a conscious effort to develop an alternative strategy for
survival."[6]

The external constraints imposed by the authoritarian structure
of the plantation system and the repressive labor legislation that
upheld it were the primary deterrents of collective self-assertion by
the indentured laborers. But their difficulties were compounded by
serious problems of organization within the nascent Indian community itself. The laborers' diverse social and cultural background,
their differing aspirations and motivations for migrating to Fiji,
their varying individual experiences on the plantations, and the
absence of institutional structures within the indentured community, which could have become avenues for mobilization, all combined to frustrate the potential for collective action.

Resistance and accommodation were not necessarily contradictory and mutually exclusive strategies. Rather, they were a part of
the same continuum. Some persons resisted more consistently and
tenaciously than others. Some individuals chose to resist at one particular moment in time but preferred accommodation at another.
As Walter Rodney has observed in the case of Guyana, "Moments
of struggle and moments of compromise appeared within the same
historical juncture."[7] This is an important though frequently misunderstood phenomenon; but it is also one that is extremely diffi-

cult to delineate with any precision. The indentured laborers have left few accounts of their experiences of plantation life,[8] while in the documentary sources, their perspective is often presented in a highly distorted fashion. These difficulties are familiar to students of subaltern history, who have increasingly turned to oral traditions, folktales, and quantification in an effort to portray the lives and experiences of unlettered peoples. Here, I too have endeavored to present a wide range of evidence to show the impediments that confronted the Indian indentured laborers and that stifled protest and promoted accommodation among them.

Indian indentured emigration to Fiji began in 1879, forty-five years after it was first started (in 1834) and five years after Fiji became a British Crown Colony (in 1874). Reluctantly acquired, the new colony was expected to become economically self-sustaining in a short period of time.[9] But neither capital nor labor, the two obvious preconditions for rapid economic development, were readily available in the new colony. Sir Arthur Gordon, the first governor, had prohibited the commercial employment of Fijian labor as part of his native policy to shield the indigenous population from the destructive effects of western penetration. And local European planters were in straitened financial circumstances, following the collapse of the cotton boom of the 1860s. With previous experience as governor of Trinidad and Mauritius, both plantation sugar colonies, Gordon turned to plantation agriculture and to outside sources for both capital and labor. He invited the Colonial Sugar Refining Company (CSR) of Australia to extend its operation to Fiji, which it did in 1882, and CSR remained in the country until 1973. For a cheap and reliable source of labor supply, he predictably turned to India, which was already an important supplier of indentured labor to colonies scattered across the globe.

Between 1879 and 1916, when indentured emigration to Fiji was finally abolished, 60,000 Indian indentured laborers were imported into the colony. Most of them chose to remain there after their contracts expired, and in the end only 40 percent returned to India.[10] Three-quarters of Fiji's Indian migrants came from North India, and there especially from the impoverished eastern districts of the United Provinces, such as Basti, Gonda, Faizabad, Gorakhpur, and Ballia. Recruitment in South India began in 1903. Most of the migrants were a part of an already uprooted peasantry in search of employment in urban industrial centers such as Calcutta, or in Bihar coal mines or the Assam tea gardens. Economic hardships caused by British penetration of the Indian countryside that restructured rural relationships and impoverished local industries provided the primary impetus for migration. The recruiters' suitably embellished tales of opportunity and fabulous wealth awaiting the recruits in Fiji and other colonies propelled others into migrating. Most of the recruits did not intend to migrate permanently; on the contrary, they probably expected to return to their villages after they had acquired enough wealth in the colonies. They were thus sojourners, a part of circular migration that was on the increase in the Indian subcontinent in the late nineteenth century.[11] Their sojourner mentality manifested itself in several ways, most significantly in the yearly remittances to the families they had left behind in India. From Fiji alone between 1889 and 1912, the returning migrants deposited £62,773 for transmission to India, while taking with them jewelry and other items worth £111,962.[12] The need and the pressure to save, to abide by the rules, and to return to India at an early date—these were the primary and understandable considerations in the sojourners' calculations; and they acted as powerful disincentives to any involvement in time-consuming, potentially costly, and futile struggles against the plantation authorities—whom they knew from personal experience to be all-powerful, more so than the colonial government itself.

The diverse cultural and social background of the Indian laborers also hindered the development of common perceptions, interests, and values among them. The laborers came from all strata of rural Indian society. Over three hundred different castes of varying status, from the lowly Chamars to Brahmins, were represented in the emigrating population, originating in over 250 districts in Northern India alone.[13] Speaking a host of different tongues, worshiping a multitude of different gods, and occupying widely varying positions in the Indian social structure, they were brought together for the first time at the emigration depots in Calcutta and Madras, on the long sea voyage, and on the plantations where they lived and worked together for five years, at the very least. Emigration across the *kala pani* (dark seas) was a traumatic experience for a primarily landlocked people, which fragmented the values of the "old world," especially those that emphasized the importance of tradition and group solidarity. New values, forged in the crucible of indenture, stressed new goals: individual achievement and personal survival. This transition from an emphasis on community to self-preservation was gradual, not precipitous. Many elements of the old world survived the trauma of plantation work, but the fundamental change in world view of the immigrants was unmistakable.

Most of the migrants were young: 87 percent were below twenty-six years of age.[14] This is not surprising. The planters wanted young workers, and in India, as elsewhere, these were among the most mobile and adventureous members of society. But like their counterparts in other traditional societies, these young people did not enjoy a high status, being generally untutored and unskilled in deeper political and cultural matters and unprepared for leadership roles. The disruption of the institutions of religion, caste, and community that occurred on the outward voyage and on the plantations further exacerbated the problem of disorientation. The laborers had little or no formal education, certainly not in English, which placed them at a great disadvantage in articulating their grievances to the colonial officialdom in Fiji. The new kind of leadership that emerged on the plantations focused on the *sirdar*, or Indian foreman, and it did not serve, nor indeed was it designed to serve, the interest of the indentured workers. The new leadership was, in Hugh Tinker's apt phrase, "lackey leadership,"[15] created and sustained by the plantation management to achieve their goal of exercising a tight control over the labor force, and extracting the maximum amount of work out of them. As we shall see later, the *sirdars* were chosen for their loyalty to the goals of their masters and for their "ability to get the immigrants to complete their tasks in the field by methods which were anything but diplomatic."[16] They often turned out to be the indentured laborers' worst enemy. "The sirdars were never with us,"[17] recalled one indentured laborer, echoing a widespread sentiment. The absence of good leadership, then, posed a major problem for the laborers in their struggle against the planters.

However, while these internal social and political constraints to collective action were certainly important, it should also be recognized that for some immigrants at least, indenture still represented an improvement on their position in India. This was particularly the case with the lower castes, which were permanently consigned to the social and economic fringes of rural Indian society as untouchables, tenants-at-will, and landless laborers—mired in poverty and degradation. They were no strangers to strenuous physical labor, which was a daily condition of their lives in India. In Fiji at least, their individual identity was recognized, and their effort rewarded on the basis of personal achievement rather than traditional status, or other such social criteria. For them, the leveling tendencies of the plantation system heralded a welcome change that broke away from an oppressive past and promised a brighter future. Others, perhaps those who had been victims of natural calamities such as famines and droughts, or of exploitative landlords, welcomed the apparent stability and security that the authoritarian structure of the plantation system offered. Indeed, for them indenture was better than the unsettled life they encountered after they became "free." Reflecting on his indenture experience, one laborer told the anthropologist Adrian Mayer in the 1950s:

> The time of indenture was better than now. You did your task, and knew that this was all. You knew you will get food everyday. I had shipmates with me, and we weren't badly off when there was a good *sirdar* and overseer. Of course, if they were bad then you had to be careful. But now what do I do? I have cane land, bullocks and a home. Yet every night I awake, listening to see if someone is not trying to burn my cane, or steal my animals. In indenture lines we slept well, we did not worry.[18]

Such people were probably in the minority, but they had little reason to fight the system.

It was the conditions within Fiji itself, however, that posed the most serious obstacles to the development of organizational activity among the indentured laborers. On arriving in the colony, the laborers were allotted to employers on the basis of requisitions received by the colonial Immigration Department. The laborers, of course, were not given a choice of employers, nor, ordinarily, the right to change them on account of ill-treatment or any other reason. At the time of allotment, an effort was made not to disrupt families, but once on the plantations, the movement of laborers was controlled by the overseers, whose decisions, naturally enough, were guided not by humanitarian concerns but by the needs of the plantation management. There were instances of families being broken up and sent to different plantations for long periods of time.[19] On some plantations, immigrants were moved about from one place to another to prevent alleged breaches of the peace.[20] Immigration officials, at the behest of planters, split up immigrants from the same districts of origin to prevent the possibility of "ganging." Older, more experienced immigrants were often made to work with newer arrivals to help them acculturate into the new way of life. The practice of breaking up old connections and relationships and creating and fostering new social groupings rendered the laborers vulnerable and thus more amenable to plantation control.

The laborers were further immobilized by the fact that the estates on which they worked were widely scattered across the colony. The initial plantations were established on Viti Levu, Fiji's largest island of some 4,000 square miles. The CSR erected its first sugar mill at Nausori (in the southeastern part of the island) in 1882, and in the west at Rarawai (Ba) in 1883 and at Lautoka in 1901. It erected its first and only mill at Labasa, on the second largest island of Vanua Levu, in 1890. The western part of Viti Levu was divided from the south by rugged mountainous terrain and by poor communication. Thus, a very large number of indentured laborers in different parts of the island, or even within a particular region, spent their entire indenture insulated from each other, without the opportunity to develop and coordinate strategies for collective action. Limitations imposed by geography were compounded by indenture legislation that severely restricted mobility. An ordinance passed in 1886[21] and in force for much of the period made it unlawful for more than five laborers employed on the same plantation to absent themselves from employment for any purpose (such as for laying complaints) without the authorization of the employer. Violation of this rule could fetch a fine of up to £2 or imprisonment up to two months. Even laborers who had completed the required fifty hours of timework or five-and-a-half tasks per week required the permission of the overseer to leave the estate. The overseer was under no compulsion to issue the required ticket of absence. Without this piece of paper, any person wandering about could be apprehended by the police or the overseers and convicted for desertion. In practice, for the most part, the movement of the laborers depended upon the goodwill of the plantation management. The workers were their "property," and they alone decided how they would deal with them.

The employers not only enjoyed great authority over their laborers but also exercised considerable influence on the colonial government, which was required by law to act as their trustee. This was largely due to the dominance of the CSR in the colonial economy. At first only one among several sugar companies, the CSR was able, because of its more secure financial base and skilled management, to withstand economic vicissitudes and edge others out of competition. By the turn of the century, it was the dominant concern in the sugar industry, with investments in excess of £1.4 million and employing over three-quarters of all the indentured laborers.[22] From 1924 to 1973, the CSR and its wholly owned Fiji subsidiary, the South Pacific Sugar Mills Limited, became the sole miller of sugarcane in Fiji. The revenue-conscious colonial government had a keen appreciation of the role and contribution of the CSR to the economy, and the company used its dominant position as powerful leverage to obtain concessions and to "rely on government not to check illegal efforts of planters and its overseers to reduce the costs of labor."[23]

Furthermore, the colonial government consistently sided with the planters in disputes over work and compensation. In 1887, Bootan, an indentured laborer, lost his hand in a mill accident at Nausori.[24] The CSR refused to pay the injured laborer his wages and rations on the grounds that it could not be "called upon to help a man who will not help himself." Without his wages and perhaps with a family to support, Bootan would have to use up his meager savings, contract indebtedness, absent himself from work, and in consequence have his indenture extended. The Colonial Secretary endorsed the CSR position even while acknowledging that the injury had "not resulted from carelessness" on Bootan's part. He wrote, "The bare fact that a servant is injured whilst working for the master's benefit does not impose any obligation on the master." The government adopted a similar position on the question of remuneration for incomplete tasks. When indentured laborers were unable to complete tasks that were widely believed even by the colonial officials to be excessive, the CSR refused to pay them wages, even for the completed portion of work. The Attorney General of Fiji in 1886 offered an opinion that pleased the planters and sanctioned their practice: "It may be stated as a general legal proposition that if a person engages to perform a task, he forfeits all claim to the wage: for the performance of the task is the condition precedent to the payment of the wage."[25] No one thought to ask why the indentured laborers were unable to complete the tasks or if the tasks were excessive. Fortunately, with mounting evidence of overtasking and the increasing misery among the indentured laborers, the colonial government was forced in the 1890s to require employers to pay wages proportionate to the amount of work they had accomplished.

Clearly, there was an identity of interest between the colonial government and the planters, but it would be misleading to suggest a simple collusion between the two. Some governors and immigration officials were more sympathetic than others. Under Sir John Thurston's tenure as governor (1888–1897), for example, progressively more stringent controls were placed upon the indentured population. To be sure, Thurston was faced with an economic depression and a precarious financial situation in the colony, which weakened his hand in remonstrating with the CSR.[26] Nonetheless, there was also an apparent unwillingness on his part to enforce the existing laws governing indenture. He appears to have shared the view of the indentured laborers as lazy, improvident, and disinclined to work except under close supervision and strict discipline,

and therefore sanctioned new repressive legislations that stipulated severe penalties for even minor breaches of the labor laws.
Ordinances passed in 1896 [27] imposed fines of up to 3 shillings per
day or imprisonment for three months with hard labor if the laborers were convicted of "unlawful absence," "lack of ordinary diligence," and "neglect." It was made unlawful for laborers visiting
their employer's office or house or the office of a Stipendiary Magistrate or any other public official to make complaint. The punishment for this offense: a fine of up to £1 or one month's imprisonment with hard labor. The indentured laborers could smoke outside
their dwelling houses only at the risk of being fined 10 shillings or
one week's imprisonment with hard labor. Further, it was during
Thurston's tenure that the "block system" was introduced, whereby indentured laborers were allotted and indentured to different
groups or blocks of plantations with liability to serve on any of
them as desired by the employers. This practice brought much
hardship to the laborers, for it meant longer working hours with no
additional pay, separation from friends and family, and general
uncertainty. Sensitive officials said as much, but Thurston did not
pay heed.

But perhaps the governor's most serious act of disregard for the
welfare of the laborers was retrenching the office of the Agent
General of Immigration (AGI) in 1888 and amalgamating it with the
office of the Receiver-General. This reduced the services of the
Immigration Department at a time these were most needed. Henry
Anson, the sympathetic AGI whose persistence in enforcing the
regulations had irritated the planters and forced the government to
act, resigned and left the colony. Inspections became infrequent
and abuses went unrectified; and when some of these came to light,
they were not mentioned in the Annual Reports, which themselves
were not forwarded to India and London for fear that a knowledge
of the conditions on the Fiji plantations might lead to the cancellation of indentured emigration to the colony.[28] Other governors,
however, were slightly more sympathetic to the plight of the indentured laborers. Sir Henry Jackson, governor in 1900, showed a
greater tenacity in resisting the planters' demands. Ahmed Ali has
argued that Jackson was "not prepared to succumb to CSR pressure."[29] Sir Everard im Thurn, his successor, stiffened the penalty
for abuse of the indentured laborers. In general, however, the colonial government did not take seriously its role as trustee of the
indentured laborers' rights, and this, as much as anything else,
aggravated the laborers' demoralization and engendered a lack of
confidence in obtaining redress for their grievances.

In theory, the conditions of employment and the general provisions of indenture were clearly laid out. The "Form of Agreement for Intending Emigrants," which outlined the details, was distributed by recruiters and subagents in the districts of recruitment in India.[30] Among other things, it stipulated that indenture would be for five years, and the immigrants would be required to do work relating to the cultivation of soil or the manufacture of products; and that they would work five-and-a-half days a week (Sundays and holidays being free) at the rate of 1 shilling for men and 9 pennies for women. Further, the laborers would be given the choice of either timework (nine hours daily) or taskwork, the latter being defined as the amount of work an able-bodied adult could accomplish in six hours of steady work. The employers were to provide free accommodation as well as rations for the first six months at a daily cost of 4 pennies for each person over twelve years of age. Finally, the indentured laborers could return to India at their own expense at the end of five years, or at government expense at the end of ten years of "industrial residence" in the colony.

By the standards of the nineteenth century, when the very notion of a contract between an employer and an employee was largely unknown, the Agreement was a truly remarkable document. Students of British imperial history have been at pains to point this out in an effort to provide a more sympathetic appreciation of the imperial position.[31] The critics of indenture, on the other hand, dismissed the document as inadequate and deceptive. For them, the real significance of the Agreement for the indentured laborers lay not in what was stated on paper, but rather in what was left unsaid—especially about the social and economic realities they would encounter in Fiji and the actual conditions of employment. Gopal Krishna Gokhale, the moderate Indian nationalist leader, provided a scathing critique of the indenture system:

> Under this system, those who are recruited bind themselves, first to go to a distant and unknown land, the language, usage and customs of which they do not know, and where they have no friends and relatives. Secondly, they bind themselves to work for any employer to whom they may be allotted, whom they do not know and who does not know them, and in whose choice they have no voice. Thirdly, they bind themselves to live there on the estate of the employer, must not go anywhere without a special permit, and must do whatever tasks are assigned to them, no matter how irksome those tasks may be. Fourthly, the binding is for a certain fixed period, usually for five years, during which time they cannot voluntarily withdraw from the contract and have no means of escaping from the hardship, however intolerable. Fifthly, they

bind themselves to work for a fixed wage, which invariably is lower and in some cases much lower, than the wage paid to free labor around them. And sixthly, and lastly, and this to my mind is the worst feature of the system, they are placed under a special law, never explained to them before they left their country, which is in a language they do not understand and which imposes on them a criminal liability for the most trivial breaches of the contracts, in place of civil liability which usually attaches to such breaches. Thus they are liable under this law to imprisonment with hard labor, which may extend to two and in some cases to three months, not only for fraud, not only for deception, but for negligence, for carelessness and will the Council believe it?—for even an impertinent word or gesture to the manager or his overseers.[32]

The thrust of Gokhale's argument is the deception of the indentured laborers. In Fiji, some of the laborers recruited under false pretenses struck work and even committed suicide because the conditions of employment they encountered in the colony differed so greatly from those they were promised at home. One of the main reasons for the Labasa strike of 1907 was the laborers' refusal to do manual work because, they complained, they were promised nonmanual jobs in India. Deception was certainly present in the system, as it is bound to be in most systems of labor recruitment. But it was not the prime mover of people; the deteriorating economic condition of the Indian countryside was.[33] Would the emigrants have migrated had they been apprised fully of the conditions that lay ahead in Fiji? We can only speculate at the answer, but the fact that many saw emigration as a temporary strategy to alleviate some plight at home may suggest that a fuller knowledge of Fiji probably would not have played a decisive role in their decision.

A more serious breach of trust occurred after the indentured laborers arrived in Fiji, where there was a great discrepancy between legislative enactment and its enforcement by the officials of the Immigration Department. In the first decade of indenture, a number of ad hoc legislative measures were passed to govern indenture. These were consolidated into a single statute for the first time in 1891.[34] With minor subsequent amendments, this legislation provided the basic framework of indenture in Fiji. It was a very thorough piece of legislation (of 64 pages) that defined all aspects of plantation life, from general powers of immigration officials to the release of prisoners and their delivery from public institutions. In practice, however, the ideals enshrined in the legislation varied greatly from the realities that confronted the laborers in the field.

One of the most important provisions of the Ordinance was the creation of the Office of the Agent General of Immigration, known in some other colonies as the Protector of Immigrants. His duties included requiring any immigrant on the plantation to be produced before him for the purposes of examination, examining the state and condition of any dwelling or hospital on plantation and of any rations supplied or kept for the purpose of being supplied to any immigrant, inquiring into any complaint that the employer might bring against any indentured immigrant or any immigrant against the employer, looking into any breach of the peace or offense against the provisions of the Labour Ordinance not already adjudicated upon by a magistrate that may have taken place on any plantation, cross-examining any person under oath or otherwise who might be able to give evidence touching such matter, granting a warrant for the immediate arrest of any person guilty of breach of the peace, and finally, examining all pay-lists and other books of any plantation and all returns and documents to be kept under the provisions of this Ordinance.

The Agent General of Immigration was to be assisted in his duties by a subagent and a number of Inspectors of Immigrants. While the list of the AGI's responsibilities was impressive enough, many factors combined to limit his effectiveness. The retrenchment of his office following its amalgamation with the office of the Receiver General was a serious blow. Much depended on the energy, diligence, and attitude of the person who occupied the office. A Henry Anson or a John Forster was insistent that the planters honor their obligation to provide proper housing and medical facilities to the immigrants, even at the risk of inviting the wrath of their superiors. Most others, however, were less diligent in carrying out their duty as the trustees of the immigrants. Many of them shared, with the planters, a deeply derogatory view of the indentured laborers. A. R. Coates, the AGI in 1911, penned in a portrait of the laborers that reflected the racist ethos of the times. The Indian indentured laborers, he wrote, were a people "of emotional temperament [who] have low moral standards, [are] prone to trickery, and under certain excitement to crimes of violence, even under the discipline of continuous labour."[35] Over and over again, the indentured laborers were apportioned a large part of the blame for the social and moral ills of indenture. Indian mothers, consistently described as "indolent" and "careless," were held responsible for the astonishingly high infant mortality rate, despite evidence to the contrary.[36] Indian women and their supposedly "immoral character," rather than the destructive effects of the plantation system, were universally though erroneously seen as the primary cause of suicide in the Indian community.[37] Then there was the tendency to see problems from the point of view of the planters. When in May 1886 indentured laborers in Koronivia struck because their tasks had been increased from seven to ten chains, J. C. Carruthers, the subagent, did not ask why the tasks had been increased but argued, as the planters did: "The men certainly had . . . not the shadow of a right to leave their work en masse and rush to Suva to complain, without so much as putting shovel to ground to see whether they could do the 10 chain task."[38] Among the remedies Carruthers recommended for stopping future strikes were a liberal use of corporal punishment ("would no doubt have a marvellous effect upon habitual idlers"); infliction of heavier fines than the maximum 3 shillings provided by law; limiting the option of fine in favor of imprisonment; making prison work tougher ("let hard labour be hard labour"); and instituting a system of random checks for leave of absence tickets. The planters could not have asked for a better friend than Carruthers.

The AGI's main contact with the indentured laborers in the field were the District Medical Officers (DMOs) and Inspectors of Immigrants. The Labour Ordinance required all planters to provide adequate medical facilities to the laborers under their control. The DMOs were expected to visit the plantation hospitals regularly and to inspect "the supply of water for and the rationing of any such immigrants in such hospital and the supply of clothing, bedding, furniture, medicine, medical comforts and medical and surgical appliances." They were empowered to impose a penalty of up to £10 on planters who failed to provide full amenities to the laborers.[39] The reality was somewhat different. The *Annual Report* of 1894 noted: "The efficiency of the medical care required by the law is somewhat hampered by the expense (a serious one on small plantations) of maintaining a competent hospital attendant, the want of a working standard of competence and the absence of any control over this class of estate official by the DMO."[40] Some women died in maternity confinement because of negligence on the part of the plantation management.[41] How effective were the occasional inspection visits of the DMOs? One indentured immigrant recalled:

We were never told about the arrival of the big doctor. Once or twice a year, a new *sahib* would suddenly appear, peep into our rooms, shake his head, lift his nose to smell something, point to the overgrown grass to the accompanying *sahib*, talk very fast gesturing at our toilets, and then walk away smartly. Sometimes he would ask us whether we liked the place. We would complain about the overcrowded room, about theft, about heavy work, and other hardships. Once he was gone, our complaints remained only complaints and nothing came out of them.[42]

The inspectors, whose job was to enforce the employers' compliance with the provisions of the Labour Ordinance, also functioned with partial effectiveness. In the early years, roving inspectors visited, or were required to visit, plantations every six months, but by the turn of the century, resident inspectors were appointed in the main plantation areas of the colony. Some, such as Russell and Harpur in Labasa, were admirably persistent in investigating the laborers' complaints. Many, however, were not. They came from the ranks of the CSR overseers and generally shared their values and interests. K. L. Gillion, the historian of Fiji indenture, writes that the Indians had "no confidence in most of the inspectors." He goes on to argue that the

> inspectors shared much of the outlook and attitudes of the overseers, and it is not surprising that the employers were generally satisfied with their work. Although there were several instances of friction between employers and inspectors, usually the Europeans of a particular locality belonged to the same social circle, the government officers being dependent on the employers and overseers for fellowship, and on the companies for meat, ice, and transport for their families. The road to ease and even promotion did not lie along the way of trouble-making.[43]

Even the more assiduous inspectors conceded the hopelessness of their task. As one official noted in 1892, it was only a few cases in which the inspector could "induce the employer to give relief, knowing even then that the relief will only be temporary and his interference be bitterly though silently resented."[44] And there were many means by which the employers could pressure their employees to withhold evidence from the inspectors. One indentured laborer recalled:

> Before the inspector arrived, the *kulambar* (overseer) would come to us in the field, assemble us all, look at us with red fiery eyes, stare at some complaining type of people and begin: "The inspector will be here one of these days. He will ask you some questions and then he will be gone. If you report anything against your bosses (meaning the *sirdar* and himself), we will come to know of it. You know, we white men can find out things quickly. But before we can find out, your *sirdar* will find out about your reports. He has got friends among you. You should be able to guess what the outcome will be if you pinch the serpent. You

have to work under me all the time. Don't spoil your chances of survival in five minutes talk." After he was gone, the *sirdar* would begin his harangue: "You heard the *sahib*. He is right. You have to live and the only good way to live is to obey your superiors. The inspector will just write down your report, but who has got the key to your future? WE. Now move to your sections of work."[45]

The key to an untroublesome future lay in complying with the wishes of the overseers and *sirdars*, not in creating trouble for them.

Even magistrates tended to favor the planters.[46] They deliberately stuck to the letter of the law, even in circumstances where the evidence was far from conclusive. In Lautoka in 1903, some South Indian laborers refused to work because, they argued, nothing about taskwork was mentioned in the Tamil version of the agreement they had signed in India. The magistrate refused to listen to their complaints, evidently rigid in his belief that the laborers had been fully apprised of the terms of their service: "I can only assume as before stated that they, one and all, perfectly understand the terms and conditions of their contract and the pains and penalties attaching thereto for non-fulfilment."[47] This was an expedient assumption in a sticky situation, but the magistrate was wrong. The agreement the laborers signed in India had mentioned the possibility of both time- and taskwork. A change had been effected in Fiji that practically abolished the alternative of taskwork from the very beginning,[48] but this was not brought to the attention of potential recruits in India. The laborers, therefore, were entitled to seek legal clarification of their understanding. Instead, they were fined for absence from work and had their indentures extended. Excessive penalties were common. A. R. Coates wrote in 1910 that "instances have been brought to the notice of His Excellency the Governor of excessive or improper penalties being awarded for minor breaches of the Ordinance that by some magistrates' views are held in regard to the position and liability of an indentured immigrant, that are not in accordance with the letter or the intention of the law."[49] The system of colonial justice, even government officials were forced to conclude, was double-faced for the indentured laborers.

Further evidence of this is provided in the startling discrepancy between the extent to which the indentured laborers and their employers were able to use the courts to enforce the Labour Ordinance. Table 1 gives an indication of the nature and volume of charges the indentured laborers brought against their employers.

Several things stand out. The first is the extremely small number of complaints that the laborers laid against the employers. Indeed, there were some years in which the laborers were unable to lay any complaints at all. The paucity of the complaints, however, was no indication of the plight of the indentured laborers. As one official noted in 1892, "That there are no or few complaints is no more an indication of perfect satisfaction than the paucity of departmental prosecutions of employers is an indication of a careful and conscientious observance of the law and their obligation by the latter."[50] Laying a complaint against an employer was a serious "offense" and entailed great risks for the indentured laborer. It involved absence from work and therefore loss of pay, the extension of indenture by the number of days the immigrant was absent, and the wrath of the overseers. There were also instances of laborers being prevented from reporting abuses to the inspectors. These were especially common in isolated areas such as Labasa where, wrote Sergeant Mason in 1897, "it is a usual thing for Indians to come to the police station between the hours of 9 and 12 at night to complain of the treatment they get on some of the plantations."[51] But perhaps a more important reason why the laborers reported so few complaints was the "uncertainty of relief":[52] after taking all the risk, to see accused overseers discharged or fined lightly, or to witness the reluctance of the Immigration Department to press charges even in the face of solid evidence against the employers. Another striking feature illustrated by Table 1 is the surprisingly low conviction rate of the employers, which is in marked contrast to the conviction rate of the laborers, as we shall see later. The main reasons were the laborers' ignorance of the law, inexperience in conducting their cases, frequently without any assistance from the Immigration Department, and, as mentioned above, the prejudice of the colonial judiciary in favor of the planters. Their cases also broke down because the overseers were able to bribe or coerce other laborers to give testimony in their favor.[53]

Assault and battery were the major complaints of the laborers, accounting for 61 percent of all the charges. To some extent, this is not surprising, since violence, coercion, and control are an integral part of the plantation system. As Eric Wolf has pointed out, a plantation is "an instrument of force, wielded to create and maintain a class-structure of workers and owners, connected hierarchically by a staff-line of overseers and managers."[54] Race also contributed its share, not so much in causing violence on the indentured workers as in blunting sensitivity to it. All the planters and overseers were white, while the laboring force was black, members of an assumed

Table 1. Indentured Laborers' Complaints against Employers

Complaints	1886	1890	1891	1892	1893	1894	1895	1896	1897	Total
Assault and Battery	29	18	6	7	8	28	14	36	40	186
Nonpayment of Wages	—	3	6	7	12	10	25	15	19	97
Not Providing Tools	—	—	—	—	2	—	—	2	—	4
Not Supplying Rations	—	—	—	—	—	—	—	—	1	1
Not Providing Work	—	—	—	—	1	2	—	—	3	6
Using Insulting Language	—	—	—	—	—	1	—	—	1	2
Requiring Work Illegally	—	—	—	—	—	—	2	1	—	3
Falsifying Paylist	—	—	—	—	—	—	3	—	—	3
Failure to Take Delivery of Discharged Prisoner	—	—	1	—	—	—	—	—	—	1
Miscellaneous	1	—	—	—	—	—	—	—	—	1
Overtasking	—	—	—	1	—	—	—	—	—	1
Withdrawn	—	—	—	1	1	—	1	—	3	6
Dismissed	18	8	5	14	19	34	29	36	29	192
Convicted	12	13	7	1	3	7	15	18	32	108

SOURCES: *Annual Reports* 1886:16; 1890:30; 1892:28; 1894:33; 1895:35; 1896:31; 1897:30.

NOTE: Similar complaints have been grouped together.

inferior race whose own best interests were served by being kept under white tutelage. Employer violence was rampant in Fiji at the turn of the century. Governor Sir Everard im Thurn noted in 1907 that "the habitual attitude of many of the overseers towards the immigrants under them is, to put it plainly, brutal,"[55] and he legislated stiffer penalties for the "ill-use" of immigrants. A few years earlier, an official had noted that if "assault convictions mean that a man is a bad character then nearly all the officers of the company [CSR] are bad characters."[56] Some areas were especially notorious for overseer violence. Labasa and Ba had the worst reputation of all, and there retaliation against overseers and *sirdars* was correspondingly greater. On the smaller plantations, where the planters and overseers took a greater personal interest in the welfare of their laborers, life was generally better. It was on the larger mechanically run CSR plantations that conditions were especially bad. There the overseers competed with each other to produce more with less, and when production declined (because of bad weather, deteriorating soil conditions, and other reasons), the burden to meet quotas fell on the indentured laborers.

The overseers were assisted by *sirdars*, described by one scholar as the "lynchpin of the system."[57] As already mentioned, the *sir-*

dars were chosen for their unquestioning loyalty and willingness to serve the plantation management. The relationship between the overseer and his *sirdar* was one of mutual self-interest. An effective overseer needed a loyal and strong *sirdar*, while the *sirdar* needed the ear of his master. The threat of dismissal or relegation to field labor was a powerful incentive to please the management. The position also brought power and influence in the indentured community, and the *sirdars* used these effectively to enhance their own interests. They were allowed to own stores on the plantations, and as one official pointed out, "to those who have a knowledge of the conditions on the plantation, it is unnecessary to state how pressure can be put on an immigrant by the sirdar to compel him to deal at his store."[58] They extorted money from immigrants and even forced laborers in their charge to work free for them on Saturdays and Sundays.[59] Many interfered with indentured women and some even engaged in sexual trafficking.[60] There was at least one instance when a *sirdar* participated in the murder of a man who had caused trouble for his overseer.[61]

The indentured laborers found it extremely difficult to obtain convictions for assault and battery, despite clear evidence of physical injury inflicted by overseers. A case in point is overseer H. E. Forrest's assault on his cook, Thermadu, because the breakfast curry had not been prepared to the overseer's satisfaction.[62] Thermadu had evidence of physical injury on his body: a black eye and a long cut over his left temple. Forrest, of course, denied the assault, and was able to call two *sirdars* to support his evidence. His counsel "practically put the words into the mouths of all the witnesses for the defence who were not cross-examined nor examined by the court." The charge was dismissed despite the fact that Thermadu's testimony was "sound and practically unshaken by rigid cross-examination by counsel and court." This was not an isolated incident. As John Forster had correctly remarked seven years earlier in 1900: "There have been too many cases where the complainants have the evidence of violence on their bodies and yet could not prosecute successfully. It is obvious that where violence is resorted to on a plantation and goes lightly punished, the victims might often be unable to prove their case and are under strong inducement to abstain from complaint or to withdraw complaints made."[63] While the assault and battery cases were very difficult to prosecute, it was especially difficult to convict European overseers. As figures in Table 2 show, less than one-third of the overseers were successfully prosecuted.

Even when the overseers were convicted, the penalty was light.

They usually escaped with small fines, hardly ever imprisonment, especially before the turn of the century. Some employers even thought the transfer of the offending overseer to another estate was a sufficient punishment in itself.[64] The *sirdars*, too, got away with light fines, and some even with a history of previous convictions were reemployed, in spite of the remonstrance of the Immigration Department.[65] In an open defiance of the law, some overseers publicly returned their *sirdars'* fines, and the "fact of such return has been made known to the immigrant laborers,"[66] further undermining their confidence in the efficacy of colonial justice and reminding them, if, indeed, any reminder was needed, of who had the last laugh.

The nonpayment of wages constituted the second largest ground for the laborers' complaint against the employers. The agreement the indentured laborers had signed in India had promised a daily wage of 1 shilling for men and 9 pennies for women. This was the maximum pay the laborers could make under ideal conditions, but these hardly existed on the Fiji plantations. In fact, it was not until 1908, twenty-eight years after indenture had begun in Fiji, that adult males were able to earn an average wage of 1 shilling per working day.[67] Sickness, absence, noncompletion of tasks, and other such factors explain why the indentured laborers were unable to earn the maximum pay for such a long period of time. The

Table 2. Conviction Rate for European Overseers

Year	No. of Charges	Convictions	Withdrawn/Dismissed	% Convicted
1897	64	32	32	50
1898	7	4	3	57
1899	10	4	6	40
1900	2	—	2	0
1901	5	1	4	20
1902	14	6	8	43
1904	37	9	28	24
1905	20	4	16	20
1907	35	8	27	23
1908	30	8	22	24
1909	29	7	22	24
1910	39	5	34	29
1911	48	14	34	29
1912	62	22	40	35

SOURCES: *Annual Reports* 1902:27; 1904:24; 1905:24; 1907:25; 1908:26; 1909:24; 1910:18; 1911:19; 1912:16.

greed of the planters also played a part. Some of them devised their own tactics to retain a portion of the laborers' wages as punishment for absence without their approval. In some places, the planters used the practice of "double cut," by which they docked two days' pay for each day the laborer was away from work. Others disregarded the rules for the time when the wages had to be paid. The Immigration Ordinance required the payment of all wages on the Saturday of each week after noon, or if this was not possible because of bad weather or public holidays, on the first available working day of the following week after working hours.[68]

In practice, however, different estates paid their workers at different times, at the convenience of the employer. On some estates, only fully earned wages were paid weekly, the rest being paid at the end of the month. Thus an immigrant who had a dispute about the amount of work completed in the first week had to wait until the end of the month before he could take any action. His disadvantage vis-à-vis the employer increased with each day. Inspector Hamilton Hunter noted: "This delay confuses the immigrant as to time, and he has merely his own vague recollection of day and date to lay before the courts, whereas the employer has his fieldbook and paysheet to produce, and these are taken in evidence that the task was either badly done or not completed."[69] Once again, the powerlessness of the indentured laborers was starkly underlined.

It is remarkable that for the years included in Table 1, only one charge for overtasking was laid against the employers. This is especially surprising in view of the universal complaint of overtasking among the laborers.[70] Overtasking was the major cause of the Koronivia strike of 1886. A task, it will be recalled, was supposed to be the amount of work an able-bodied laborer could accomplish in six hours of steady labor. In practice, as the Immigration Department officials themselves conceded, tasks were frequently set by overseers on the basis of the amount of work a few handpicked men could do.[71] As one official wrote in 1886, "I believe that they are being pushed too hard and think that a proper man should pass the greater part of his time on the river and go about constantly and examine all tasks and assist Indians in prosecuting employers."[72] Yet prosecutions for overtasking were virtually nonexistent, mainly because of the absence of a precise definition of what constituted an acceptable amount of taskwork. "The indefinitiveness of the legal definition obviously leaves the limits of a fair task entirely an open question and a matter of opinion," wrote one official, "and supposing a prosecution for overtasking can be and is instituted, the weight of opinion is found on the side of skilled evidence the employer can bring forward in his favour as against the evidence of an ignorant coolie. The court has to decide on evidence not on the private opinion of the presiding magistrate."[73] The "ignorant coolie," of course, was fully aware of this reality. When redress could not be obtained for overtasking, it was pointless to complain.

In contrast to the indentured laborers, the planters enjoyed astounding success in prosecuting a very high percentage of their workers. Every year, as Table 3 shows, they laid complaints against a very large proportion of the indentured population. Both men and women were complained against, though in a number of years, proportionately more women were complained against than indentured men. Women constituted a more vulnerable segment of the indentured population. They absented, or were forced to absent, themselves from work more often than men on account of the pressure of domestic work, pregnancy, child rearing, and sickness, while in the field, they were frequently unable to complete their tasks in time. Hence the higher rate of complaints against them.[74] Some areas were worse than others. Labasa stands out, as it did in many other aspects also. The situation there was especially bad during the 1890s. In 1895, 96 percent of the total indentured population was complained against, in 1896, 68 percent, in 1898, 90 percent, and in 1899, 68 percent. Most of the complaints were successfully prosecuted. Ba was a distant runner-up. The highest per-

Table 3. Percentage of Male and Female Workers Complained Against

Year	No. Males and Females/100 Adults		No. of Total Charges	
	Males	Females	Males	Females
1897	73.8	26.2	65.2	34.8
1898	74.3	25.7	69.4	30.6
1899	73.9	26.1	67.3	32.7
1900	73.9	26.1	61.7	38.3
1901	72.8	27.1	56.2	43.8
1902	72.7	27.3	70.3	29.7
1903	74.2	25.8	75.4	24.6
1904	72.0	28.0	76.8	23.2
1905	73.5	26.5	76.0	24.0
1906	73.4	26.6	73.5	26.5
1907	74.7	25.3	70.5	29.6

SOURCES: *Annual Reports* 1897:20; 1902:26; 1904:25; 1905:26; 1908:24; 1910:17.

centage of complaints ever laid there was in 1895: 50 percent.[75] The reasons for the differences among the different districts are not explained in the Annual Reports, but they are not difficult to guess. Labasa was far away, on the island of Vanua Levu, inspections of plantations were less frequent, and the planters exercised much greater control over their laborers.

The Labour Ordinance provided a very large number of offenses for which the employers could prosecute their laborers. The offenses for which the most number of convictions were obtained are shown in Table 4. In the light of what has been said above, most of the charges are unexceptional, though the fact that the employers were able to convict 82 percent of all the charges they laid, while the indentured laborers were able to obtain convictions in only 35 percent of their cases, starkly underlined the inescapable conclusion about who held sway on the plantations. In many instances, the indentured laborers were convicted for trivial offenses, many of which they probably had not heard of until they were brought before the courts. The prosecution for "committing a nuisance" (defecating) underlines the point well. The Labour Ordinance provided that any person "who shall commit any nuisance [i.e., defecate] within sixty yards of any stream running through or any thoroughfare running through or adjoining any plantation shall on conviction in a summary way forfeit any sum not exceeding ten shillings [more than a week's pay] or be imprisoned for any term not exceeding one month."[76]

Unlawful absence by far comprised the largest source of planter complaints against the indentured laborers. It was defined as absence by truancy, detention in jail, and even attendance at court. It is doubtful that the indentured laborers deliberately absented themselves from work, knowing that it meant loss of pay, a fine, or even imprisonment. It is more likely that they did so because they were unable to work due to sickness, debility, and hospitalization. Perhaps many were also marked absent because they turned up for work late: a normal working day began at the crack of dawn. Desertion was narrowly defined as absence "without lawful excuse for three whole days exclusive of Sundays or lawful holidays."[77] A deserter could be arrested without warrant wherever he or she was found and faced a penalty of fine up to £2 or imprisonment for up to two months. The alternative to imprisonment was abolished in 1912, though a person convicted three times for desertion could face a fine of up to £5 for three months of desertion.[78] Immigration officials always saw desertion as a deliberate act of defiance. It was seen in 1885, for example, as a tendency "on the part of a limited

Table 4. Employers' Complaints against Indentured Laborers

Years	Non-performance of Tasks	Unlawful Absence	Damaging Property	Want of Ordinary Diligence	Desertion	Committing a Nuisance	Breach of Hospital Discipline	Others	Total	Convicted	%
1885	344	3,536	30	774	94	63	24	207	5,101	4,140	81
1886	52	7,121	15	1,200	272	10	40	143	8,853	6,680	75
1887	358	1,814	41	1,308	162	6	9	122	3,820	3,124	82
1890	376	969	25	29	106	5	9	50	1,569	1,282	82
1891	596	955	16	5	87	18	2	118	1,797	1,602	89
1892	793	1,125	4	6	82	10	0	133	2,153	1,825	85
1896	734	973	18	6	454	167	114	234	2,700	1,952	85
1897	83	802	45	694	29	41	83	134	1,911	1,603	84
1898	1,080	813	75	35	36	60	142	386	2,627	2,046	78
1900	881	625	91	20	92	59	151	258	2,177	1,895	87
1901	1,007	567	111	59	152	146	137	289	2,468	2,202	89
1902	717	619	110	30	120	111	113	314	2,134	1,814	85
1904	820	947	112	171	243	90	115	304	2,802	2,404	86
1905	625	671	54	216	342	42	76	326	2,352	1,875	80
1906	305	378	74	164	226	106	47	277	1,577	1,315	83
Total	8,771	21,915	821	4,717	2,497	934	1,062	3,295	43,614	45,759	83

SOURCES: *Annual Reports* 1885:16; 1886:17; 1887:11; 1888:35; 1889:29; 1890:38; 1891:46; 1892:29; 1893:33; 1894:32; 1896:30; 1898:34; 1899:36; 1900:39; 1901:42; 1902:37; 1905:39; 1906:37.

NOTE: Some similar complaints have been grouped together.

number of dissipated, dissatisfied and vicious coolies to desert from indentured service in order to indulge in gambling, prostitution, or seclusion and idleness."[79] Desertion was, indeed, a strategy some indentured immigrants used when other means of seeking redress had failed. And they did so openly. As the Immigration Inspector in Ba reported in 1900, "The intention of desertion has usually been avowed beforehand, at the time of making the complaint in the most stubborn and determined manner. . . . It was planned and systematized protest against assault [by overseers and *sirdars*]."[80] Whether the desertions were deliberate or unintended, the fact that they were so pervasive made a serious indictment of the indenture system. The immigration officials, of course, rarely thought that the indentured laborers might have genuinely good reasons to abscond from the plantations.

Indentured laborers convicted of breaching the Labour Ordinance could either be fined or imprisoned. Neither, however, was the end of punishment for the indentured laborer, for the planters were legally entitled to recover lost work by extending the contract of workers by the number of days they were absent from the plantation. The extent to which employers were able to use this provision is shown in Table 5. The indentures of both men and women were extended, though there was a decline in the proportion of extensions over the years. Around the turn of the century, proportionately more women than men had their indentures extended; however, the extensions for men were for a much longer period. These extensions pointed to the same general conclusion as the high prosecution rate of the indentured laborers, the violence of the overseers, and the indifference of the colonial government: protest did not pay. Were the indentured laborers ever notified about this provision? What was the nature of the "evidence" presented in court? But even if we accept that the laborers heeded the call of nature in the open fields, it is still astonishing that the employers were able to lay complaints under this category while the indentured laborers were unable to lay charges for overtasking. It is difficult to believe that planters who laid such complaints were serious in their intent to prosecute and obtain charges. More likely, it was a tactic to harass and intimidate those indentured laborers whom the overseers could not prosecute under any other category.

Many other provisions of the Labour Ordinance were also vague and placed a great deal of power in the hands of the planters. The indentured laborers could be prosecuted, fined, or imprisoned for using "Threatening and Insulting Language," for "Threatening Behaviour," and for "Refusal and Neglect to Go to Hospital." Many

Table 5. Extension of Indenture for Males and Females

	Women				Men		
Year	Total Expired	Extended	%	Days	Total Expired	Extended	%	Days
1898	145	67	46	1,213	363	169	47	6,705
1899	183	69	38	1,690	491	176	36	6,835
1900	277	127	46	1,984	739	315	43	16,417
1901	251	141	56	3,293	660	252	43	11,454
1902	309	148	48	2,976	798	321	40	11,208
1904	211	111	53	2,399	564	169	30	11,085
1905	524	276	53	6,097	1,254	500	40	32,404
1906	587	169	29	2,336	1,385	539	39	31,581
1908	358	125	35	1,577	1,014	386	38	31,851
1909	204	40	20	779	630	178	28	22,925
1910	577	129	22	2,581	1,648	437	27	19,917
1911	445	88	20	1,819	1,379	303	22	22,859
1912	407	76	19	2,814	1,314	333	25	36,748
1913	415	45	11	291	1,259	316	25	8,694
1916	682	38	5	783	2,061	287	14	20,516

SOURCES: *Annual Reports* 1898:19; 1899:11; 1900:12; 1901:13; 1902:12; 1904:13; 1905:12; 1906:12; 1908:13; 1909:12; 1910:8; 1911:10; 1912:9; 1913:6; 1916:6.

were convicted for "The Want of Ordinary Diligence," which was decided by the overseers who set excessive tasks in the first place. The planters viewed the laborers who did not complete the assigned work as malingerers, who needed to be worked through firm discipline. Were these workers really as lazy as the planters made them out to be? One official noted in 1882:

> The fact remains that cases were brought to notice, and more possibly escaped notice, in which immigrants were committed to prison for the non-performance of acts that they were not capable of performing, and in more than one instance the condition of the immigrants was found to be such on entering gaol as to necessitate their speedy relegation to hospital.

This chapter has attempted to highlight those factors that stifled protest in the Indian indentured community. It has focused especially on the actions of the planters and the colonial government, which were the primary deterrents of collective action in the Indian community. But to leave the impression that there was no protest whatsoever would be misleading. Powerless as they were, the indentured laborers did resist, in their own ways, undue pressures that were put on them. Some attempted desertion, as we have seen, hoping perhaps to submerge themselves in the free Indian community slowly emerging on the fringes of the plantations. Some vented their rage on the crops and tools of the employers, feigned illness, and absented themselves from work "as deliberate retaliatory devices against the system."[81] A few sought redress by direct petitions to the government, though this practice was short-lived. In 1912, Karim Bux from Waiyevo, Taveuni, telegramed the governor in the most humble and respectful terms— "You are our Father and the Lord and the Coolie Agent is our Mother"—complaining of violence.[82] The AGI disliked the immigrants' resorting to the practice of using telegrams, and instructed the wireless operators not to translate messages sent from Indians to various government departments. And he refused to investigate the complaints because the "staff of inspectors is too small, and the difficulties of communication are too great to permit an officer being sent without sufficient grounds." Anyway, he concluded without any investigation that "the complaints are vague and apparently not probable as to the alleged facts." A handful on the Rewa showed a "spirit of combination" in the nineteenth century by contributing to a mutual assistance fund to pay fines or to fight court cases. In 1917, they burned an effigy of the indenture system, "this old friend of the European planters, and the enemy of Indian national self-respect, national honor, national name and fame—a hideous monster preying on Indian womanhood and torturing its victims into a life of misery and shame, and bringing up its offsprings in sin and filth."[83] The use of such language was a sign of the increasing maturity and confidence in the indentured community.

But there were many other areas where the indentured laborers were less fortunately placed. Tight control by the plantation management rendered them immobile and vulnerable. And when no respite was in sight and redress seemed impossible, some were left with few alternatives but to resort to violence. Murder became an instrument of revenge, and habitually brutal overseers and *sirdars* who interfered with Indian women paid with their lives.[84] What was particularly disturbing to the officials was that these took place in the open, and the perpetrators were fully aware of the consequences that followed: execution or life imprisonment. Others, unable or unwilling to retaliate, took their own lives through suicide. This was especially the case with the new arrivals, the South Indians, who were also the victims of North Indian cultural prejudices, and others who found the plantation routine too demanding.[85] Such acts of violence might strike temporary terror in the hearts of the overseers or raise concern in official circles. But these were aberrations that posed little threat to the planters and the colonial officialdom.

During the nineteenth century, the outside world knew little about the plight of the Indian indentured laborers on the Fiji plantations. That changed after the turn of the century, as the writings of Rev. J. W. Burton and C. F. Andrews described for the first time the conditions in Fiji. An emerging nationalist movement in India seized upon their writings to bolster their struggle for the abolition of the indenture system.[86] Indentured emigration to Fiji ceased in 1916, and all indentures in the colony were canceled on 1 January 1920. At last "free," the Indians immediately (in 1921) mounted a concerted challenge to both the CSR and the colonial government, to retrieve their *izzat* (honor), to obtain better wages, and to secure a measure of equality with the Europeans.[87] The vigor and energy with which the Indians took to these challenges showed them to be anything but a docile and submissive people. It was their struggle, perhaps more than anything else, that led to the independence of Fiji in 1970 and to the departure of the CSR in 1973.

Abbreviation

CSO   Records of the Colonial Secretary's Office, National Archives of Fiji, Suva.

Notes

A version of this chapter was published in the *Hawaiian Journal of History*, 20 (1986), 188–214.

1. K. L. Gillion, *Fiji's Indian Migrants: A history to the end of indenture in 1920* (Melbourne, 1962), 93.

2. C. F. Andrews and W. W. Pearson, *Indian Indentured Labour in Fiji: An independent enquiry* (Perth, 1916), 29

3. K. O. Laurence, *Immigration into the West Indies in the Nineteenth Century* (Kingston, 1971), 74. For other comparative studies of Indian indenture, see Alan H. Adamson, *Sugar Without Slaves: The political economy of British Guiana, 1838–1904* (New Haven, 1972), Walter Rodney, *A History of the Guyanese Working People, 1881–1905* (Baltimore, 1981), and several studies in Kay Saunders (ed.), *Indentured Labour in the British Empire, 1834–1920* (London, 1984).

4. Kusha Haraksingh, "Indian Leadership in the Indenture Period," *Caribbean Issues*, 2:3 (1976), 31.

5. Hugh Tinker, *A New System of Slavery: The export of Indian labour overseas, 1830–1920* (London, 1974), 226.

6. Eugene D. Genovese, *From Rebellion to Revolution: Afro-American slave revolts in the making of the modern world* (Baton Rouge, 1979), 7.

7. Rodney, *Guyanese Working People*, 151.

8. The only published account of the indenture period by an indentured laborer is Totaram Sanadhya's *Fiji Dvip Men Mere Ikkis Varsh* [My Twenty-One Years in the Fiji Islands], 4th ed. (Benaras, 1973; first publ. 1914); Totaram Sanadhya, *My Twenty-One Years in the Fiji Islands and The Story of the Haunted Line*, ed. and transl. by J. D. Kelly and U. K. Singh (Suva, 1991).

9. Introductions to Fijian history are Deryck Scarr, *Fiji: A short history* (Sydney, 1984); and Bruce Knapman, *Fiji's Economic History, 1874–1939: Studies of capitalist colonial development* (Canberra, 1987).

10. The background to indenture is treated fully in my *Girmitiyas: The origins of the Fiji Indians* (Canberra, 1983).

11. For a fuller discussion of this, see my "Peasants on the Move: Circulation and the migration of Indian indentured labor," unpublished paper presented at the 15th Pacific Science Congress, Section C (Geography), 2–8 February 1983, Dunedin, New Zealand.

12. The amounts annually remitted to India are noted in the *Annual Reports* of the Agent General of Immigration. For a summary of the amount sent to India for the period 1907–1917, see Fiji Legislative *Council Paper* CP 48/1917. See also James McNeill and Chimman Lal, *Report of the Government on the Conditions of Indian Immigrants in Four British Colonies and Surinam*, Command Paper 7744-5 (Simla, 1914), Appendix.

13. For a complete list of all the castes and districts of origin, see my "Leaves of the Banyan Tree: The origins and background of Fiji's North India indentured migrants, 1879–1916." Ph.D. dissertation, Australian National University (Canberra, 1981), Vol. II, Appendixes IV (pp. 12–28) and VIII (pp. 82–100).

14. Lal, *Girmitiyas*, 103.

15. Tinker, *A New System of Slavery*, 226.

16. Adrian Mayer, *Peasants in the Pacific: A study of Fiji India rural society*, 2nd ed. (London, 1973), 5.

17. Ahmed Ali (ed.), *The Indenture Experience in Fiji* (Suva, 1979), 10.

18. Mayer, *Peasants in the Pacific*, 5.

19. Minute Paper 1050/86, Suva, National Archives of Fiji.

20. CSO 351/1914; CSO 457/1913.

21. Ordinance XIV of 1886.

22. See Agent General of Immigration *Annual Report* 1899:9; 1900:10; 1902:10; 1905:10. See also Jay Narayan, *Political Economy of Fiji* (Suva, 1984), 42.

23. Michael Moynagh, *Brown or White?: A history of the Fiji sugar industry, 1873–1973* (Canberra, 1981), 59.

24. CSO 1591/1887.

25. CSO 443/1887.

26. This is the view of Thurston's biographer. See Deryck Scarr, *Viceroy of the Pacific* (Canberra, 1981), 192.

27. Ordinance XIV of 1896.

28. Gillion, *Fiji's Indian Migrants*, 89.

29. Ali (ed.), *Indenture Experience*, xviii.

30. Lal, *Girmitiyas*, Appendix 1.

31. See, for example, I. M. Cumpston, *Indians Overseas in British Territories, 1834–1854* (London, 1953), 174; George R. Mellor, *British Imperial Trusteeship, 1783–1850* (London, 1951), 223.

32. Quoted in Dharam Yash Dev, *Our Countrymen Abroad* (Allahabad, 1940), 14. This view has been reasserted by modern scholars. See Bridget Brereton, "The Experience of Indentureship," in John La Guerre (ed.), *From Calcutta to Caroni: The East Indians of Trinidad* (Port of Spain, 1974), 29: "Perhaps the most obnoxious feature of the system was the fact that labourers who infringed the immigration laws, even quite trivially, could be persecuted as criminals and sentenced to jail terms."

33. For a fuller discussion, see my *Girmitiyas*, ch. 3.

34. Indenture Ordinance No. 1 of 1891 (an ordinance to amend and consolidate the law relating to Indian immigration).

35. CSO 3027/1911.

36. This is discussed at length in my "Kunti's Cry: Indentured women on Fiji plantations," *Indian Economic and Social History Review*, 22:1 (1985), 55–72. The question of infant mortality is also dealt with by Ralph Shlomowitz, "Infant Mortality and Fiji's Indian Migrants, 1879–1919," *Indian Economic and Social History Review*, 23:3 (1986), 289–302.

37. For an extended discussion of this see my "Veil of Dishonour: Sexual jealousy and suicide on Fiji plantations," *Journal of Pacific History*, 20:3 (1985), 135–155.

38. Minute Paper 3481/86.

39. Indenture Ordinance 1 of 1891, Part XI, section 141.

40. *Annual Report* (1894), 29.

41. CSO 5800/1909.

42. Shiu Prasad, *Indian Indentured Workers in Fiji* (Suva, 1974), 9–10.

43. Gillion, *Fiji's Indian Migrants*, 111.

44. CSO 1955/1892.

45. Prasad, *Indian Indentured Workers*, 19.

46. Ali (ed.), *Indenture Experience*, xvii, as well as studies in Saunders (ed.), *Indentured Labour in the British Empire* (for example, M. D. North-Coombs' contribution on Mauritius at page 97).

47. CSO 4705/1903, cited in my "Veil of Dishonour."

48. On many estates the change to taskwork, without the option of timework, took place in the early 1880s. The *Annual Report* for 1882, §41, noted that "the change has not been brought about without the expression of great dissatisfaction on the part of the immigrants."

49. Im Thurn to Colonial Office, No. 277, 23 December 1910.

50. CSO 1955/1892.

51. CSO 1315/1897. See also CSO 4224/1895, CSO 4215/1899, CSO 3237/1900.

52. CSO 1955/1892.

53. CSO 5579/1914.

54. Eric Wolf, "Specific Aspects of Plantation Systems in the New World: Community sub-cultures and social class," in M. M. Horowitz (ed.), *Peoples and Cultures of the Caribbean* (New York, 1971), 163, cited in Ali

(ed.), *Indenture Experience*, xii. See also George Beckford, *Persistent Poverty: Underdevelopment in plantation economies of the Third World* (New York, 1972), for an extended discussion.

55. CSO 800/1907. Penalties for assault on indentured laborers were stiffened by Ordinance VI of 1907, which provided for a fine of up to £3 or two months' imprisonment with hard labor.

56. CSO 5064/1899.

57. Ali (ed.), *Indenture Experience*, xxiii.

58. CSO 1045/1906. This practice was barred by Ordinance II of 1912.

59. Ali (ed.), *Indenture Experience*, xxiii.

60. Andrews and Pearson, *Indian Indentured Labour*, 36; *Annual Report*, 1909:23.

61. Walter Gill, *Turn North-East at the Tombstone* (Adelaide, 1970), 44. Gill was an overseer in Fiji during the last days of indenture, and his book is a moving and vivid account of the conditions under which the indentured laborers lived.

62. CSO 4012/1907.

63. CSO 3491/1900; CSO 3237/1900; *Annual Report*, 1901:25.

64. CSO 4412/1902; *Annual Report*, 1901:25.

65. CSO 4412/1902; CSO 2555/1893.

66. CSO 3121/1897.

67. Wadan Narsey, "Monopoly Capitalism, White Racism and Superprofits in Fiji," *Journal of Pacific Studies*, 5 (1979), 86.

68. Indenture Ordinance I of 1891, Part VIII, Section 116.

69. CSO 2315/1888.

70. CSO 2315/1888; CSO 511/1886; CSO 1955/1892; CSO 3481/1886.

71. *Annual Report*, 1886: 15.

72. CSO 1800/1886.

73. CSO 1955/1892.

74. For a fuller discussion, see my "Kunti's Cry."

75. *Annual Reports*, 1897:20; 1902:26.

76. Indenture Ordinance I of 1891, Part VII, Section 92.

77. *Annual Report*, 1882:30.

78. *Annual Report*, 1912:17.

79. *Annual Report*, 1885:13.

80. CSO 3237/1900.

81. Ali (ed.), *Indenture Experience*, xvii.

82. CSO 3280/1912; CSO 4100/1912.

83. Ali (ed.), *Indenture Experience*, xxvi.

84. CSO 4050/1908; Gillion, *Fiji's Indian Migrants*, 166 ff.

85. For a fuller discussion, see my "Veil of Dishonour."

86. On the abolition of indenture, see Gillion, *Fiji's Indian Migrants*, 164–189.

87. This is discussed in K. L. Gillion, *The Fiji Indians: Challenge to European dominance, 1920–1946* (Canberra, 1977); and in Ahmed Ali, *From Plantation to Politics: Studies on the Fiji Indians* (Suva, 1980).

7

Hegemony and Repression in Rural Guatemala, 1871–1940

David McCreery

The shift to large-scale production and export of coffee in late nine-teenth-century Guatemala prompted an enormous increase in the demand for cheap agricultural labor, a demand the indigenous majority of the population was reluctant to meet. Centuries of experience had taught them that only abuse, disease, and death awaited them on the *fincas* (plantations) of the *boca costa* (pied-mont) and lowlands. In order to overcome this resistance and to ensure a smooth flow of workers to coffee *fincas*, the neo-Liberal regime that came to power in 1871 with coffee planter support put into place and enforced a variety of coercive labor measures, including *mandamientos* (corvée labor) and debt peonage.[1] While vestigial regional and personalist conflicts persisted for a time, as did upheavals provoked by the occasionally quixotic ambitions of a succession of *caudillos*, all but the rare skeptic among the elites quickly came to accept the coffee export economy as central to national development.[2] The Liberal state generally served these interests faithfully, but it was a state that rested increasingly on simple force, on coercion and violence, rather than on the shared culture and values of consensus. This chapter will examine the changing ideology and instruments of state control in nineteenth-century Guatemala and the modes of Indian peasant resistance to state demands, and particularly to demands for forced coffee labor.

To function, the state must have hegemony. By hegemony is understood the ability to guarantee at least minimal compliance with a policy or goal of the government throughout the sovereign territory.[3] It is important, in this context, to identify and differenti-ate political from ideological hegemony. Political hegemony rests directly on the threat or use of force or coercion, whereas ideologi-cal hegemony implies the achievement of policy ends based on will-ing, or apparently willing, compliance, on shared ideas and values. This split is possible, and in premodern societies usual, because ideological hegemony is not the exclusive, or even the normal, property of the elites or the state they control, but resides rather in civil society:

> The term "civil society" connotes the other organizations in a social formation which are neither part of the process of material production in the economy, nor part of state-funded organizations, but which are relatively long-lasting institutions supported and run by people outside of the other two major spheres. A major component of civil society so defined would be religious institutions and organizations, apart from entirely state-funded and state-controlled organizations.[4]

In the case of Guatemala's indigenous population, the institutions of civil society were, and are, the product of a long history of struggle and are embodied in *costumbre* (custom), the most important arbiter for that group of what is acceptable or valuable, and socially and morally correct. In illiterate and largely unschooled societies not only is such *costumbre* rarely under the control of Althusser's "ideological state apparatuses," but it commonly remains quite beyond the ken of most of the state's representatives.[5] As a result, it may not, indeed rarely does, share in all of the values, goals, and orientations—the ideological project—of elite-dominated "national" culture. By language, moral assumptions, and socially esteemed activities the masses are a "race wholly apart."[6] This was particularly so for rural Guatemala, with its *ladino*[7] elite and indigenous majority. In Guatemala's situation, in fact, it would be more accurate to speak of civil societies in the plural, for there were, and are, great differences between even adjacent communities, as well as within and between language and culture groups. Where political and ideological hegemony coincide, as is typical of modern, industrial societies, they reinforce each other and the power of the state, allowing that institution to maintain an illusion, if rarely the reality, of working in the general interest. Where ideological hegemony is lacking, the state typically relies on threatened or real force, on political hegemony, if it is to maintain control. Few modern peasantries anywhere have expected the state to be anything but oppressive, exploitative, and immoral, an expectation rarely disappointed.

Over the course of the long nineteenth century (1760–1930), the political hegemony of the Guatemalan national government increased precisely as its ideological hegemony declined. The Bourbons in the late eighteenth century proposed a course of economic and political change at odds with what the indigenous population, probably correctly, saw as its own best interests.[8] New taxes, free trade, export agriculture, and, above all, attempts to weaken the secular power of the church tended to alienate the rural Indians. From the first years of the Conquest, the language of the church and of "Catholicism" had provided the ideological glue of the empire, the idiom that allowed the state and peasantry to speak the same language and to agree on morals, values, and priorities. To be sure, the popular or folk catholicism(s) in the villages was a far cry from what the pope imagined Catholicism to be, but in practice such differences mattered little at the local level. If the institutional church did make occasional and largely unsuccessful efforts to reconcile these multiple visions, more generally the priests were content to collect their fees and accepted that too zealous an examination of the religious activities of their parishioners would put at risk their income and perhaps even their lives.[9] The result was an illusion of consensus, but a very important illusion nevertheless, remembering that individuals and groups act not on reality but their perception of reality. Catholicism and the role of the church in everyday life provided the basis for an apparent societal consensus and a sense of shared values. Support for the church gave the state some claim to ideological hegemony and to the consent of the ruled. This was important precisely because neither the Hapsburgs nor the Bourbons ever achieved dependable political hegemony over the countryside. If the state, usually slowly and with great effort, could assert its power at most points within the *Audiencia* (High Court) of Guatemala, its day-to-day hold on the rural areas always remained extremely tenuous. Nothing illustrates this better than the regime's inability to control contraband trade or to enforce tax reform after 1800.[10] For the same reason, the Bourbons largely failed to shake the hold of the church on the communities and of popular catholicism on their inhabitants. The political weakness of the state meant that it could not penetrate or orient popular society to serve its goals of change, and so, and inadvertently, could not threaten the basis of apparent consensus and of its own ideological hegemony over the peasantry.

The Independence Liberals also sought to substitute their vision of development for *costumbre*,[11] and they too encountered opposition from among the mass of the population. This resistance was strong enough by the end of the 1830s to result in the overthrow of the existing regime and its replacement with one more congenial to peasant priorities. What the reforms of the late 1820s and 1830s and the violence they provoked did accomplish, however, was to seriously weaken the church in the countryside by stripping the religious orders of their properties and driving out much of the regular and secular clergy. This assault on "religion,"[12] in turn, seriously weakened, in the popular eye, any claim by the state to participation in the existing popular ideological consensus. In their attack on the church, the Enlightenment Liberals of the 1820s and 1830s succeeded where the Bourbons had failed. But they could not construct a state strong enough either to impose political hegemony over the countryside or to substitute and make hegemonic their new ideological project, one of nationalism, and secularism, for the old. The Conservatives who replaced the Liberals after 1839 did not try. Theirs was a minimal state, the sort peasantries prefer. President Carrera did not so much protect or advance peasant interests as leave most of the population to its own devices.[13] Conservatives sought no extensive changes in rural society, and the export crop of cochineal dominant in these years (1840–1860) made even less labor and land demands than had the production of indigo, the cash crop staple that preceded it. While the Conservative regime did allow the church to regain some of its former power and presence in these years, it did not return most of the properties the orders earlier had lost, and the church continued to suffer from a shortage of priests, disruptions in the hierarchy, and a lack of funds. It was this period, for example, that witnessed the loss of the church and state control over the *cofradias* (brotherhood) and the transformation of these into "Indian" institutions.[14] Indeed, what is most characteristic of the half century from the 1820s to the 1860s, and whether Liberals or Conservatives controlled the central government, was the ebbing of the state's political and ideological power, or, more directly, of the state itself, in the countryside.

The second Liberal generation, or neo-Liberals, of 1871 came to power with a development agenda that required, to a far greater extent than had that of the Bourbons or the Enlightenment Liberals, the penetration and partial disintegration of existing indigenous rural society. Coffee production gave the motive and the wherewithal to shake land and labor out of existing rural structures and to make these available for the rapid expansion of export agriculture. Such an effort clearly conflicted with the perceived self-interests of indigenous rural society and could in no way be made to fit the existing, if at least partially illusory and greatly decayed, popular ideological consensus. The Liberals did little to seek such agreement. Despite much rhetoric about schools, education, and uplifting of the masses, most of which was, in any event, the work of Liberal intellectuals[15] rather than practical men of government and business, what is striking about the neo-Liberals is the absence of any serious attempt to indoctrinate the peasantry in its vision of development or modern society. Whereas the Enlightenment Liberals had seen the Indian as a block to national development but a block that could and should be overcome by education and integration, by raising the Indian to equality, the neo-Liberals of the post-1871 *Reforma* viewed the Indian as probably essentially, and certainly in the short run unalterably, inferior:

> The Indian . . . is always rebellious, malicious and distrustful by nature, a swindler by routine without any stimulus to work or needs to satisfy who would undoubtedly improve his social condition [if made to work]; happy to vegetate in indolence and slovenliness . . . this immense quantity of human beings does not wish to understand what is in his own good. [I]t is today an insuperable obstacle that tenaciously opposes the production, the wealth and the general wellbeing of the country.[16]

Instead of schools, the dictator Manuel Estrada Cabrera, "Protector of Studious Youth," built temples to Minerva. All agreed, nevertheless, that the labor of the Indian, unless and until he could be replaced or "bleached out" by a superior immigrant population, was a necessary and vital ingredient for export agriculture and for elite prosperity. It mattered less, if at all, what the Indian thought than that he or she should be readily and cheaply available for labor in the coffee groves. The growers wanted their bodies, not their minds.

The neo-Liberals, secure in their racist assumptions and in a rush for "development," effectively abandoned efforts to construct ideological hegemony and fell back on coercion, on political hegemony in its most bald and undisguised forms. Greatly facilitating this was the increase in resources available to the state. Export and import taxes grew rapidly after 1870, as did spending on agencies and instruments of domestic control (see Table 1).

With the repeating rifle, the telegraph, and new bureaucratic forms of organization funded with a rapidly expanding budget, the state was sufficiently powerful that it did not need, nor did the neo-Liberals think it worth the resources to attempt to build, an ideological consensus among the indigenous majority. Force, the coffee planters reasoned, was what the Indian understood, force would do, and force was what the Indians got.

Control of the countryside after 1871 rested on the twin pillars of the army and the militia,[17] and tying the state and this repressive apparatus together was the telegraph. Although the Conservatives had flirted with the new technology, Guatemala's first telegraph

Table 1. State Revenues/Expenditures

Year	Revenue (in pesos)	Expenditure (in pesos)	
		Administration	War
1870	1,130,449	132,816	327,779
1880	3,844,413	—	—
1890	7,500,142	817,534	4,389,158
1900	8,860,947	1,664,621	3,392,824
1910	51,571,441	2,812,703	3,739,657

SOURCES: The figures for 1870 and 1880 are taken from Ignacio Solis, *Memorias de la Casa de Moneda de Guatemala y del desarrollo economico del pais,* tomo 4 (Guatemala, 1979), 1333; those for 1890 and subsequent years are available in the annual reports *(Memoria)* of the Ministerio de Hacienda.

line came into operation in 1873, between the Pacific port of San José and Guatemala City.[18] The events of their own successful revolt made the Liberals acutely aware of the importance of rapid communication for political control, and they spread the telegraph quickly to the rest of the country. The lines extended to the *cabeceras* (far reaches) of Chimaltenango, Totonicapán, and Jalapa in 1874, Chiquimula and Huehuetenango in 1875, and Cobán and Zacapa by 1876. The various government ministries, the *jefes politicos* (political leaders), and the local officials such as *alcaldes* (mayors), *comandantes* (commandants), and *comisionados politicos* (political commissioners), kept the wires alive overseeing the population. Local telegraph operators complained that they worked late into the night after the end of public hours, receiving official messages and circulars and transmitting local news to the central authorities. Government telegrams increased from an average of less than a thousand a month in the early 1870s to 15,000 a month in 1898, and more than 20,000 in the peak labor mobilization month of August; in that year the system handled more than 182,000 official messages.[19] The telegraph played a vital role, too, in repressing revolts such as Julian Rubio's attempt to raise Quiche against the government in 1877 or the *levantamiento* (revolt) at San Juan Ixcoy in 1898.[20] After 1893 a modest telephone network supplemented the telegraph. As the judge remarked in Miguel Angel Asturias' novel *El Senor Presidente:* "What was the telephone invented for? To see that orders were carried out! To arrest the enemies of the government!"[21] Given the difficulty of building adequate roads or railways in the highlands, no innovation did more to help to assert state control over the interior than the prosaic telegraph and telephone.

Armies and militias had existed in the areas from the colonial period, but generally these had remained ad hoc, ragtag affairs based in the towns and officered largely by amateurs; in times of conflict the state filled out the ranks with press-ganged and poorly trained and armed *ladino* and Indian peasants.[22] None of the governments before 1871 could bear the expense of maintaining any substantial number of regular troops under arms for any period of time, so they and their opponents put armed bands together as needed and disbanded these as quickly as possible when the immediate fighting ended. Without a regular armed force at its disposal, state control over the countryside was limited, as, for example, Santa Catarina Ixtahuacan's defiance of the central government for several decades after 1839[23] and the activities of the various *facciones* (factions) that plagued Guatemala in the 1840s and 1850s made clear. The Liberals' own triumph in 1871 in a series of small, if hard fought, battles impressed upon them the weakness of the state's available defense and control apparatus. Among the new government's early reforms was the creation of the *Escuela Politecnica* (Polytechnic School) in 1983 to professionalize the officer class.[24] The regime moved too to put the militia on a regular footing. It was now possible to differentiate the regular army, which in peacetime rarely numbered more than 2,000 to 4,000 men and was used to garrison a few urban centers and the frontiers, from the militia, which in theory included all *ladino* males between the ages of eighteen and fifty not otherwise exempt.[25] The army's principal task was national defense, whereas the militia, although it acted as a reserve for the army and might be called upon in time of war, served, in the absence of a rural police force comparable to El Salvador's National Guard, as the state's chief instrument of control and repression in the rural areas; by the turn of the century, there were 173 militia *comandancias* (command areas) based throughout the country in *ladino*-controlled towns and *aldeas* (small villages).[26] If still far from adequately trained, the militia was now armed with modern repeating rifles, giving the minority *ladino* population in the labor recruitment areas for the first time a decisive advantage over even large masses of Indians, typically armed only with rocks, sticks, and an ancient shotgun or two. Militia soldiers under the command of the *comandante local,* usually a retired regular army officer or noncommissioned officer, mustered conspicuously for drill in the central plaza and acted to guard land surveyors from irate contestants, put down conflict arising out of disputes between communities or *fincas,* transported recalcitrant workers to the *fincas,* and generally intervened in any problem or disturbances in the neighborhood of interest to state authorities.[27]

The indigenous population abhorred regular army service and did

all they could to avoid it. Service in the army meant brutal treatment under *ladino* officers, poor food, harsh and humiliating living
conditions, and often years away from the individual's home and
his plantings or trading interests. "Please exempt us from military
service," begged the Indians of San Pedro Carcha (Alta Verapaz) in
1872, "for we know the carrying trade, the machete and the axe,
and we don't like being soldiers."[28] A decade later, the indigenous
inhabitants of Santiago Sacatepequez (Sacatepequez) agreed, protesting "we don't know how to take up arms as soldiers."[29] In contrast to the regular army, into which they might be drafted against
their will, Indians were not encouraged or often allowed to join the
militia units,[30] for readily apparent reasons. In the aftermath of the
general mobilization for the 1890 war with El Salvador, for example, President Barillas ordered the reorganization of the armed
forces, "eliminating completely all of the Indians who have become
part of the army, who for their conditions of unskillfulness and
ineptitude lack the capacity to understand what it is to be a soldier
and the very important mission he has to fulfill in the defense of
the country and in the maintenance of internal order."[31] Only those
well on the way to becoming *ladinos* might be allowed to remain.
The role of the army was largely to fight foreigners, and for this
Indians would serve, but the job of the militia was to uphold *ladino*
and state authority in the countryside, and any substantial component of Indians in the units might put into question its reliability for
this task.

Coffee producers relied on state-enforced coerced labor systems
to mobilize their work force. Although a few Indian families settled
as *colonos*, or permanent residents, on the *fincas*, most of the laborers mobilized under various forms of forced labor for sixty to ninety
days a year, migrating from their highland villages to the coffee belt
to clean the groves and work the harvest. Violence dominated labor
relations, and *finquero* (planter) control on the properties was by
no means always assured.[32] On the *fincas*, the owners and administrators treated harshly workers they assumed to be their racial infe-
riors, "almost always addressing them as 'chucho,' a word used to
call a dog."[33] Foremen and owners beat Indians with their fists,
with whips, and with the flat of a machete; they set dogs on them,
and placed them in stocks or in *finca* jails, and drunken bosses shot
workers and assaulted women with impunity. Labor recruiters and
state agents in the villages jailed workers, beat and defrauded
them, kidnapped their wives and children, and burned their
houses.[34] The pervasiveness of state power and the readiness of the
state to use this power to aid coffee production greatly restricted

how the Indians could respond to such demands and abuses. Premeditated attacks by the workers on administrators or recruiters were rare and even spontaneous outbursts uncommon. Although the Indians always outnumbered the administrators and his underlings on the *fincas* and in the villages, the latter went about armed and mounted and could count on assistance from the authorities and militia in nearby towns. Unplanned outbursts did occur, as, for example, when one or a group of workers threatened or attacked a *caporal* (foreman) over an unfair *cuerda* (unit of area for work) or *caja* (unit of quantity for harvest) or in response to egregious blows.[35] On the *finca* Ona, for example, the workers "rioted," which is to say they complained in a group, in July of 1901, because they had not received their food ration and because the owner jailed a municipal official who, in this instance, sided with them. Militia soldiers from the nearby town of La Reforma quickly put down the protest, and remanded the leaders to hard labor on public works.[36] Similarly, when the administrator of *finca* Lorena reprimanded a group of Indian workers for, he said, pulling down and breaking the branches of coffee bushes instead of using the ladders provided, they "tried to hit [him] with [his] own club" and chased him out of the field. The *comandante local* dispatched a militia unit to the *finca*, which arrested the few workers who had not fled, and the judge sentenced them to fifteen days in prison.[37] But such instances are rare precisely because, as these demonstrate, violence against *finca* authorities was not an effective means to righting wrongs. Open resistance brought immediate and effective repression by the state, and while such opposition was occasionally a useful and even necessary bargaining tool, the price for the Indians was usually disproportionately high for the results likely to be obtained.

For the same reasons, large-scale uprisings or *levantamientos* became almost nonexistent after 1880.[38] The few that occurred the state crushed quickly and bloodily. The Alta Verapaz, perhaps because of its isolation and the very small *ladino* population, seems to have been particularly prone to such outbreaks, or so the local *ladinos* felt. According to one source, an Indian "revolt" broke out in 1885 and "[t]here are allusions to many other altercations in the late 19th century, usually led by Indian religious practitioners and having the aspect of natavistic movements involving gatherings of several different Indian groups."[39] While this may have been the case, and the subject awaits more intensive examination,[40] a closer look at one of these purported uprisings suggests, at least in that particular instance, a rather more pedestrian explanation. In 1897, *jefe politico* noted rumors of a conspiracy among the Indians of Cobán lead by the *principal* (community elder) Juan de la Cruz against "local *ladinos* and foreigners."[41] De la Cruz, it was reported, for some time had been "exploiting the Indians with guile and promises, taking advantage of their tendency to oppose anything that is progress" and collecting money for a supposed revolt: "It is said that the object of the revolt is to end [export] agriculture and attack the right of private property, and one should also note the resistance of the Indians to working on the plantations." Juan de la Cruz fled Cobán before he could be captured, but upon investigating, the *jefe politico* concluded: "It may well be the case that Juan de la Cruz from the beginning had no other object than to obtain money from those gullible enough to believe him, promising the Indian that for which they hope." As the *jefe* acknowledged, the Indians of the Alta Verapez certainly dreamed of expelling the foreigners attracted to the area by coffee and of reclaiming control over their land and labor, but the possibilities for doing this, as most must have realized, were slight. In this case hope may simply have made them the victims of petty fraud. The most famous *levantamiento* to actually erupt in these years was that which broke out at San Ixcoy in July of 1898, resulting in the murder of all but one of the *ladinos* in the village.[42] Even in the high Cuchumatan mountains, however, militia units from neighboring *ladino*-dominated municipalities repressed the uprising the following day, with the loss of hundreds of Indian lives and the forced migration of hundreds more out of the community to the northern lowlands. Because the state now had the ability to deliver effective and immediate violence to the countryside—and it is noteworthy in this regard that, in contrast to the 1830s and 1840s, there was very little rural banditry in Guatemala after the 1870s—open violence, whether against the individual employers or the state, made little sense.

   This is not to say that the Indians subjected to forced labor drafts or caught in peonage contracts or faced with loss of access to vital lands did not resist. But because they were not foolish they chose those modes of resistance most effective to their purposes and least likely to bring down upon themselves the devastating violence of the state. James Scott in *Weapons of the Weak* defines "resistance" as "*any* act(s) by member(s) of a subordinate class *intended* either to mitigate or deny claims . . . made on that class by superordinate classes . . . or to advance their own claims . . . vis-à-vis those in superordinate classes."[43] In the broadest sense, Indian opposition to elite and state exploitation and oppression in Guatemala from the sixteenth century to the present has taken the form of denying *ladino* superiority, even given the reality of political and economic subordination.[44] While accepting useful elements of the conqueror's material culture and adopting a version of his religion modified to fit their needs, the Indians also perfected the well-known "*cargo* system" and the "closed corporate community,"[45] the manifest function of which was to keep the *ladino* and *ladino* culture at arm's length. The expansion of coffee in the late nineteenth century brought *ladino* immigrants to the highland Indian municipalities as never before, but the result was not usually acculturation (ladinization) of the dominated population but rather the constitution of dual or parallel societies that interacted as little as possible.[46] Popular folktales, dances, and religious ceremonies and rituals took elements of *ladino* culture and used these to define and reinforce the barriers between Indian and *ladino* "worlds."[47] If this Indian world itself was, of course, a colonial construct, it nevertheless provided the basis for a separate Indian identity morally superior, in the Indians' perception, if not in terms of secular power, to that of the *ladino*. The Liberals, however, and to repeat what has been indicated above, cared little what the Indians thought so long as they were able to work under conditions favorable to the coffee producers. It was this very favorable manifestation of *ladino* power with which the Indians had to deal.

Resistance of the weak must necessarily be oblique if it is to succeed at all. The daily norm of peasant life, in fact, is grudgingly acquiescent to an existing if unjust system, mixed with minor but largely anonymous opposition to its particularly egregious manifestations: the Indians' "silent but inflamed hostility toward the owners."[48] Their effort was, in Eric Hobsbawm's words, to work "the system to their minimum disadvantage."[49] Lying deception, for example, played a prominent role in Indian-employer relations. Employers manipulated their records to keep the Indian in debt or to avoid having to pay him for his work: "Ten pesos I give you, ten I write in your book, and ten you owe me makes total debt of thirty pesos," went the *finqueros'* old joke.[50] Any "miserable tramp," on the other hand, a newspaper exaggerated, could get an exemption from forced labor by bribing a local official or by pretending to be a traveling merchant or on the basis of, for example, health problems or landownership.[51] Agricultural laborers struck deals with employers to work for lower wages for fewer days than the labor law demanded in return for being credited with what they needed. The Indians, too, accepted *habilitaciones* (wage advances) from various employers, violating the law that limited them to one *patron*, and often in amounts or under conditions with which they could not possibly comply. If the law required that they work for the *fincas*, the Indians would turn this to what advantage they could. Some took multiple advances brazenly in their own name, and others used fictitious names and falsified papers.[52] The bolder ones, the *jefe* of the Alta Verapaz reported in 1896, for example, accepted *habilitaciones* to gain exemption from military service and, that secured, then tried to give these back.[53] The eagerness of *habilitadores* (labor agents) to sign up workers and their willingness to ignore the law and overlook the debts these individuals already owed other employers greatly facilitated the task of the Indian seeking money. The anthropologist Juan de Dios Rosales reported in the 1940s watching the *habilitadores* in the market of Aguacatan attempting to steal workers from one another.[54] Indians regularly pushed their debts far above the amounts the law demanded.[55] This was not "debt slavery." If the law required them to work and to be in debt, it made the best sense to force from their employers as much money as possible. Owners and the state were reluctant to accord such "fraud" the status of resistance, because it bespoke their failure to achieve ideological hegemony and because it granted to the Indians both an understanding of their real self-interest and an agency the *fiqueros* were loath to admit. It was more comforting to continue to see the Indian as stupid, lazy, brutish, and drunken and easily deceived by competing labor recruiters; at best, or worst, he was deceitful and sly. The Indian could not be granted the consciousness that systematic resistance necessarily entailed.

Forms of self-help resistance common among coerced labor populations are that of theft, tool breaking, and sabotage. In the case of Guatemala, however, and at least while at work on the *fincas*, the indigenous population seem to have been either remarkably honest or exceptionally skillful in their crimes. There are only scattered reports of theft of coffee from the *fincas*.[56] A willingness to work steadily while on the *finca* undoubtedly was related also to the common system of payment by a piecework (the *tarea*, or task) system, which linked wages and food rations to work completed.

By far the most common mode of resistance was the undramatic but often effective petition of rights and grievances, of which the indigenous population filed thousands with various public authorities in three years. Because most Indians were illiterate, or at best could only sign their names, the documents themselves were usually the work of local *secretarios* (scribes) or *guisaches* (untitled lawyers). What resulted ranged from the most unintelligible in content and writing to elaborate presentations citing law, moral philosophy, and current events in Guatemala and other countries. These cannot in any real sense be said to represent the Indians "speaking for themselves," but the sheer repetition of complaints made clear the population's concerns. Petitions inevitably took the familiar "good Tzar" tack: the problems they were experiencing were not, the Indians suggested in most cases, the result of the system itself, although on occasion they did compare their situation unfavorably to that of the colonial period or with that of urban workers, and certainly had remained unknown to the *jefe politico* or the president, their "father" or "grandfather" *(tata)*. The distress they suffered was, rather, the result of the activities of corrupt and abusive lower officials and of exploitative employers, and especially foreigners—"gringo jews"[57]—who took advantage of the isolation of rural areas and of the Indians' own lack of familiarity with Spanish and with the law to abuse them. Once the *jefe politico* or the president understood the situation, the applicants were certain he could only decide for justice, that is, in favor of their claims. Such confidence was not entirely misplaced. There were corrupt officials whose activities so exceeded those accepted as customary that they threatened the functioning of the labor recruitment process as a whole, and on occasion these were disciplined or removed. But forced labor was by definition abusive of the Indians' natural, if not legal, rights and needs, and tinkering with the details of its operations could do little to ameliorate their situation. If some of the rural population was naïve enough to anticipate relief from *tata presidente*, their leaders were less so, and the *guisaches* they hired to frame the petitions certainly knew better. Yet the documents took, or claimed to take, laws at face value. This was not stupidity or ignorance. It represented, instead, purposefully suspended disbelief, or perhaps better, temporarily and artificially assumed belief. To argue their case before the authorities, the Indians had to accept the rules and categories laid down by these authorities, however detached from reality and weighted against them these might be. To deny these legitimacy, given the political hegemony of the state, was the bureaucratic equivalent of a *levantamiento* and simply short-circuited the complaint process. By agreeing at the onset to "play by the rules," the Indians could hope to get a hearing.

For the authorities who received these petitions, they presented no small problem and could rarely be ignored with impunity. A number of towns after 1900, for example, refused to send men ordered to fill *mandamiento* drafts because, they said, not only had an 1893 decree ended such practices but many of the villages involved enjoyed individual exemptions granted them on past occasions by the president himself.[58] Since the 1897–1898 collapse of coffee prices, however, the *jefes politicos* had been under orders from President Cabrera to supply forced drafts to the struggling planters, regardless of laws or exemptions. "We are free individuals," protested the Indians of Santa Maria de Jesus (Sacatepequez) in 1906 and under the constitution could not be compelled to work against their will.[59] Of course, they could be and were compelled to work against their will, and they understood this well enough, but appeal to the law provided a point from which to bargain and something of a shield from state violence. The reality that without Indian workers there would be no harvest, whatever the ultimately superior and coercive power of the state, forced the authorities to note and to attempt to resolve complaints put forward by the workers. The *jefe politico* was under intense pressure from his superiors and from the coffee growers to ensure that there was labor available when needed, but he was also charged with maintaining peace in the countryside. The resources available to him for this, if striking by comparison with a half century before, were still quite limited. Moreover, armed coercion was not ultimately a viable, or at least not a cost-effective, means of mobilizing the tens of thousands of workers needed each year. What was called for was negotiators.

Commonly, disgruntled workers opened negotiations by fleeing the *finca*. To where could or did they escape? Most returned to their home communities, which was also the first place the *finca* administrator and *habilitador* looked.[60] Thus it is clear that flight in these cases meant less to effect genuine escape than to strengthen the workers' bargaining position or hasten the negotiating process. To apply for redress while on the *finca* itself was possible, but it exposed the protesters to the vengeance of the employers and limited the workers' room to maneuver. They could also go to the capital to appeal directly to the president, but that was expensive and usually undertaken only as a last resort. The overwhelming number of petitions and complaints to the authorities came from Indians in the highland towns, written for workers who either had escaped what they argued were oppressive conditions on the *fincas* or who were resisting having to go on *mandamientos* or to work off wage advances. This strategy gave them a number of advantages. Most obviously, the support and solidarity of the community greatly aided the individual or group in their struggle with the employers. The administrator or *habilitador* and even government agents remained outsiders who operated in the village at disadvantages quite in contrast to the upper hand they held on the *fincas*. By staying at home, the Indians could also continue to look after their crops and other interests. Too, they petitioned the *jefe politico* of their own jurisdiction, who had a vested interest in his population and was more likely to view their problem sympathetically than was the governor of a coastal department, whose first concern was the *fincas*. The pattern, then, was for one or a group of workers to flee a *finca* and immediately upon arriving at their own town to formulate and send to the *jefe* of the department, and perhaps the president as well, a long and detailed exposition of their employer's abuses, which, they regretted, had forced otherwise obedient workers to flee for their lives. This drew the authorities into the bargaining process and usually forced some concessions, or at least the promise of concessions, from the *finqueros*, who as they themselves admitted,[61] and the Indians understood, needed the Indians more than the Indians needed the *fincas*, at least before the 1930s.[62] Offered a few more cents a day or a more honest measure of the *cuerda* or perhaps better food, and threatened now by the *jefe* with dire consequences if they failed to comply, the Indians returned to work. Both sides understood that nothing prevented the workers from initiating the process again, after a time.

Individuals or families slipped away from the *fincas* easily, but where a large group sought to escape, the situation became more complicated. Around the turn of the century, for example, and according to popular tradition in the community, a large number of families from the town of Aguacatan (Huehuetenango) were working on the *finca* Santa Augustina (San Marcos?). They became increasingly angry, however, about the abuses they experienced there, particularly the use of excessively long cord to measure each day's cleaning tasks.[63] They resolved to take the offending cord to Huehuetenango to argue their case before the authorities there. A *zahorin* (ritual specialist) told them they must escape at night and indicated the proper day "according to the old calendar." He took the bones of a long-dead *aguacateco* (water animal) who had been buried in the *finca's* fields and spread these about so that "all of those in the *finca's* living quarters would sleep and not realize that the Indians were escaping." The owner of the property kept two large dogs to intimidate the workers and to patrol the grounds: the *zahorin's* magic killed them. At midnight the *aguacatecos* slipped out of their huts and climbed into the mountains. "The zahorin did his rituals for the journey, asking the hills to give permission for the families to pass, and they said they agreed and assured the Indians that they would do them no harm." After two days of walking, the *aguacatecos* reached Huehuetenango and went to the *jefe politico* to show the cord. He fined the owner and the administrator, or so the Indians remembered, and allowed them to work off what they owed building the road from Aguacatan to Huehuetenango.

As the activities of the *zahorin* make clear, local leaders played key roles in the struggle over labor mobilization and control. Civil and religious leadership activities in the community overlapped, a result of the *cargo* system, in which individuals alternated munici-pal government offices with service in the *cofradias*, exiting at the top, if successful, into the category of elder or *principal*, the ulti-mate authorities among the indigenous population. *Zahorines* or *brujos* (sorcerers) on the other hand, were religious specialists called individually, usually through dreams and life crises, to serve the saints and the ancestors and typically remained apart from the organized civil-hierarchy.[64] The state, however, tended to lump all manner of local leaders together under the rubric of *"justicias"* and to make them collectively responsible not only for the behavior of the community but for delivering to the state or the landowners whatever taxes, materials, or labor these required. It is hardly to be wondered that often the identity of the *principales* was kept secret.[65] And few men admitted openly to being *brujos*. As a result, state demands fell almost directly on the *alcaldes* and *regidores* of the municipal government (the *cabildo*) and the heads of the *cofra-dias* (religious brotherhoods). They were required to see the filling of *mandamiento* orders[66] and to provide *alguaciles* (policemen) to help labor recruiters and round up the men said to owe them money. Some used this intermediary position to exploit their fel-lows,[67] but most appear to have sought to deflect or resist the demands of the state and the employers, or at least resist those evi-dently excessive and not *costumbre*. For this they sometimes paid with beatings, fines, and months in jail. The state, in any event, whenever it encountered opposition, and in the manner of Robert Cobb's Paris policeman, "found" and punished "leaders" *(cabecil-las)*, whether these were the *justicias* or simply the first half-dozen Indians the troops encountered.

Large groups of Indians who fled the *fincas*, unless they simply moved to another property and took their chances there, had little choice but to return to their home villages, but a single family could escape the demands of coffee altogether, if at no small cost. Some sought refuge in the *mote* (bush).[68] A traveler reported shortly after the turn of the century: "There are people living in the woods here and there, though you could not see them until you come on them; Indians mostly from the settled parts of the Alta Verapaz, running away from plantation work and the oppressions of Government or the authorities."[69] In the 1930s, the anthropologist Jackson Lincoln was warned of "dangerous" Indians who, it was said, lived in the hills around Nebaj evading road taxes and labor duty and preying on travelers.[70] Life on the run and cut off from regular interaction with the community, however, was precarious and *triste* (sad).[71] *Habilitadores* and village *alguaciles*, pressed to deliver labor, also increasingly took their search for *brazos* (literally "arms," i.e., workers) into every corner of the municipality, making evasion more difficult. Another possible response to labor pressure was to migrate to a *departmento* (department) where demands were less or state control was less effective. This was common, for example, among the residents of the Alta Verapaz, where many escaped into the almost uninhabited and unpoliced mountains north of Lake Izabal, the neighboring department of Izabal. Indeed, something of a border war broke out between the two departments. The *jefe politico* of the Alta Verapaz claimed that agents from Izabal encouraged this emigration and even evaded the jurisdiction of Cahabon to assault a municipal secretary trying to halt the flow.[72] The *jefe* of Izabal, for his part, blamed the migration, and the Alta Verapaz's governor was forced to agree, on the heavy labor requirements of coffee and on the effects of *ladinos* invading the Indian towns and depriving the inhabitants of their lands. By the late 1890s, he reported hundreds of Indians living in the mountains between the lake and the Sarstoon River who resisted "anything that might be called a wage advance."[73]

Much more worrisome than internal migration was emigration to neighboring Mexico and Belize.[74] In Mexico the Indians found work with which they were familiar, on the developing coffee plantations of Chiapas and Soconusco, and at better wages than available in Guatemala. The Brazilian agronomist Agusto Ramos, touring the region in the early 1900s, found that not only was the labor force on the Mexican plantations predominantly Guatemalan, but so was the money, as soaring paper inflation in Guatemala drove silver out of circulation.[75] Peonage was difficult to enforce along the notoriously porous border, and the inhabitants of the Occidente moved back and forth at will. The *jefe politico* of San Marcos department sought to enforce pass laws, ordered officials in the border municipalities to check the papers of all individuals traveling through, and jailed *habilitadores* thought to be recruiting workers for Mexican employers, but to little avail.[76] When explaining in 1901, for example, why he could not fill the *mandamiento* order for fifteen workers for *finca* Porvenir, an *alcalde* of Tajumulco (San Marcos) pointed out that when he attempted to cite local residents, they slipped over the border; inhabitants of Jacaltenango to the north rather more inventively took *mandamiento* advances and then fled to Mexico.[77] In its more candid moments, the government admitted that they had cause. The *Ministerio de Formento* lamented that in 1902:

> The old laws which [the government] enacted to protect and promote the planting of coffee, with detriment to the liberty of the Indian, provoked the emigration of a great number [of Indians] to the neighboring countries of Mexico and British Honduras, with the result that . . . the shortage of workers . . . will be one of the great obstacles that the coffee planters will have to overcome in their cultivations.[78]

For years the Indians of the Alta Verapaz had fled that department not only to Izabel but across the frontier into British Honduras. The attraction in this case was not wages but land, as well as freedom from labor and military pressures. In an effort to counteract this outflow, the government offered land at Chisec in the Alta Verapaz and a two-year exemption from taxes and military or labor services to the Indian refugees "to attract the return of our fellow nationals who have emigrated to neighboring countries, fleeing unnecessarily severe laws which obligated them to work on the *fincas.*"[79] The labor laws, however, it is worth noting, remained essentially unchanged. Whether the offer of land prompted any to return is unclear, but in the early 1920s the *jefe politico* of the Peten reported "more than eighty thousand families from Cobán, San Pedro [Carcha] and San Juan [Chamelco]" on the other side of the Sarstoon River.[80]

Guatemala's rural indigenous population has resisted state and landowner control and exploitation from the Conquest. The specific forms this resistance has taken have depended on the nature and intensity of elite demands, on the ability of the state to enforce these, and on the conjunctural capacity of the indigenous population to resist. The Indian communities held the Bourbon reformers at bay and destroyed the hopes and the government of the first generation of Independence Liberals, and they forced from the Carrera Conservatives an edgy independence. Coffee changed the situation. Rapid expansion of the new crop beginning in the 1860s required the planters to seek new supplies of cheap labor, and the income from coffee exports gave the government, for the first time, the resources necessary to create a strong centralized state to assist with labor mobilization. The Bourbons and the Enlightenment Liberals had been unable to impose their development projects on the rural population, but advances in the technology of repression tipped the balance in favor of the state after 1871. Rather than attempt to convert to their vision of national development a rural population deemed fundamentally inferior, the neo-Liberal regime came to rely above all on simple force and coercion, on political hegemony. The Indians and the Indian communities shifted their resistance strategies too, away from the often open confrontation characteristic of the century before 1870 to more subtle forms of avoidance and mitigation. Oppressed peoples have no obligation to act in ways academics find dramatic or exciting, but rather to survive and endure and to ensure the survival of their families and communities in the face of what threaten to be literally overwhelming pressures. Guatemala's Indians adopted modes of resistance—from deception to evasion and escape to the forcing up of debts and the resort to violence on a rare occasion—after 1871 best calculated to extract from a system into which they had been drawn involuntarily the best result at the least risk, and they waited for a better day.

Abbreviations

AGCA   Archivo General de Centro América, Guatemala City.
MG   Ministerio de Gobernación.

Notes

A version of this chapter appeared in *Peasant Studies*, 17:3 (1990), 157–177.

1. David McCreery, "An Odious Feudalism: *Mandamientos* labor and commercial agriculture in Guatemala, 1861–1920," *Latin American Perspectives*, 13:1 (1986), 99–117, and "Debt Servitude in Rural Guatemala, 1876–1936," *Hispanic American Historical Review*, 63:4 (1983), 735–759.

2. On the role of a "dominant ideology" in unifying a ruling elite (much more so than the "nation" as a whole), see Nicolas Abercrombie et al., *The Dominant Ideology Thesis* (London, 1980), esp. chs. 1 and 3.

3. The effort here is to extract "hegemony," at least for the moment, from under the accumulating weight of Gramscian scholarship, which threatens to squeeze the life from a very useful concept. The following paragraph relies heavily on Robert Bocock, *Hegemony* (London, 1986); Joseph V. Femia, *Gramsci's Political Thought* (Oxford, 1981); James Scott, "Hegemony and the Peasantry," *Politics and Society*, 7:3 (1977), 267–296, and *Weapons of the Weak* (New Haven, 1985), ch. 8.

4. Bocock, *Hegemony*, 33–34.

5. This is quite clear, for example, in the correspondence of the parish priests *(curas)* with Guatemala's archbishop throughout the nineteenth century. See "Cartas," 1821–1920, Archivo Eclesiastico de Guatemala, Guatemala City.

6. Femia, *Gramsci's Political Thought*, 32, quoting Engels.

7. In Guatemala a *ladino* is an individual of European/North American or "national" culture, whatever his or her "racial" makeup; the term is not to be confused with *mestizo*, which refers to mixed blood and is not commonly used in Guatemala.

8. On the impacts of the Bourbon Reforms in Guatemala, see Miles Wortman, *Government and Society in Central America, 1680–1840* (New York, 1982).

9. See, for example, Ann Cox Collins, "Colonial Jacaltenango, Guatemala: The formation of a corporate community," Ph.D. dissertation, Tulane University (New Orleans, 1980), chs. 6 and 9.

10. On the failure of tribute reforms, for example, see Manuel Fernandez Molina, *Los tributos en el Reino de Guatemala, 1786–1821* (Guatemala, n.d.).

11. A good, short treatment of this period, with an extensive bibliography, is to be found in R. L. Woodward, *Central America* (New York, 1976), ch. 4.

12. The battle cry of Carrera's troops was "Long Live Religion and Death to Foreigners." See J. L. Stevens, *Incidents of Travel in Central America, Chiapas and Yucatán* (New York, 1941), I:225.

13. A sympathetic portrait of Carrera as "pro-peasant" is available in E. Bradford Burns, *The Poverty of Progress* (Berkeley, 1980), 97 ff.

14. David McCreery, "Rural Guatemala, 1750–1940," Part II, ch. 6: The Communities. Forthcoming, Stanford Univ. Press, 1993.

15. For example, Antonio Batres Juarequi, *Los indiod: Su historia y civilizacion* (Guatemala City, 1893).

16. *El Guatemalteco* (Guatemala City), 13 September 1886.

17. There was also an urban police force in Guatemala and a small enforcement arm attached to the Ministerio de Hacienda called the *Montada*, or mounted police, which patrolled the countryside to control smuggling and contraband alcohol.

18. Pedro Barreda, *Geografica e historia de correos, y telecomunicaciones de Guatemala* (Guatemala, 1960), 202 ff. See Ministerio de Fomento, *Memoria-1924* (Guatemala, 1924), 251 ff for a list of stations and when they were established.

19. Fomento, *Memoria-1898* (Guatemala, 1898), 159 ff.

20. MG 28663/74, 14 September 1877, AGCA; David McCreery, "Land, Labor and Violence in Highland Guatemala: San Juan Ixcoy (Huehuetenango), 1890–1940," *The Americas* 45:2 (1988), 237–249.

21. Miguel Angel Asturias, *El Senor Presidente*, trans. Francis Partridge (New York, 1982), 91.

22. There are no historical studies of the army; for an introduction, see M. McClintock, *The American Connection* (London, 1985), ch. 1.

23. McCreery, "Rural Guatemala," Part II, ch. 5.

24. *Recopilacion de las leyes de Guatemala*, tome 1 (Guatemala, 1881), 157–169.

25. William T. Brigham, *Guatemala: Land of the Quetzal* (New York, 1887), 296; James Boddam-Whettham, *Across Central America* (London, 1877), 205–206.

26. Ministerio de Guerra, *Memoria-1905* (Guatemala, 1905), 56.

27. For example, MG 28638/216 and MG 28762/1334, AGCA.

28. Indians of San Pedro Carcha to subjefe Verapaz [Coban], 17 July 1872, Papers of the Jefe Politico Alta Verapaz, 1870–1872, AGCA.

29. Indians of Santiago Sacatepequez to President, May 1884, Jefe Politico Sacatepequez, 1884, AGCA.

30. There were exceptions: see, for example, Robert Carmack, *Historia social de los Quiches* (Guatemala, 1979), 271 ff. on the activities of the warlike *momostecos*.

31. President–Jefe Politicos (circular), 13 October 1890, Jefe Politico Solola 1890, AGCA.

32. *Finqueros* strongly resisted Guatemala's signature on the Washington Convention of 1923, limiting the use of physical coercion for labor mobilization and control. See *El Imparcial* (Guatemala City), 14 May 1925.

33. Helen Sanborn, *A Winter in Central America and Mexico* (Boston, 1886), 83.

34. On abuses in labor recruiting and on the *fincas* see the articles cited in note 1, above.

35. For example, L. J. *[finca* Altamira]–Jefe Politico Solola, 20 June 1890, Jefe Politico Solola, 1890, AGCA; 1a Instancia Penal San Marcos, leg. 18/exp. 54, AGCA; B. I. companieros–Jefe Politico Alta Verapaz, 19 September 1930, Jefe Politico Alta Verapaz, 1930, AGCA.

36. Juez de Paz La Reforma–Jefe Politico San Marcos, 27 July 1901, Jefe Politico San Marcos, 1901, AGCA.

37. Comisionado Politico San Rafael Pie de la Questa–Jefe Politico San Marcos, 27 October 1915, Jefe Politico San Marcos, 1915, AGCA.

38. On *levantimientos* related to land conflicts, see David McCreery, "State Power, Indigenous Communities, and Land in Nineteenth Century Guatemala," in C. Smith (ed.), *Indian Communities and the State: Guatemala, 1520–1988* (Austin, Tex., 1990), 96–115.

39. Arden King, *Coban and the Alta Verapaz* (New Orleans, 1978), 34.

40. A review of the archive of the Ministerio de Gobernacion, of the papers of the Jefe Politico of the Alta Verapaz, and of the 1a Instancia Penal, Alta Verapaz does not reveal particularly rebellious behavior.

41. Jefe Politico, "Concejo [sic] Consultativo," [n.d. 1897], Jefe Politico Alta Verapaz, 1897, AGCA.

42. McCreery, "Land, Labor and Violence in Highland Guatemala," 237–249.

43. Scott, *Weapons of the Weak*, 290. Scott's emphasis.

44. Particularly useful on India-*ladino* relations and on the nature of subordination is Kay Warren, *The Symbolism of Subordination* (Austin, 1978).

45. Although associated with Eric Wolf, the paradigmatic description of

the "cargo system" and "the closed corporate [peasant] community" is found in Frank Cancian, "Political and Religious Organization," *Handbook of Middle American Indians*, vol. 6 (Austin, 1967), 283–298.

46. See, for example, Benjamin Colby and Pierre L. van den Berghe, *Ixil Country* (Berkeley, 1969).

47. The classic study is by E. Michael Mendelson, *Los escandalos de Maximon* (Guatemala, 1965); see also John Hawkins, *Inverse Images* (Albuquerque, 1984).

48. *El Imparcial*, 1 November 1922.

49. Quoted in Scott, *Weapons of the Weak*, 301.

50. *Diario de Centro America*, 3 May 1919.

51. *El Norte* [Coban], 5 June 1937; *El Imparcial*, 13 August 1943.

52. Ministerio de Fomento, letterbook 14865, 3 November 1897, and 14866, 28 January 1898, AGCA.

53. Fomento letterbook 14862 [n.d. 1896], AGCA.

54. University of Chicago microfilms, Middle America Ethnographic Notes, No. 24, "Aguacatan," 147.

55. On debts, see artices cited in note 1, above.

56. For example, Jefe Politico, Libro de Sentencias Economicas, March 1881, Jefe Politico Sacatepequez, 1881.

57. Residents fina Conception–Jefe Politico Escuintla, 16 April 1934, Jefe Politico Escuintla, 1934.

58. *Dairio de Centro America*, 25 October 1893; *El Republicano* (Heuheutenango), 31 October 1893; Fomento letterbook 14935, 1 and 3 September 1900.

59. Indians of Santa Maria de Jesus–President, 14 March 1904, Jefe Politico Sacatepequez, 1906 [sic], AGCA.

60. See, for example, Appendix 3 of Maud Oakes, *The Two Crosses of Todos Santos* (Princeton, 1951).

61. For example, B119.21.0.0 47789/9, AGCA.

62. On labor changes in the 1920s and 1930s see David McCreery, "Wage Labor, Free Labor, and Vagrancy Laws: The transition to capitalism in Guatemala, 1920–1945," in William Roseberry (ed.), *Coffee, Class and Change in Latin America*. Forthcoming.

63. Chicago microfilms, "Aguacatan," 82 ff.

64. For a life history of *zahorin*, see Benjamin and Lore Colby, *The Daykeeper: The life and discourse of an Ixil diviner* (Cambridge, Mass., 1981).

65. Ruth Bunzel, *Chichicastenango* (Seattle, 1952), 186.

66. See, for example, the report of the "Comicion conciliadora en los cuestiones que surjan entre jornaleros y patrones," July 1877, Jefe Politico Alta Verapaz, 1877, AGCA.

67. Jackson Lincoln, "An Ethnographic Study of Ixil Indians of the Guatemalan Highlands," Chicago microfilms, 88.

68. MG 28669/117; Comisionado Politico San Pedro las Huertas–Jefe Politico Sacatepequez, 9 June 1884, Jefe Politico Sacatepequez 1884, AGCA; Fomento letterbook 14914, 13 December 1902; Batres Juaregui, *Los indios*, 178–180.

69. Robert Burkett, "Explorations in the Highlands of Western Guatemala," *The Museum Journal* (Philadelphia), 20:1 (1930), 45.

70. Lincoln, "Ixil," 53.

71. For a recent analogous situation, see Ricardo Falla, "Struggle for Survival in the Mountains," in Robert Cormack (ed.), *Harvest of Violence* (Norman, Okla., 1988), 235–255.

72. MG 28699/117, AGCA.

73. MG 28673/131 and 28757/225, AGCA; Fomento letterbook 14865, 3 November 1897.

74. There is no mention of emigration to El Salvador, where the working conditions were perhaps worse. In the Oriente some individuals fled to Honduras around the turn of the century to avoid being drafted for the building of the Northern Railroad. See Fomento letterbook 14855, 21 November 1893, AGCA.

75. Agusto Ramos, *O cafe no Brasil e no entrangeiro* (Rio, 1923), 295–296.

76. AGCA, Juez Municipal Tacana–Jefe Politico San Marcos, 11 September 1901, Jefe Politico San Marcos, 1901; Juez Comitancillo–Jefe Politico San Marcos, 22 February 1903, Jefe Politico San Marcos 1903; R.R.–Jefe Politico San Marcos, 7 May 1906, Jefe Politico San Marcos 1906.

77. Alcalde Tajumulco–Jefe Politico San Marcos [n.d. 1901], Jefe Politico San Marcos 1901; Fomento letterbook 14895, 12 October 1900, AGCA.

78. "Agricultura," Ministerio de Fomento, *Memoria-1903* (Guatemala, 1903), 41–42.

79. Fomento, *Memoria-1903,* 46; Fomento letterbook 14873, 12 June 1902, AGCA.

80. Fomento letterbook 14909, 24 May 1922, AGCA.

8

Structure of Domination and Forms of Resistance on Yucatecan Estates during the Late Pofiriato, ca. 1880–1915

Allen Wells and Gilbert M. Joseph

> Si son concientes. . . . Deep within them the indebted peons feel hatred and a desire for revenge against their masters and the authorities who helped their matters to enslave them!
>
> —Tomás Pérez Ponce

A sensational interview with radical labor organizer and public defender, Tomás Pérez Ponce, printed in mid-1911 in an ephemeral Mérida newspaper, *El Ciudadano*, must have troubled Yucatán's henequen *hacendados (henequeneros)*, accustomed to dominating the terms of debate both in the local press and on the estates.[1] Pérez Ponce, labeled by the planter-financed press, "el líder del peonaje," provocatively demanded the end of debt peonage, freedom of movement for agricultural workers (*jornaleros de campo*[2]), and wage increases for henequen workers. The upstart attorney, who had spent the better part of the last years of the Porfiriato in Mérida's Juárez Penitentiary for mobilizing urban and rural workers and for this unbridled criticism of the oligarchy, now took advantage of the Maderista *apertura* to publicize his visits to haciendas in the *zona henequenera*.[3]

Pérez Ponce astutely observed in the interview—appropriately entitled "La gravisima cuestión agrícola"—that the culprit of this heinous labor regime was the unceasing demands of henequen-fiber monoculture. Arguing for diversification, he warned *henequeneros* that unless some of the relentless pressure of monocrop production was tempered with the cultivation of corn and beans and the reintroduction of cattle raising in the *zona henequenera*, *peones acasillados*, tired of being tricked and taken advantage of by local landowners and politicians, would in all likelihood resort to violence against their former masters *(amos)* and flee the estates for nearby villages, *cabeceras* (district or *partido* seats) and the state's

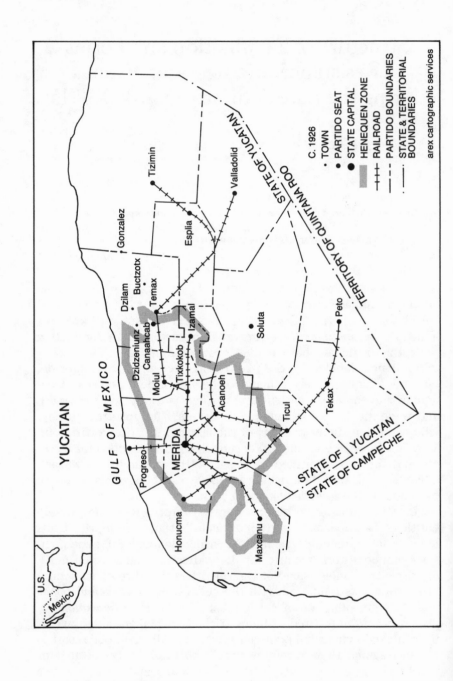

YUCATAN

GULF OF MEXICO

STATE OF YUCATAN

TERRITORY OF QUINTANA ROO

Tizimin

Gonzalez

Valladolid

Esplia

Dzilam
Buctzotx
Temax

Izamal

Soluta

Peto

Dzidzeniunz
Canaahcab

Tixkokob

Moiul

Acanoeh

Ticul

Tekax

MERIDA

Progreso

Honucma

Maxcanu

STATE OF YUCATAN

STATE OF CAMPECHE

C. 1926

• TOWN
• PARTIDO SEAT
● STATE CAPITAL
▨ HENEQUEN ZONE
┼┼┼ RAILROAD
–·–·– PARTIDO BOUNDARIES
—·—·— STATE & TERRITORIAL
        BOUNDARIES

arex cartographic services

U.S.

Mexico

principal urban centers: Mérida, Progreso, and Valladolid. More-over, Pérez Ponce went on to urge henequen *jornaleros* to forget about their "illegal" debts and abandon their estates. If district prefects *(jefes politicos)* arrested and returned them to the hacienda, their relatives should contact him and he would make sure that "the authorities upheld the laws."[4]

Such blatant rabble-rousing would have been unthinkable a decade earlier. The publication of the interview in June 1911 is sig-nificant for several reasons. From a political perspective, it demon-strates the willingness of a disgruntled elite faction, known locally as the Morenistas (after their standard-bearer, journalist Delio Moreno Cantón) to reach out to progressive elements in Meridano society.[5] Previously frustrated in their bids for state power, the Morenistas now felt they had no choice but to embrace organizers like the articulate Pérez Ponce, who had important contacts with both the urban and rural working classes.

Moreover, the interview signaled a recognition by elites that the rural "labor problem" was something that needed to be addressed —at least rhetorically—as a quid pro quo for the support of persua-sive hingemen or agitators like Pérez Ponce during the first months of the Maderista period. Interestingly, *El Ciudadano* gave Pérez Ponce ample space to present his radical views, but the Morenista paper distanced itself from the agitator's remarks. It was one thing to enlist the assistance of rabble-rousers, quite another to affirm the kind of radical changes and class struggle that Pérez Ponce espoused.

Clearly, the issues raised by the interview were designed to frighten landed interests throughout the *zona*. The power-seeking Morenistas stood to gain, at least in the short-run, by fomenting chaos in the northwest portion of the peninsula: their gain was to engineer "manageable" unrest that would trigger the ouster of the Maderista governor, José María Pino Suárez.

The freedom of movement that Pérez Ponce advocated for hene-quen workers in 1911, however, would have threatened the essence of henequen monoculture by effectively undermining social relations of production in the *zona*. If *jornaleros* left their estates for nearby villages and towns and became in the vernacular of the day, "hombres libres," the world the *henequeneros* knew would be in serious jeopardy.

That this issue could be publicized in such an open and dramatic fashion attests to the cleavages that existed in late Porfirian Yuca-tán. Widespread rural discontent from 1907–1911 enabled disgrun-tled elite factions, like the Morenistas, who struggled for political control of the state, to enlist hacienda *peones* and marginalized villagers to participate in the mosaic of jacqueries, riots, and rebellions that rocked the peninsula during the first years of the Mexican Revolution. It is beyond the scope of this chapter to attempt to unravel the entire bundle of political, economic, and social contradictions that promoted waves of rebellion throughout the state during the first years of the Mexican Revolution—veritable "seasons of upheaval" that only with great difficulty were defused by local elites and state authorities by mid-1913. Rather, we will examine the prospects for, and limitations of, mobilization within one of Yucatán's most important subregions, the *zona henequenera*.[6]

Interestingly, perhaps owing to comparisons with other Mexican regions that experienced more violent revolutionary pasts, modern historians have consistently underestimated the resistance of Yucatán's *campesinado* prior to the overthrow of oligarchical rule by General Salvador Alvarado's Constitutionalist Army of the Southeast in March of 1915. Particularly misunderstood is the social behavior and *mentalidad* of the *peones acasillados:* unlike the Porfirian *hacendados* and their nemesis, Tomás Pérez Ponce, modern writers have dismissed the *peones'* capacity to oppose or protest the demands of their masters. To be sure, *henequeneros* effectively utilized both the carrot and the stick, blending paternalist incentives and a measure of security with restrictive mechanisms of coercion and isolation. It is hardly surprising, therefore, that their *sirvientes* lacked the revolutionary potential—or, as Eric Wolf has put it, the "tactical mobility"[7]—manifested by the *comuneros, vaqueros, mineros,* and *campesinos serranos* who made up the revolutionary armies of central and northern Mexico. By the same token, however, the *peones acasillados* who toiled on Yucatán's henequen estates were not inherently passive. Their characterization in the historical literature as a lumpen mass of docile retainers is a profound exaggeration. Ironically, such a portrayal harkens back to the contemporary stereotypes put forward (for very different reasons) by foreign muckrakers and *henequenero* apologists alike.[8]

Our own research in the Ramo de Justicia of the Archivo General del Estado de Yucatán prompts a very different characterization of the late Porfirian *campesinado*, and particularly recasts prevailing notions regarding the inability of *acasillados* to resist their masters. We contend that while henequen monoculture's characteristic structure of domination restricted the potential for self-generated insurrection on the estates, it frequently could not prevent *acasillados* from joining the revolts that originated on the periphery of the *zona henequenera* during the early years of the revolutionary era. Moreover, the fact that Yucatecan *peones* were not as overtly rebellious as *comuneros* outside the henequen zone does not mean that they did not resist the monocultural regime. On the contrary, their personal testimonies as well as other local documentation suggests that they partook of "quieter," "everyday forms of resistance," that were safer and more successful over the long run in contesting, materially as well as symbolically, the stepped-up work rhythms and other exploitative aspects of henequen monoculture.

The goals of this chapter, then, are several: (1) to make explicit the structure of domination of planter and state that hampered the prospects for mobilization in the *zona henequenera* during the *auge* (fiber boom); (2) to relate the wide range of "routine" and "everyday" responses that *peones* made to the demands of monoculture; and (3) to brush in broad strokes the intensifying social contradictions that fueled repeated cycles of unrest throughout the Yucatecan countryside during the final years of ancien régime. First, however, let us briefly characterize the henequen estate, focusing upon the special problem that monoculture presented for labor relations in the *zona*.

The Henequen Estate: A Hybrid Institution

The largely self-contained world that Yucatecan *hacendados* and *peones* fashioned on henequen estates during the *auge* was rife with inconsistencies. The late nineteenth-century boom was so sudden that the henequen estate was unable to erase vestiges of the cattle-and-maize hacienda that had dominated the northwestern *zona* from 1750 to 1850. A brief comparison of henequen monoculture and its more progressive Caribbean counterpart, the sugar complex, will confirm that henequen estates never evolved into modern commercial plantations during the boom.

Whereas the Caribbean sugar industry utilized new techniques to separate the agricultural and industrial sectors during the late nineteenth century and thereby increased productivity and efficiency, Yucatecan *henequeneros* never substantially altered the means and factors of production during the boom. The agro-industrial sugar complex was characterized by corporate ownership, ample capitalization, profit maximization, and a variegated yet dependent labor force organized to supply a distant yet substantial market—all characteristics of the prototypical plantation. By contrast, Yucatán's syncretic henequen estate fell under the rubric of family ownership rather than corporate management; it featured an entrepreneurial class that had a desire to follow in the footsteps of its forebears, entertaining lavish status aspirations that inhibited the reinvestment of profits; and it faced chronic shortfalls in capitalization brought on by volatile price fluctuations in the world market price of fiber. The result was an estate geared to modern commercial operations but lacking a farsighted, sophisticated ownership and management class that could complete the transformation.[9]

While the henequen hacienda may have physically resembled a modern commercial plantation—with modern machinery, narrow-gauge tramways, and land-intensive cultivation of the staple crop—family ownership, management, and *mentalité* continued to imbue the institution with characteristics of the prehenequen hacienda. Symptomatic of a rural society in the midst of a complex process of transition, the henequen estate is best viewed as a hybrid institution, illustrating some of the traits of its predecessor, the corn and maize hacienda, but reflecting inevitable adjustments in land, technology, labor, and infrastructure. Moreover, it appears that the emergence of a full-fledged plantation society was inhibited by lingering vestiges of the earlier institution, particularly the way in which *hacendados* confronted their labor problems.

In the classic plantation model, corporate managers take advantage of a free (or relatively free) labor market and pay cash wages.[10] Unlike the hacienda, the modern plantation does not have to rely on debt or cultivate a paternalistic relationship. Henequen estates, on the other hand, more closely resembled their poorly capitalized ancestors—the corn-and-cattle haciendas—that "[bound] labor by means other than money wages."[11] Landowners continued to rely on traditional labor practices like debt peonage and the *tienda de raya* (hacienda store) to inhibit mobility and ensure a dependent labor regime. Moreover, coercive techniques such as naked force and the manipulation of debt were coupled with more subtle, indirect methods like the parceling out of subsistence plots and the "institutionalization of personal relationships between employer and employee,"[12] to secure at least an outwardly compliant labor force.

Given this preference for dependent labor, as well as the rapid take-off of the boom, Yucatecan *henequeneros* were faced with the unenviable task of converting the traditional labor relations of the hacienda to the iron-clad discipline required for henequen production. With North America demand driving local production, exports grew from 30,000 bales of raw fiber in the 1870s to more than 600,000 bales a year in the 1910s. Inevitably, as the henequen economy became more engaged by, and ultimately subordinated to, the demands of international capitalism, labor conditions became increasingly rigorous. Overseers *(mayacoles)* drove work gangs of peons, who cleared, planted, weeded, and harvested fiber throughout the year. Moreover, the *desfibradora*, henequen's industrial equivalent of the sugar mill *(inquenio)*, which was located right on the estate, required disciplined teams of workers to decorticate and press thousands of henequen leaves a day.

But interestingly, the idiosyncratic nature of henequen production did not necessitate a battery of trained specialists to care for specific agricultural and industrial tasks. Although technological advances would increase production capabilities, the henequen industry never demanded a skilled work force. Rather, a large and mobile group of laborers was needed to work in all phases of production as circumstances warranted. Unlike sugar, the fibrous agave was harvested throughout the year, requiring a permanent labor supply, ruling out—from management's point-of-view—more flexible, cost-efficient, seasonal employment schemes.

In essence, henequen monoculture produced social relations that in some ways resembled a rural proletariat (routinized work gangs, piecework, daily wages, and a quasi-cash economy) but in other ways continued to embody many of the characteristics of the paternalistic cattle-and-corn hacienda. Although working conditions on the henequen estate were certainly more onerous than those of its labor-extensive predecessor, the hybrid fiber hacienda continued to offer an essential measure of security to *peones acasillados*—at least, as we shall see, until the final years of the Porfiriato. The new land and labor arrangements meant that the Maya were now totally dependent on the *amo* and his *encargado* for food, firewood, water, famine relief, justice, protection, medical care, and other services. For Yucatecan *peones*, their life on the henequen estate represented a compromise of sorts. The monocultural institution offered minimal security, but at the cost of the *campesinos'* autonomy.

Yet if resident peons were most concerned with the sense of stability and security that the new social relations fostered, *henequeneros* pursued a much different agenda, one that diminished direct contact between themselves and their *sirvientes*. Unlike their parents and grandparents, this generation of landowners saw their estates as business investments first and patrimonies second. Speculation in rural real estate—particularly during bust cycles in the regional economy—forced entrepreneurs to consider henequen haciendas as liquid assets. In this free-wheeling investment climate, where fortunes were made and broken with regularity, *peones* had difficulty maintaining a close relationship with their *amos.* While, as we shall see, patron-client relationships did exist in the *zona henequenera*, they never approached the level of sophistication that prevailed in the antebellum U.S. South, simply because most henequen producers never invested the time and effort required to nurture such a complex relationship. As the *auge* wore on, and as business investments multiplied and mortgage credits were swapped at a dizzying pace, clientelistic bonds on the estate inevitably suffered.

A flagging paternalistic ethos and the incomplete transformation of northwestern cattle-and-corn haciendas into modern commercial plantations during the fiber boom ensured that social tensions would run high among state authorities, *henequeneros,* overseers, and *peones* within the *zona.* Although the mechanisms of social control that estate managers and local authorities utilized throughout the Porfiriato kept a lid on collective action, these restrictive measures did not prevent *peones acasillados* from demonstrating their unhappiness with the new order. Before we can examine the responses workers made to this hybrid institution, we must first characterize the cooptive and repressive environment that workers had to contend with on Yucatecan henequen estates.

The Idiom of Power: Isolation, Coercion, and Security

Why did henequen workers obey their masters even to the degree that they did during the Porfiriato? Force (or the threat of force) alone would not have kept the *peones* subservient.[13] Furthermore, armed might cannot be everywhere at all times. While the fear of retribution is an essential ingredient of all coerced labor systems, force must be combined with an underlying obedience to authority to ensure compliance. The ultimate success of such asymmetrical systems has historically depended on the ability of elites to convince their workers that such an existence was, on balance, in their best interests. Even the most heinous forms of servitude had elements that at least paid lip service to the "rights" and needs of dependent workers.[14]

The subordinate classes must somehow be led to believe that such a repressive system is part of the logical, immutable order of things, and, hence, that collective action is not a viable option. Moreover, as long as the subordinate classes do not compare their own lot with that of their masters (implicitly accepting their own inferiority), the prospects for rebellion are greatly diminished.[15]

To appreciate how obedience to authority was conveyed to *peones acasillados* by *henequeneros* during the Porfiriato, let us briefly focus on "idioms of power," that is, on the ways in which the coercive aspect of domination has been represented to subordinate classes throughout history.[16] In many premodern, traditional societies, force was presented directly, in face-to-face relations, most often in transparent fashion. Although this personalistic idiom of power was at times camouflaged by such contrived strategies as *compadrazqo* (fictive kinship), paternalism, and asymmetric gift exchanges, there was little doubt on the part of the superordinate or subordinate elements that domination was the essence of the relationship.

More recently, a second, materialistic idiom of power emerged in which the unequal relationship was represented as power over commodities, rather than individuals. As modern capitalism evolved, labor-management relations became increasingly depersonalized, as both parties grew more detached from the commodities they produced. Dependency was "disguised under the shape of social relations between the products of labor."[17] The resulting power relationship, diffused by the growing alienation of the workplace, was no longer portrayed as power over persons, but instead concealed as power over commodities.

The social relations of henequen monoculture represent something of a halfway point along the continuum joining the personalistic and materialistic idioms of power. Just as the syncretic henequen estate combined characteristics of both the traditional hacienda and the commercial plantation types, so too did its labor relations represent an amalgam of the two modes of coercion. Indeed, many of the constitutive elements of the personalistic idiom found in traditional societies—such as a thinly veiled paternalism, *compadrazqo*, and the ever-present agency of human force, exemplified, for instance, in floggings meted out by overseers—were carried over to the new plantation-style society.

To more fully illustrate the manner in which henequen monoculture's idiom of power adapted some of the characteristics of traditional labor relations to the requirements of a more "proletarianized" regime, it is helpful to conceive of three complementary mechanisms of social control: isolation, coercion, and security. Underwritten by the assistance of the state political apparatus, these mechanisms allowed *henequeneros* to maintain the disciplined work rhythms of monocrop production. Isolation, coercion, and security worked in unison to cement a structural relationship that not only suited the production requirements of management but also served the subsistence needs of workers, at least until the eve of the Mexican Revolution.[18]

Isolation

Obedience to authority was inculcated by *henequeneros* through their deft management of physical and social space. Utilizing four reenforcing spheres of isolation, Yucatecan landholding elites and their ally, the state, effectively channeled and tied *jornaleros* to the haciendas, while closing off other options and avenues of escape.

At the core of the isolation mechanism was the henequen hacienda itself. A series of positive and negative inducements ensured that peons remained entrapped on the estate. Even as workers were constrained by debt, the *tienda de raya*, corporal

The Isolation Mechanism

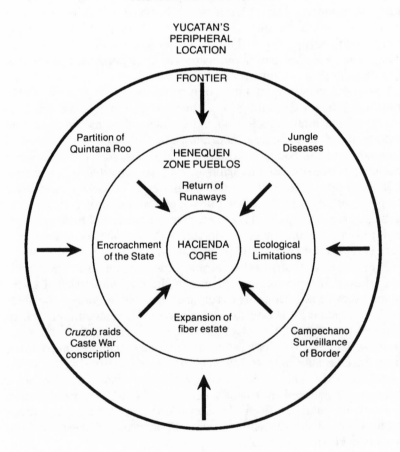

punishment, and identification papers, their lot was made more tolerable by piecework wages, medical care, and funds advanced for weddings, deaths, and baptisms. Nevertheless, despite its ameliorative, paternal aspects, the *henequeneros'* double-edged strategy ultimately rested on coercion, for *peones acasillados* could only leave the estate with the express written consent of their *henequeneros* or overseers.

Henequeneros consciously promoted the isolation of their workers, regulating communication with the world outside the hacienda. Although individual villagers were periodically granted access to the estate as part-time workers, landowners discouraged fraternization or intermarriage between their *peones* and residents of other estates and villages. As a general rule, *hacendados* kept city visitors and peddlers off their estates and restricted *peones* to them. Yucatán's primitive internal road system impeded such relationships and exchanges anyway. Finally, the heterogeneous composition of the work force augmented the prevailing climate of isolation. To meet escalating fiber demand after 1880, *henequeneros* cobbed together diverse conglomerations of human beings that diluted majorities of indebted Maya peons and village part-timers with admixtures of ethnic and linguistic strangers: Yaqui deportees, indentured Asian immigrants, and central Mexican *enganchados*. Thus, the *henequeneros'* strategy of "importing" Mexican contract workers and prisoners of war, and what they euphemistically called "foreign colonists," not only addressed an endemic labor shortage but also dampened possibilities for the creation of viable workers' communities.

Of course, *henequeneros'* power and influence transcended the physical boundaries of the estate. Proprietors regularly engaged bounty hunters and *jefes políticos*—the latter typically members of their extended families and clienteles—to return runaway peons seeking a haven in urban areas or, more likely, within villages or on neighboring estates within the henequen zone. During the early years of the fiber boom, *henequeneros* placed advertisements in Mérida's newspapers that offered rewards for the return of escaped *sirvientes*. Only when Mexico City journalists cited the advertisements as clear proof of the existence of slavery in the peninsula did these notices disappear.[19]

So precarious was life in the zone's pueblos that they themselves collectively constituted a second, complementary sphere in the landowners' mechanism of isolation. That the *henequeneros* could exercise their power so freely within the henequen zone and the runaways had such bleak prospects for refuge there supports something of the villages' miserable existence during the late Porfiriato. As economic options narrowed and the social fabric of village life in the zone frayed, ever greater numbers of *campesinos* were channeled to the great fiber estates.

Quite simply, the logic of monoculture's expansion consigned traditional peasant life in the zone to oblivion. As the boom intensified, henequen cultivation required ever-greater amounts of labor and land. Planters bought out neighboring smallholders as much to acquire their labor as to expand production on their fields.[20] *Heneplaneteros* were aided in their absorption of these *pequeñas propiedades* by the agrarian policy of the oligarchical state. Under the provisions of the 1856 reform laws, all corporate forms of landownership had been abolished and *ejido* lands were to be divided among village heads of families, each receiving a small donation. With the fiber boom under way, *hacendados* wasted no time in buying up *ejido* lands from individual heads of families throughout the zone. Sixty-six *ejidos*, totaling 134,000 hectares, were purchased by encroaching estates and given over to henequen cultivation between 1878 and 1912.[21] Sometimes disputes and acts of violence accompanied the surveying and parceling of *ejidos*, particularly on the less-controllable fringes of the henequen zone. Ultimately, however, *comuneros* and smallholders were powerless to stop the expansion of the large estates.

Sooner or later, most of the villagers based in the zone went to work on the henequen estates as tenants, as part-time workers, or, most commonly, as *peones acasillados*. Late nineteenth-century records reveal that the number of *peones* increased from 20,767 in 1880 to 80,216 in 1900, an increase of 386 percent in two decades.[22]

As the estates waxed, the *zona*'s pueblos waned. The *auge* not only transformed the social relations of production on the henequen estate but also altered the composition and role of villages and *cabeceras* within the *zona henequenera*. The hacienda's assault on the village's land and labor resources was intensified by the state's increasing involvement in rural life beginning during the last half of the nineteenth century. Increased taxes, frequent military levees, and arbitrary *corvée* labor obligations (*faqina*) further undermined corporate village bonds and weakened whatever symbiotic relationship had previously existed between the great estates and neighboring villages. With some notable exceptions—particularly Hunucmá, a *cabecera* 25 kilometers from Mérida that consistently (and often violently) opposed both hacienda encroachment and state exactions—peasant villages in the henequen zone had been reduced to a shadow of their former, semiautonomous existence by the end of the Porfiriato.

Those *campesinos* who opted to stay in pueblos now bereft of land and with only the most fragile sense of community found it increasingly difficult to sustain themselves. Most were forced to accept irregular work on the neighboring henequen estates, where they were paid at "piece rate" for cutting henequen leaves *(pencas)* or clearing new fields *(planteles)*. Otherwise, *zona* villages offered meager employment opportunities for their residents: a lucky few found work at the railway station loading bales of henequen fiber onto the rolling stock; others took jobs in the state militia (national guard) returning runaway peons to neighboring haciendas, or enforced tax collection and road *corvée* for the *jefe político;* still others worked in local mercantile establishments or wove henequen *hamacas* and bags to be sold in Mérida. Sometimes villagers had to combine several of these tasks as they struggled to maintain a marginal existence as "free men" in the zone's pueblos.

Conditions for villagers in the henequen zone were further exacerbated by the subregion's ecological limitations.[23] Because the northwestern quadrant of the peninsula has a foundation of porous limestone rock, water has historically been a more precious commodity than land. With no rivers and only a few scattered sinkholes *(cenotes)* to draw upon, water has been tapped from deep, underground canals with the assistance of costly pumps and windmills. The growth of large landholdings during the *auge* concentrated scarce water resources into the hands of aggrandizing *henequeneros*, who now had one more tool at their disposal to recruit villagers for their work force.

Often, even where water was not a problem, the dearth of topsoil was. Low corn yields have been a persistent problem for Maya *campesinos* residing in the calcareous northwestern districts. The combination of scarce water, poor soil quality, recurrent locust plagues, and the expansion of *latifundia* cut deeply into the resourcefulness of local *campesinos*. By the end of the Porfiriato, many villagers, now lacking the collective human infrastructure that the traditional *ejidos* had formerly provided, could not individually overcome the uniformly poor quality of land throughout the barren, rocky northwest.[24]

Faced with meager employment opportunities and an inhospitable social and physical environment in the surrounding pueblos of the henequen zone, most *campesinos* made an economically rational decision in opting to reside on the haciendas. Not only did the estate provide work and life's basic necessities (corn, water, and firewood), it also offered legal exemptions from military levees and *faqina*. Moreover, the henequen zone presented no other alternatives. Even flight to Mérida or Progreso, the principal urban centers, offered runaway peons or beleaguered villagers little advantage. *Campesinos* might temporarily find refuge in the capital's or port's rapidly growing slums. But with little knowledge of Spanish or of the city, with no job, and all the while exhibiting the gnarled hands of a *penca* cutter, the Maya *jornalero* would ultimately prove an easy target for the authorities to discover.[25] If he was lucky he might be returned to his estate to be whipped or jailed; otherwise, he might find himself dragooned into the state militia by a district prefect in search of able bodies to fulfill his quota *(la leva)*.

A third reenforcing sphere of isolation further induced the possibility of escape from henequen cultivation. Unlike their ancestors in previous generations, these henequen workers did not have the luxury of "voting with their feet," of finding a zone of refuge in the open southeastern jungle.[26] The brutal midcentury Caste War had effectively precluded that possibility by partitioning the peninsula into two distinct zones: the northwestern zone of commercial agriculture dominated by white Yucatecan elites, and the dense, disease-infested jungle of the southeast, the last refuge of the rebel Maya. Indeed, even at the turn of the century, much of the southeast was still held by the implacable, marauding *cruzob* and the somewhat more tractable, semiautonomous *pacíficos*.[27] Moreover, even if the acculturated northwestern Maya could somehow make their way to the jungle past all of the obstacles imposed by state authorities, they would find little in common with their now–culturally distinct brethren, the *cruzob* and the *pacíficos*. It is not surprising, then, that despite vast extensions of cheap fertile land, the southeast remained undeveloped, a virtual no-man's-land during the second half of the nineteenth century.

One of the major preoccupations for peninsular authorities during the *auge* was to keep the northwestern districts, given over to henequen production, safe from *cruzob* incursions. From a Maya villager or peon's perspective, however, *mala suerte* meant being taken to the frontier to serve on the state militia's first line of defense against the *cruzob* in remote hamlets like Xoccén or Chemax.

When President Porfirio Díaz sent a federal army to root out the *cruzob* once and for all, in 1901, he insisted that Yucatecan troops *(colonias del sur)* participate in the "pacification" campaign.[28] More than two thousand Yucatecans would die in the military expedition, many of them Maya villagers from the *zona henequenera*.[29]

The *levas* for the *colonias* and the state national guard were uniformly abhorred by all recruits, many of whom attempted to flee to neighboring Campeche at first sight of the despised *jefes politicos.* Males between the ages of fifteen and sixty were required to serve in the militia each year. Exemptions could be secured in three ways: if the individual could pay a set fee or find a replacement to serve for him; if one had a debilitating illness (an affidavit was required from a doctor as proof); or if a *campesino* was classified as a permanent laborer attached (read "indebted") to a hacienda.[30] Since the first option was typically out of the reach of many *campesinos,* and securing a medical certificate also cost dearly, villagers who wished to avoid military service in the *colonias* or the state militia often found refuge (in debt) on the hacienda.

Thus, the encroaching state and the local interests it often served had created a mechanism that served the federal government in its campaign to eliminate the *cruzob,* even as it ensured *henequeneros* a permanent labor force. Significantly, the conscription statutes gave villagers the illusion of a choice by making it appear that it was they themselves who chose servitude on the estate over "freedom" in the *pueblos.*

Villagers' "choices" were further circumscribed in 1902, when the national government subsequently partitioned the peninsula, creating the federal territory of Quintana Roo.[31] The pacified interior was now transformed into a vast federal prison for Mexican criminals, recalcitrant deserters, and Yaqui Indian prisoners-of-war. Opponents of the Díaz regime served their sentences in malarial-infested labor camps patrolled by the federal army, cutting mahogany and cedar trees or working on government-sponsored chicle concessions. The southeastern jungles, traditionally a refuge for dissident Maya, no longer remained a viable option in the late Porfiriato.

Yucatán's southern neighbor, Campeche, provided no better prospects. Campeche was essentially a mini-Yucatán during the Porfiriato, complete with henequen haciendas, a *leva* of its own, the same hostile frontier, and a state militia that patrolled its border with Yucatán with a keen eye for wayward *jornaleros.*

In short, the peninsula's outer rim acted as a vice that hemmed in villagers, making them aware that their best option was steady work on henequen haciendas. The *leva,* the Caste War, Quintana Roo labor camps, jungle fevers, and a southern neighbor that offered no safe haven—all conspired to teach *campesinos* that "voting with their feet" was fraught with too many risks.

Finally, Yucatán's location, remote from the rest of Mexico, constituted a fourth circle of isolation that buttressed the idioms of power conveyed by *henequeneros* and state authorities. For all intents and purposes, as far as localistic Maya *campesinos* were concerned, Yucatán was an island unto itself. Even if *campesinos* had wanted to leave their homes and their *patria chicas* entirely (and they did not for any number of culturally sound reasons), it would have proven very difficult. Railway or road service with the rest of the nation would not be realized until well after World War II. In the interim, the only way to reach the Mexican mainland was an irregular steamship service with the port of Veracruz. Geography reenforced the overarching control mechanism by constricting freedom of movement for Yucatán's laboring classes.

If a lack of mobility tied the Yucatec Maya to the *zona henequenera*, Yucatán's distance from the capital diminished the *campesino*'s prospects for social justice. Existing documentation suggests that the federal government turned a deaf ear toward *campesino* complaints during the Porfiriato. Until his celebrated visit to the peninsula in 1906, President Porfirio Díaz remained very much an abstract concept to most people in the Yucatecan countryside. Among the tens of thousands of letters written to Díaz during his tenure, only a handful came from Yucatecan villagers (and fewer still from *peones acasillados*).[32] Although the Porfirian polity was hierarchical and based on patronage, it seems clear that the Maya felt excluded from the hierarchy. They seldom took the opportunity to appeal directly to the *Gran Patrón* in Mexico City—though, no doubt, many peons would have felt too constrained to even think about appealing. Moreover, even when *campesinos* did find a sympathetic lawyer to set down their grievances in a letter, their chances of receiving a meaningful response from the dictator were slim. Díaz generally answered mail from petitioners and governors alike in a perfunctory and cryptic manner. Many letters appear to have elicited no response at all.[33] The dictator's apparent disdain for immersing himself in purely local matters appears to have been matched only by his constituents' unwillingness to petition their grievances to a higher authority. Yucatecan *campesinos* during the Porfiriato did not exhibit the same feeling that their colonial ancestors had for their sovereign, the king of Spain. Reverence for the presidential office was conspicuous by its absence, and (particularly after 1910) the Maya would look to local institutions such as the judicial system to seek redress for unjust treatment.

In sum, four concentric circles of isolation acted like magnetic poles, creating a force field that propelled *campesinos* (and wayward *peones*) away from the *pueblos*, Mérida, and the southeastern interior and toward the fiber estates. At times this was accomplished through coercion, but more often by diminishing the attractiveness and feasibility of other choices. The mechanism of isolation not only predisposed *jornalero* obedience to hacienda authority and facilitated the coercive aspect of power relations; it also gave villagers "on the outside looking in" a distorted view of the henequen estate itself. Given their druthers, many beleaguered villagers within the zone came to prefer resident status on the hacienda to their own emiserated existence in the *pueblo* (see the discussion of security below). It is important to keep in mind that the social control mechanism of isolation did not completely take away the ability of independent villagers to make decisions about their own (and their families') future. Such circumscribed choices, whether it be a personal preference to accept resident status, hire oneself out as a part-timer worker, or stay put in the declining villages and *cabeceras* of the *zona henequenera*, gave Maya *campesinos* the sense that they were independent actors involved in a decision-making process—even if the options framed by the landholding elite and state amounted to little more than a "lesser-of-two-evils" choice. Those who opted, however, to select permanent status on the henequen estate came up against a brand of coercion that exploited their labor through an effective combination of the personalistic and materialistic idioms of power.

Coercion

If the isolation mechanism gave the northwestern Yucatec Maya a somewhat illusory set of choices, the messy business of coercion reminded resident peons on the estates that naked force could and would be used against them if disciplined work rhythms were not maintained. But, even coercion was a variegated mechanism that conveyed diffused images of power to *jornaleros*. It was this ambiguous quality—with coercion imparted in some instances with cathartic fury and at other times conveyed more subtly to intimidate or "make an example"—that invested this mechanism of control with such versatility.

Like the sugar plantations of the Caribbean, the henequen estate was a world unto itself. With the blessing of the state, *hacendados* fashioned a powerful extra-legal system of justice to reenforce the relative isolation of the estate. Overseers—they are variously referred to as *encargados, mayordomos,* or *personeros* in the haciendas' account books *(libros de cuenta)*—had only to answer to their employers, who on average visited their properties one weekend a month.[34] As a result, until the national Madero rebellion

opened up greater political space, peons rarely registered complaints of *encargado* brutality or other abuses to unsympathetic
state authorities "on the outside."

The personalistic idiom of power was most transparently represented by corporal punishment. Whipping was believed to be an
overseer's most effective tool in controlling his laborers. The most
graphic representation of this venerable practice is found in a photograph that appeared in Henry Baerlein's muckraking account,
Mexico: Land of unrest (1914). The photograph showcases the
scarred back of a Maya *sirviente* on the hacienda of Don Rogelio
Suárez, Olegario Molina's son-in-law. *Encargados* inflicted the lash
for a variety of reasons: insubordination, shirking, dissimulation,
flight, and as the standard antidote against the generalized violence
that was so much a part of the daily routine on henequen haciendas. Whipping, after all, did not cost the *hacendado* the daily labor
of his *jornaleros;* placing them in the hacienda jail *(calabozo)* for
any length of time did.[35] The Yucatecan aphorism, "The Indian
hears only with his backside," was more than just black humor; it
reflected a widespread belief among the dominant class that discipline needed to be reenforced by stern example. Governor Santiago
Méndez reaffirmed this arrogant sentiment when he noted:

> As a result the Yucatecan Indians are regarded as being meek,
> humble and not easily stirred to ire and cruelty . . . the most cus-
> tomary punishment among them was a whipping applied with
> moderation. This kind of punishment did not offend them, if they
> were informed of the reason why it was meted out to them, nor
> did they consider it degrading.[36]

Although we now recoil at the idea, the notion that flogging "in
moderation" reenforced discipline at the workplace, and was an
effective tool of social control, was a view shared by many *encargados.*[37] Maya *jornaleros,* another observer commented, were
"docile, obedient and compliant with their masters's orders, the
application of a few strokes being, in the last resort, invariably
effective."[38] The implications of the *henequeneros'* rationalization
was an unmistakable message to the laboring classes: *jornaleros*
were subordinates who required, indeed expected, the lash, just as
children expect to be disciplined by their parents. In Yucatán, this
explicit dehumanization took on a racist character, since *mayordomos* and other managers invariably had Spanish surnames, while
henequen workers were for the most part Maya Indians.

One of several cases found in the judicial records in which a whipping was contested by resident peons demonstrates why such cases were rarely brought before the state's judiciary. In 1912, the stepparents of a sixteen-year-old *sirviente* took the *mayordomo* of a small hacienda in Sotuta district to court for the *flagelación* of their stepson for insubordination. As the case unfolded, however, the plaintiffs retracted their original story and said that the *mayordomo* was only disciplining their unruly son, who sorely needed it. The stepmother, in fact, went on to contradict her own earlier testimony, denying that the boy had been whipped, despite a medical affidavit corroborating the corporal punishment that had been filed by the plaintiffs. Needless to say, the charges were dropped. While we may never know exactly what happened, the implications are clear: the *hacendado* (in this case, Hernando Ancona Pérez) had so many ways to "persuade" his servants of the importance of settling their affairs outside the courts that few *acasillados* on the estates took advantage of their rights as Mexican citizens.[39] This minimal recourse to legal redress by resident peons stands in sharp contrast to the attitude of free villagers on the fringes of monoculture, who regularly petitioned judicial authorities with their grievances.

Hacienda "justice"/coercion was further dispensed through the use of *calabozos* designed to lock up peons temporarily for infractions. Since alcoholism was a pervasive problem, many peons who created disturbances while intoxicated found themselves sleeping off their binges in the *calabozo*. On Hacienda Tamanché, on the road to Progreso, the jail attached to the *tienda de raya* was so small that it would appear to have been little more than a cell used for solitary confinement. With only two small windows and low ceilings to counteract the stifling tropical sun, even a short stay in Tamanché's *calabozo* must have been a hellish experience.

Two indispensable components of the *henequeneros'* personalistic idiom of power adapted from the traditional cattle-and-corn hacienda were the debt mechanism and its complement, the *tienda de raya*. Together, these institutions reenforced the immobility of henequen workers. An 1882 state law, the *Ley agrícola industrial del estado de Yucatán,* reiterated earlier peonage laws, stipulating that the *peón* who left work without paying the sums that he owed would be legally prosecuted. In addition, if an indebted servant escaped and took refuge on another estate, the landowner who hid the servant could be arrested.[40] By contrast, Article V of the 1857 constitution—which stated that "nobody shall be obliged to render personal service without proper compensation and full consent"—a clause that should have rendered debt servitude illegal, was ignored by elites, who breached the spirit of the federal law by euphemistically justifying peonage as a "simple contract convenient to both parties."[41]

Debts were, after all, an established labor practice dating back to colonial rule, and historically had permeated social relations throughout the peninsula. The peon had two running debts, *chichán cuenta* and *no hoch cuenta;* the former was a small debt for daily purchases and weekly wages, while the latter was a large account (used for important rites of passage, such as marriages and baptisms) from which a servant rarely extricated himself. In fact, debt peonage was so ingrained in local labor practices that Yucatecan railway entrepreneurs utilized the institution to limit the mobility of their own scarce labor force. Moreover, local legal statutes accepted the valuation of workers's debts as a legitimate expense, granting further legitimacy—not to mention a financial advantage —to debt peonage. Specifically, Article 2030 of the state's Civil Code stipulated that peons' debts should be included in a property's inventory. Nelson Rubio Alpuche, a prominent local lawyer, put it bluntly in a legal brief written in 1895 when he said, "In Yucatán, loose teams of peons do not exist. . . . A hacienda that does not have persons obligated by personal contract . . . is not worth anything."[42]

If debt labor, then, was an established practice, what changed during the *auge* was the compelling need of *henequeneros* to make sure that their *jornaleros'* debts were not paid off.[43] As a result, the relatively mild relations that had characterized the prehenequen labor regime quickly hardened. *Henequeneros* now enlisted the aid of the burgeoning state bureaucracy to ensure that debts were not repaid. Local *jueces de paz* and district-level *jefes politicos* cooperated with proprietors to make it difficult for debtors to find relief in the court system. In many cases, collaboration was cemented by kinship ties between *hacendados* and state officials.

The case of Juan Bautista Chan provides a useful illustration of such collaboration among elites.[44] By 1912, Chan, a *peón acasillado* on Hacienda San José in the *municipio* of Hoctón, had run up a debt of about 400 pesos. He fled San José (for unspecified reasons) and took up residence in a distant *pueblo*, Tahmek. According to Chan, he tried to pay off his debt by giving the sum to the justice of the peace of Tahmek, Segundo Sergio Echeverria.[45] Nevertheless, the judge refused to accept payment because the matter was out of his jurisdiction and advised the peon to go back to Hoctón and pay the *cuenta* at the business establishment of San José's politically powerful *hacendado*, Aurelio Gamboa. Chan's lawyer pointed out that his client had a legal right to repay the debt at his choice of venue, and petitioned the tribunal Superior to make the judgment accept the payment. For his part, Justice Echeverria denied that he had ever seen Chan and paraded witnesses before the court who supported his contention. Not surprisingly, the higher court found for the judge. The case implicitly illustrates the collegial relationship between *henequeneros* and the court system and indicates how difficult it was for *acasillados* to repay their debts, even in those rare instances when they were able. Moreover, it suggests that Chan was unwilling to return to Hoctún, because he might be snared by local authorities in cahoots with Gamboa and returned to Hacienda San José.

If authorities worked hand in glove with *henequeneros* to preclude repayment of debts, *tiendas de raya* afforded *henequeneros* an excellent opportunity to increase the size of *jornaleros'* credits. Most *peones acasillados* purchased their goods at these *tiendas*, either because of the availability of credit, or simply for convenience, since, as we have seen, peons often found it difficult to travel off the estates to purchase goods elsewhere. Sometimes owned by the *henequenero*, sometimes leased to local merchants, the *tiendas de raya* were instrumental in reenforcing the immobility of henequen workers. *Hacendados* paid wages in *vales* (scrip) redeemable at the hacienda store.[46] The use of scrip varied from hacienda to hacienda and only in rare cases was negotiable at shops in nearby towns. Although many *jornaleros* were paid by the piece rather than a set daily wage, in order to boost production levels, it would be wrong to infer that these "rural proletarians" were wage earners in the modern sense of the term. There was never any pretense on the part of the *henequeneros* that these wages were meant to foster either a money economy or a free labor market in the *zona*. Henequen workers had little access to "real" money; the wages they earned were deducted from a *cuenta* that for all intents and purposes was never repaid.[47] Finally—as if to add insult to injury—some hacienda records tell us that proprietors enhanced the indebtedness of their *jornaleros* by charging for breakfasts.[48]

Conceivably, the peon never had to leave the inner sanctum of the hacienda. In terms of convenience, the store offered a wide variety of goods, including food, unfinished cloth, household goods, shotguns and gunpowder for hunting, and last—but by no means least in popularity—liquor. *Libros de cuenta* from the Compañía Agrícola del Cuyo y Anexas, S.A., a multifaceted enterprise in the eastern hinterlands, testify to the considerable income that a *cantina* attached to the *tienda* generated for its owners.[49] While some estates did not have bars on the premises, most haciendas sold cane liquor *(aquardiente)* to their workers. Social critics, fond of needling Governor Olegario Molina for his attempts to infuse Yucatán's laboring classes with a pristine, positivistic morality that forcefully advocated temperance for the laboring classes, gleefully pointed out this apparent hypocrisy.

This, a flexible mechanism of coercion that combined elements of corporal punishment, a propietary, extra-legal system of justice, debt peonage, and the *tiendas de raya,* was a powerful weapon in the *henequenero's* arsenal of social control. It was the juxtaposition of such personalistic institutions—singularly and in combination— with the materialistic mode of production enforced on the estates that fostered henequen monoculture's greatest social contradictions. Since the ultimate goal remained the maximization of fiber production, however, *hacendados* understood that heavy-handedness could be overdone and force was often most effectively applied "in moderation." That is why coercive institutions such as debt and the *tienda* were closely linked to another more amenable mechanism, security.

Security

Resident peons had to be made to think that the estate's social relations, as exploitative as they were, offered them something more than an occasional whipping and a lifelong debt. The fact that *henequeneros* could effectively employ institutions like debt peonage and the *tiendas de raya* to coerce *acasillados,* while simultaneously restricting their mobility and providing them with a measure of security, accentuates the multiple roles these institutions played in Porfirian Yucatán. The cooptive security mechanism was the flip side of isolation and coercion, since it provided something tangible —a fundamental security of subsistence[50]—to resident peons in return for their labor. As one student of peonage explains: "peons stayed on the plantations (maltreatment notwithstanding) because it was a rational, as well as a customary thing to do."[51]

By providing a continuous supply of imported corn, beans, and meat at the *tienda, henequeneros* ensured that they alone controlled the keys to subsistence in the *zona henequenera.* After 1880, with monoculture's gaudy profits plainly in view, *henequeneros* eschewed self-sufficiency and rapidly converted their estates' corn plots to henequen. Notwithstanding the higher prices of imported corn, it made good business sense to purchase the basic foodstuffs needed to feed *peones acasillados* from grain merchants in Mérida. Interestingly, these enterprising importers were the very same *casas exportadoras* who bought the *henequeneros'* fiber.[52]

More importantly, *henequeneros* no longer had to provide plots of land for *milperos*, since corn was available at the *tienda*. Even though henequen workers complained bitterly about the poor quality of imported corn—some called it rancid and mistakenly believed it led to outbreaks of pellagra, a nervous disorder caused by a vitamin B deficiency—imported maize had replaced homegrown as a mainstay of the Yucatecan *campesino*'s diet by 1910.[53] The losers, again, were the villagers of the henequen zone, who, for over a century, had enjoyed an appreciable domestic market, providing their surplus corn and beans to nearby haciendas.

Monoculture's infrastructure—four railway lines that articulated agricultural zones with Mérida and the port of Progreso—eroded the economic viability of the zone's pueblos, transporting imported corn, *frijoles*, and beef from the port throughout the peninsula. By 1910, corn cultivation in the zone was in a state of irreparable decline. Statistics (see Table 1) tell the tale: production levels for zone *partidos* such as Maxcanú, Mérida, Motul, and Tixkokob demonstrate the extent to which the tentacles of monoculture had usurped *milpa* lands in the northwest. When we contrast the zone's cultivation figures with those of the "fringe" *partidos*, we find that cultivation in zone districts pales in comparison with that of the peripheral *partidos*. Indeed, in districts where henequen did not overrun the countryside and where peasant communities remained relatively intact (for example, *partidos* such as Tekax, Valladolid, Ticul, Espita, Tizimin, and Sotuta) the average cultivation of maize more than doubled what it was in the henequen zone (5,683 hectares versus 2,343 hectares). Whereas a half century before, the northwestern *partidos* had supplied the product of primary necessity to peripheral districts, the former now imported Mexican and foreign corn. The parasitical henequen economy, by siphoning off labor and land from *zona* villages and by denying those communities a meaningful role in the wider regional economy, effectively condemned the embattled pueblos to a slow and painful economic death during the Porfiriato. Obviously, with the staff of life no longer plentiful in village communities, the *tiendas de raya*'s abundant reserves took on greater importance as a tactic in the *hacendado*'s overall strategy to lure (and tie) workers to the estate.

Also figuring into the mechanism of security was the tepid brand of paternalism practiced on henequen estates. As we have seen, Yucatecan *hacendados* were, for the most part, absentee landlords, who left the daily operation of their haciendas to their overseers. The *jornalero* typically directed his problems, complaints, and requests to a hired employee. Conscientious planters occasionally

Table 1. Corn Cultivation in the Henequen Zone, 1896–1912

Partido	Average Hectares Cultivated
Acanceh	2,332
Hunucmá	4,515
Izamal	4,225
Maxcanú	3,461
Mérida	579
Motul	1,800
Progreso	313
Temax	2,792
Tixkokob	1,073
Average for henequen zone	2,343
Average for nonhenequen zone	5,683
Average for Yucatán	3,805

SOURCES: For 1896–1901, *Boletín de Estadística*, passim; data on corn cultivation for 1907–1912, *El Diario Yucateco*, 22 February 1910, *El Agricultor*, 1907–1908, passim, and *Boletín de Estadística*, 1908–1910, passim.

NOTES: Reliable data exist for individual *partidos* for 1897–1901, 1908–1910, and 1912. Data on Yucatán's average cultivation include all of the above years plus 1896 and 1907. The *partido* Las Islas, which was part of Yucatán prior to the partition of the peninsula in 1902 and the subsequent creation of the federal territory of Quintana Roo, was factored out for years 1896–1901.

Partido averages computed and rounded off by authors. Care should be taken with these figures, since they represent *milpa* planted on haciendas and villages, not final production figures. Locust plagues and insufficient and overabundant rains played havoc with corn production throughout the peninsula. Unfortunately, data on statewide corn production were too erratic to prove statistically useful.

visited their estates on weekends, preferring to spend the rest of their time in Mérida, where they could monitor their myriad investments and enjoy the cultural amenities of urban life. This is not to say that certain Yucatecan *hacendados* did not take a special interest in their *peones* or that paternalistic relationships based on mutual obligations and responsibilities did not exist in Porfirian Yucatán. The institution of *compadrazqo*, or fictive kinship, for example, tied the peon somewhat loosely to his *amo*.[54]

Nevertheless, it would be a mistake to assume that such paternalism ever fostered strong bonds of attachment, let alone cultural understanding, between the acculturated Maya and their masters.[55] The Maya *sirviente* and the *dzul* (the Yucatecan master) were separated by a cultural polarity that undermined trust. A clear example of how that cultural divide weakened paternalism is reflected in the differing perceptions of medical treatment held by *dzules* and Maya during the Porfiriato.

Hacendados provided a modicum of health care on the hacienda. Since doctors were not permanent fixtures on any, save the largest, estates, most *sirvientes*, when seriously ill, were transported by train to their master's villa in Mérida, where they were treated by the family doctor until they recuperated. If *acasillados* required the services of Mérida's O'Horan Hospital, the only legitimate medical facility in the state, they needed to be accompanied by their *amo* or have the written permission of hacienda authorities.[56] But the rural Maya were reluctant to entrust themselves to their *amo*'s personal doctor, let alone to a hospital, given what they regarded as culturally bewildering, even frightening, practices on the part of city doctors.

A case in point is the unfortunate story of Eleuteria Ek, a young girl growing up on Hacienda Santa Barbara near Cansahcab in Temaz *partido*.[57] On 8 October 1907, Circiaco Santos, the *mayordomo* of Santa Barbara, demanded that the parents of the ailing twelve-year-old Eleuteria deliver her to the main house *(casa principal)*. Eleuteria was stricken with influenza, Santos testified, and it was hacienda policy to isolate such contagions. Initially, Eleuteria's parents, Crisanto Ek and Micaela López (who was from nearby Cansahcab, but now lived on the hacienda), refused to hand over their daughter, claiming that she was needed to help with household chores. According to Crisanto's and Micaela's testimony, the *mayordomo*'s men then physically hauled Eleuteria off to the *casa principal*, warning the parents that they would be beaten if they continued to object. The very next day, Eleuteria was taken by train to Mérida to receive medical treatment and to convalesce in the home of the *hacendado*, Pedro Luján.

The scene—emotionally described in the parents' court testimony—of a child wailing hysterically as her mother and father watched helplessly at the train station in Cansahcab, points up the different worlds of the rural Maya and their paternalistic masters. Although families were not, as a rule, broken up during the *auge*, Maya *campesinos* had legitimate reason to fear that once their children were taken to their master's residence in the capital, they might stay on as domestics in the *hacendado*'s urban villa. Perhaps because she was a villager who had married and then moved onto the hacienda, Micaela assertively filed the grievance with Cansahcab's justice of the peace. The judge granted the mother permission to go to Mérida to pick up her daughter and return the minor to the estate. Interestingly, the case record closes with an affidavit from Micaela that attests to the fact that Eleuteria was well treated in Mérida—a reassuring judicial epitaph for an unsettling incident.

Clearly, the court wanted it known that while the parents were to be permitted to take their child back home, the *hacendado*'s reputation also required that he be absolved from any suggestion of mistreatment.

Eleuteria's story graphically illustrates the lack of trust between Maya and *dzul.* Something as seemingly innocuous as medical treatment is reflected with cultural tensions that weaken rather than strengthen clientelistic bonding. Perhaps Nancy Farriss' observation for the colonial Yucatec Maya is apt for their more acculturated descendants of the late nineteenth and early twentieth century: "The Maya had no hospitals nor any desire for them, having come to the not wholly unfounded conclusion from their observations that these European institutions were for dying."[58] It is ironic that such suspicions were harbored by Maya families precisely when the intentions of the *amo* were so overtly altruistic (however paternal). Apologists have taken great pains to laud benevolent *hacendados* who took care of ailing servants, either transporting them to their homes in Mérida or taking them personally to Hospital O'Horan.[59] Eleuteria's case, and others like it, call for a reexamination of paternalism on henequen estates; altruism may well be in the eyes of the beholder.

While the cultural dissonance between *amo* and *sirviente* tempered patron-client relations, henequen monoculture's fundamental security of subsistence throughout most of the Porfiriato, coupled with the economic demise of nearby village communities, enlisted workers for, and harnessed them to, the disciplined work rhythms of fiber production. Skillful management by the *henequeneros* of the three complementary mechanisms of social control —isolation, coercion, and security—kept *jornaleros* tied to fiber production. Yet, although *jornaleros* found themselves very much on the defensive during the boom, they often responded to the structure of domination in a number of creative ways.

Forms of Resistance in the *Zona*

Formidable as it was, henequen monoculture's structure of domination did not—indeed, could not—completely deprive *peones acasillados* of means to protest and express their humanity. A certain amount of "play" must be built into even the most coercive labor systems. The implication of any means of social control, whether it be flogging by an overseer, the use of identification papers to inhibit worker mobility in the countryside, conscription into the despised national guard, or the workings of the criminal justice system, requires a degree of flexibility to ensure its maximum efficiency. If the mechanism of control is too rigid or stringent, if it does not allow for exceptions, its utility is likely to diminish, since those for whom it is intended will increasingly balk at its implementation. Not surprisingly, we have found that despite its coercive aspects, henequen monoculture's characteristic idiom of power still afforded resident peons a certain latitude to adapt to changing circumstances, tapping the cultural resources at their command. In the process, they became active agents in shaping the terms of their own oppression.

Here, our argument buttresses recent findings for the roughly contemporary slave regime that drove Cuba's sugar monoculture.[60] Despite profound differences on other matters, both Manuel Moreno Fraginals and Rebecca Scott argue that Cuban slaves were successful in forging a countervailing culture. Even under the brutal working conditions found on *inquenios*—where slaves would labor for up to twenty hours a day during the harvest, sleep-walking among scalding boilers and dangerous machinery that cost limbs and lives—deracinated slaves did not acquiesce to, or allow themselves to be broken by, the treatment their *amos* meted out. Scott concludes: "Slaves could be cheated, yet participate in a money economy. They could be ill-housed, yet struggle to maintain families. They could be treated worse than beasts, yet not become like beasts."[61]

Compared to the African-born *bozales* and creole slaves who toiled on Cuban *inquenios*, Yucatán's resident Maya *peones*—who always made up the vast majority of the fiber estate's labor force—enjoyed distinct advantages that enhanced their capacity to resist. As we have seen, despite their isolation and subordination by the *dzul*, the northwestern Maya still possessed a distinct culture, based upon nuclear and extended family units and reinforced by syncretic religious beliefs and practices. Unlike many slave societies, the Maya family in the *zona henequenera* was seldom broken up, even by the most despotic *amos*. In fact, the chronic labor scarcity that was a permanent feature of the *auge* worked to limit any movement of workers from estate to estate. *Acasillados* family units lived in wattle-and-daub huts, not the gender-segregated, impersonal barracoons that held as many as two hundred slaves on large Cuban *inquenios*.

Moreover, the Maya's cultural heritage was further reinforced by their language, which throughout the Porfiriato remained the lingua franca of the henequen zone. In fact, Yucatán outranked the rest of Mexico in its percentage of native speakers. While the national percentage of Spanish-speaking inhabitants in 1895 was 83 percent, in Yucatán seven out of ten citizens spoke Maya.[62] At the very least, then, Maya family cohesiveness and the persistence of language and cultural traditions offered a degree of solace and refuge against the ravages of an exploitative system.[63]

No doubt, the ability of peons to fashion a countervailing culture was impeded by the intensification of the labor regime during the *auge*. Nevertheless, *henequeneros* continued to view the potential for Maya solidarity with some alarm and, as we have seen, consciously diluted communities of Maya *acasillados* with groupings of ethnic and linguistic strangers. Cuban sugar planters adopted a similar strategy, employing Chinese contract workers. Ultimately, however, neither master class succeeded in eliminating the cultural space that legitimized challenges to its hegemony. Yet it seems clear that Yucatán's indigenous Maya had a greater capacity for resistance than Cuba's slaves, many of whom were "outsiders," uprooted and transplanted from abroad.[64]

Of course, the *forms* of resistance that Yucatán's peons were able to mount against monoculture were shaped by the relations of domination at concrete historical junctures. Until the final years of the Porfiriato, Maya peons in the henequen zone, like their counterparts in more formal slave societies, rarely took the risks attending violent collective actions against the plantocracy. Given the multitiered repressive mechanism that the *henequeneros* had fashioned in collaboration with the state, such jacqueries and insurgencies, as well as organized bandit operations, were doomed to eventual defeat, even massacre. To be sure, sporadic flare-ups did occur, signaling desperate local responses to particularly egregious *henequeneros* abuses, and typically provoking excessive planter retaliation. Yet, by and large, Maya peons, like slaves in the U.S. South and Caribbean, were not suicidal; they too appreciated that while "a man may perish by the sword . . . no man draws the sword to perish, but to live by it."[65] Moreover, there were other alternatives open to them, more modest strategies by which *peones acasillados* could deal a blow, materially and symbolically, against the exactions and domination of their masters.

Such "quieter" or "everyday forms of resistance," as James Scott terms them, might include small, self-serving acts of noncompliance, foot dragging, shirking, and flight, or, more aggressively, clandestine acts of theft, arson, and sabotage. While social scientists have given it short shrift, this "small arms fire of the class war," these "weapons of the weak" that proceed outside the bounds of organized movements, have always constituted the greatest part of peasant (and working-class) politics.[66] And, as Scott argues persuasively, they have likely accomplished what futile armed resistance could not: they have "confer[red] immediate and concrete material advantages while at the same time denying resources to the appropriating classes."[67]

In the henequen zone, such "routine" resistance was eminently suited to the highly controlled, socially heterogeneous plantation milieu: it needed little planning, required only a modicum of room to maneuver, and could be carried out secretly by *jornaleros* acting alone or in the smallest, most informal of groups. Moreover, it avoided any direct (and inevitably costly) confrontation with the *amo* or his overseers. The thrust of such resistance was not the impossible goal of overthrowing monoculture's system of domination but rather of surviving—today, this week, this season—within it. For, as Eric Hobsbawm conceives it more generally, the fundamental aim of *campesinos* has always been to work "the system to their minimum disadvantage."[68]

Such day-to-day resistance poses a challenging methodological problem for the social historian. It is little noticed in the official records of the state because it does not generate the programmatic statements, violent encounters, and public demonstrations that tend to rivet the state's attention. Indeed, the goal of the perpetrators is precisely to avoid drawing attention to themselves. Moreover, the state's bureaucrats and local landowning elites have little interest in publicizing the incidence of *campesinos'* insubordination, for to do so would be to acknowledge unpopular policies and the limits of hegemony in the countryside. Thus, with good reason, Scott argues that the historiography of class struggle has been "Statolotrous." Minor, doomed revolts that have left an impressive paper trail continue to preoccupy scholars out of all proportion to their impact on class relations, while "unheralded acts of flight, sabotage, and theft that may be of greater long-run significance are rarely noticed."[69]

While anthropologists like Scott engage in participant observation to understand such nonconfrontational strategies at first hand, we have been left an incomplete written record that systematically understates everyday forms of resistance. For example, all but a few instances of shirking, dissimulation, and insubordination—the principal forms of protest by peons on the henequen estates—have fallen through the cracks of the historical record. Agents of the state found such behavior too insignificant to document—except in rare, vexed bureaucratic asides—while *henequeneros* chose not to belabor the underside of plantation life in the lofty memoirs they commended to posterity.[70] Unfortunately, as regards the peons, few then old enough to remember have been interviewed, and few *ancianos* now remain to give testimony. Moreover, the several, incomplete oral histories that have been taken treat the past in nebulous terms that do not permit a thorough understanding of everyday forms of resistance.[71]

This methodological dilemma goes a long way toward explaining the rather monochromatic, static portrait of Porfirian life in the henequen zone that emerges in the historical literature. As we have noted, peons have been characterized as cowed and docile retainers, *amos* and their agents as omnipotent and omnipresent masters who refused to countenance the slightest rebuke to their authority, let alone a consistent challenge to the *auge's* regimented rhythms of production. Of course, such a portrait excises from the historical record not only routine forms of resistance but also, as we shall see, the more violent and coordinated acts of protest that *peones* initiated or joined during the final years of the old order.

Although they are incomplete, unorganized, and rent with official bias—problems we have taken up elsewhere[72]—Yucatán's criminal records nevertheless help to redress the methodological and interpretive shortcomings of the historical literature. Court testimonies document instances of even the most routine forms of protest on Porfirian estates (shirking, foot dragging, and insubordination)[73] and capture more aggressive forms of resistance (rustling and theft, arson and sabotage).[74] Ultimately, they provide a window as well onto the riots, jacqueries, and revolts that erupted during the Madero era. Apart from shedding light on modes of resistance, such personal court testimonies graphically—often poignantly—reveal larger snatches of working-class life during the *auge*, documenting a rigorous quotidian struggle to cope and survive. While we cannot reconstruct the texture and nuance of the larger process here, we can illustrate the broad range of resistance that ultimately culminated in waves of insurgency during the first years of the Mexican Revolution.

Despite the various forms of coercion that were brought to bear, shirking and absenteeism remained a problem throughout the *auge* for the overseers who were entrusted with maintaining production levels on estates. *Hacendados* and their personnel continually groused about the lazy, shiftless Maya. The *libros de cuenta* of haciendas regularly list peons who, for one medical reason or another, did not work on a given day. How much of this illness was feigned is impossible to say.[75] Clearly *henequeneros* had failed to inculcate a "proletarian" attitude in the *acasillados* during the boom. Then again, their hybrid labor regime was not suited to produce such a transformation in *mentalité*. Peons who were physically coerced to cut 1,500 to 2,000 spiny leaves a day under the relentless tropical sun, who existed in a perpetual state of indebtedness, whose mobility was severely constrained, found ways to avoid work. Piece wages may have been paid, but debts had become so large at the end of the Porfiriato that few had any real prospect of liquidating their accounts. Indeed, what incentive did peons have to complete their assigned tasks—let alone develop a work ethic—when all that their toil yielded was an inconsequentially smaller debt?

More likely, the *jornaleros* found escape in alcohol, with intoxication exacerbating the problem of absenteeism. Even systematic whippings failed to prevent the popular custom whereby weekends were extended to include "San Lunes." Alcohol was, of course, the principal release for resident peons; its ready availability in the *tiendas de raya* indicates that, on another level, *henequeneros* appreciated its value as a mechanism of social control.

Alcohol abuse was therefore a complex matter in social terms; alcohol likely augmented and diminished resistance on estates. Apart from its link to absenteeism, intoxication appears to have figured heavily in much of the crime and low-level violence that was endemic in rural Yucatán during the Porfiriato. In 1906 and 1907, for example, arrests for drunken disorder accounted for 55 to 70 percent of all apprehensions by state authorities.[76] While such government statistics are notoriously unreliable and reflect urban as well as rural patterns, they point up the staggering incidence of alcohol abuse. Moreover, government figures do not even include those incidents handled more "quietly," outside the state judicial system, by *hacendados* and overseers on the estates.

The criminal records abound with deaths and crimes that were found by the courts either to have been induced or mitigated by alcohol.[77] The crimes run the gamut of offenses, from simple theft and cattle rustling to arson, from domestic quarrels and assault and battery to murder. In a great many (and likely the majority of) instances, peons victimized other peons,[78] the obverse of Scott's notion of "routine" resistance. Scott argues plausibly that to the extent that *campesinos* are reduced to lashing out at, or preying upon, their fellows, appropriation by the dominant classes is added, not resisted.[79]

Yet the criminal records also contain a variety of cases in which peons robbed and rustled from (and even assaulted and killed) overseers and *hacendados*. Invariably, excessive drinking was invoked by defendants and witnesses, either as an explanation or a mitiga-

tory factor. It appears that the types of questions posed by prosecu-
tors influenced or predisposed defendants' responses, suggesting
the ability of poor rural offenders to "play the game." Because
they knew that premeditated crimes were punished more severely,
it was in the defendant's interest to claim his offense was induced
by alcoholic stupor, or merely unplanned.[80] Quite likely, then, the
criminal records overrepresent unpremeditated acts. Authorities
understood the significance of alcohol as a release for the *campesinado,* and judges appear to have indulged their stereotype of *el
indio ignorante y borracho.*[81] Yet these same judges—usually planters or their clients—also appreciated that if every infractor (drunk
or in full possession of his faculties) received a "just" sentence, the
jails would have overflowed and few *brazos* would have been left
to work in the henequen fields.[82]

In rare instances, we are presented with prima facie cases of
resistance—episodes of rustling, theft, and sabotage—in which
peons have been apprehended in the act. Arson, for example,
became a formidable mode of resistance within the henequen zone,
looming as a constant threat to production during the *auge.* In 1908,
Santiago May, a Maya *jornalero,* was caught, matches in hand,
torching 120 *mecates* of henequen (valued at 3,000 pesos) on his
hacienda, San Pedro, in Izamal *partido.*[83] Although the peon
claimed he had been drinking cane liquor, the judge, relying on the
testimony of numerous eyewitnesses, concluded that May had
acted with premeditation and never lóst control of his faculties. He
was convicted and sentenced to six years in prison, an extraordinarily harsh sentence considering that perpetrators of homicide
often received more lenient sentences in Porfirian Yucatán. The
harshness of May's sentence (and of other verdicts meted out in
cases involving the destruction of *planteles* and other hacienda
assets) can only be interpreted as a clear signal to peons by the oligarchical state that crimes against property constituted the gravest
of offenses and would not be tolerated.[84] Moreover, arson represented one of the few instances where state authorities felt a compelling need to meddle in the internal affairs of the hacienda. A
premeditated fire set in a henequen *plantel* could wipe out the life
savings of a Yucatecan *henequenero* in a matter of hours.[85]

Notwithstanding the clear-cut verdict in the May case, judges
were often hampered in ruling on arson cases by considerations of
the region's tropical climate. Accidental fires were commonplace in
the zone during the long dry season that ran from October to April,
particularly during the last two parched months of March and
April. It was during these months that *campesinos* in nearby vil-

lages, employing swidden techniques, set their *milpa* ablaze to clear land for the upcoming planting season. Occasionally, a change in wind direction blew sparks into neighboring henequen *planteles*. Once a fire was ignited, there was little that *encargados* could do to stop the conflagration, since fire-fighting capabilities were so limited in the arid northwest. One of the largest fires during the late Porfiriato took place on Hacienda Temozón, some 40 kilometers from Mérida, during the hottest part of the dry season, late in April 1908. Thirty-one thousand seven hundred *mecates* of prime henequen, valued at more than 400,000 pesos, went up in smoke.[86] Although Porfirian judges regularly ordered *diliquencias* (judicial investigations) following destruction of *planteles*, given the nature of the ecology, they were hard pressed to find culprits, let alone assess guilt.[87]

Clearly, considerations such as the peon's mental state and the region's ecology, as well as the representativeness and bias of the official documents, render conclusions regarding the incidence of "conscious resistance" problematical. James Scott argues that the social historian is unlikely to completely understand a *campesino*'s motivation. Everyday resistance depends for its effectiveness and safety on secrecy and the appearance of conformity; moreover, Scott contends, intentions may be so embedded in the rural subculture and in the day-to-day struggle for security of subsistence as to remain "inarticulate." Ultimately, in the absence of definitive evidence, the historian must assess the local setting and infer intention or motivation from the acts themselves. Where the material interests of the dominant classes are directly in conflict with those of peasants or peons (e.g., regarding access to land and subsistence or wages and conditions of work), Scott suggests that acts of rustling, theft, and sabotage may often be presumed to be resistance.[88] He would likely view the courtroom defenses that Yucatecan peons advanced—impaired mental state, accidents of the environment—as convenient alibis that masked clandestine acts of class struggle.

Scott may well overstate the case for a consistent strategy of everyday resistance, particularly where acts of *incendio* (fire) are concerned. If resistance is to be inferred from the social context, the historian should be prepared to make a compelling, if not gilt-edged case.[89] In any event, our data for Yucatecan henequen estates are sufficient to conclude that *peones acasillados* challenged the monocultural regime through a variety of forms, which escalated in terms of the damages they inflicted and the risk they entailed.

No doubt the most poignant, negative, and "total" response to exploitation was suicide. A frightening number of self-inflicted deaths—many induced by pellagra, the vitamin B deficiency that produces scaling, itching, and in its advanced stages, mental disorders—are documented in the judicial records. When coupled with the alcoholism that was endemic in the henequen zone, pellagra often had tragic consequences. That so many peons hung themselves from trees or house beams or threw themselves into wells testifies to their poor diet, inadequate medical care, and feelings of desperation—conditions that became increasingly acute during the final years of the Porfiriato.[90]

The judicial investigation of José María Eb's death provides a grisly account of the pernicious effects of pellagra and alcoholism, and points to the deterioration of paternalism and security on late Porfirian estates. Eb was a seventy-year-old Maya peon on Hacienda San José, a *finca* located in the rich henequen district midway between Cacalchén and Motul. He was discovered completely buried by *pencas* in a henequen *plantel* by his fellow workers, Isidro May and Toribio Escobarrubias. Eb's fellow *ancianos* were led to his body by a concentration of buzzards hovering above the *plantel*. What they encountered was almost too horrible to describe: the buzzards had eaten the body almost beyond recognition; identification was made possible only by Eb's characteristic clothing and *sabucán* (sack). The medical examiner found the corpse covered with the telltale lesions of an advanced case of pellagra. May and Escobarrubias testified that, owing to the disease and heavy bouts of drinking, Eb had periodically fallen down and lost consciousness during the previous few weeks. Occasionally he had even forgotten where he was and had to be led to his work in the fields.[91]

While Eb's case was not ruled a suicide, but rather ascribed to acute *conquestión alcohólica*, the mental disorders that frequently accompanied the nexus of pellagra and alcoholism often drove *jornaleros* to take their lives. Significantly, Maya peons often adhered to traditional ritual and custom, hanging themselves from trees or house beams.[92] Pre-Columbian Maya religious beliefs held that suicide victims, unlike most of the dead, who faced an arduous journey to the underworld, were rewarded with a direct trip to paradise. In effect, they would be assigned to a special heaven, where they, along with other unfortunates, such as unweaned children, would rest forever from labor and "[pass] a life of happy leisure with all imaginable delights beneath the shade of a giant ceiba tree (the sacred *yaxché*)."[93] Indeed, suicides were taken to this heaven by the Maya goddess *Ix Tab*—alternatively translated as "She of the

Cord'' and "Goddess of the Hanged," who is depicted in the Dresden Codex as hanging from the sky by a rope looped around her neck; "her eyes are closed in death, and a black circle, representing decomposition, appears on her cheek."[94] In *Relacion de las cosas de Yucatán*, Bishop Landa speculated on the motives behind such ultimate acts.

> They also said and held it as absolutely certain that those who hanged themselves went to this heaven of theirs; and thus there were many who on slight occasions of sorrows, troubles or sicknesses, hanged themselves in order to escape these things and to go to rest in their heaven, where they said that the Goddess of the Gallows, whom they call Ixtab, came to fetch them.[95]

Although archaeologists are unclear as to why the Maya accorded such special distinction to suicide victims, the fact that the cultural belief likely lingered into the twentieth century confirms our argument that the plantocracy and the Church never fully succeeded in ideologically recasting the acculturated northwestern Maya in their image. The Maya practice and perception of suicide by hanging suggest that other indigenous traditions and beliefs may have underwritten daily acts of resistance, with the Maya reshaping other cultural forms of the dominant society and investing them with a different, empowering significance.[96]

If, in many respects, the "quieter" forms of resistance that were utilized by *acasillados* were similar to those employed by slaves throughout the Americas, the cultural resources that the Maya had at their disposal made such strategies more effective. Collectively, they represented more than an escape from the daily indignities of an exploitative system: with the notable exception of suicide (which had its own cultural rationale), such forms reveal a tenacity to persevere and, when opportunities presented themselves, to contest the demands of *henequeneros* without risking open (and unequal) confrontation. Realistically speaking, until the final years of the Porfiriato, these "small arms" were the only weapons in the peons' arsenal, and permitted them to assert their own interests within a brutal regime, even if this only meant "working the system to their minimum disadvantage."

After 1907, however, new historical conjunctures facilitated more dynamic patterns of *acasillado* resistance. The consequences that the worldwide financial crisis of 1907–1908 had for Mexico—and particularly its role in eroding the Porfirian regime—have been the subject of recent debate, but there can be little question that the international panic had a profound impact at all levels of Yucatán's dependent monocrop society.[97] With fiber quotations plummeting in 1908 (and generally declining thereafter until 1912), and with credit in short supply, a chain reaction of commercial failures and bankruptcies swept the peninsula, claiming many small and medium *henequeneros*.

Their backs to the walls, the straitened planters passed the burden downward, reducing wages, restricting initial advances to peons as well as continuing credit at the *tiendas de raya*, and slashing traditional paternalist benefits and incentives. At the same time, *henequeneros* stepped up already coerced rhythms of production. Heightened labor demands left peons little time or energy to tend small traditional *milpa* plots—in those increasingly rare instances where estates had not already converted them to monoculture. In addition to the extra labor that they extracted from the *jornaleros*, according to recent research, planters may now have begun to directly appropriate the labor of *sirvientes'* family members. Peons had long augmented their household incomes through the extra *jornal* (wage) of adolescent sons and the domestic labor provided by wives and daughters. Now, to the extent that they directly tapped the labor power of household dependents, particularly the unpaid labor of wives and daughters, the *henequeneros* deprived the *acasillado* and his family of a strategic resource in their continuing struggle to maintain a security of subsistence.[98]

Increasingly after 1907, with paternalism but a faint echo and with labor demands intensifying as access to *milpa* and complementary household income dwindled, resident peons could only reflect that their calculated exchange of autonomy for security no longer provided the assurance of continued subsistence. Periodic droughts and almost annual locust plagues between 1907 and 1911 only further eroded their horizon of expectations. It is hardly surprising, therefore, that violent confrontation, previously the last and most dangerous resort, became a more frequent response after 1907.

One need not espouse a simplistic "volcanic" theory of collective violence to account for the local riots and jacqueries that occurred in the henequen zone between 1907 and 1911, or for the broader waves of rural revolt that swept the state in 1911 and 1912, enlisting substantial *acasillados'* support. Such volcanic or J-Curve models suggest that worsening economic conditions generate relative deprivation and discontent, which eventually trigger mass rage and collective action.[99] To be sure, where the peons were concerned, deteriorating, increasingly "involuted" conditions of labor and economic well-being lowered the threshold of the peon's tolerance toward his oppression and likely generated multiple local—albeit isolated—encounters after 1907. But, as we will see, worsening economic conditions never triggered a more generalized tide of protest and violence in the state, let alone gave rise to a mass movement that toppled the oligarchical order.

The regional press reported a variety of violent disputes over wage cuts, deteriorating labor conditions, and physical abuse in the face of declining fiber prices during the 1907–1911 period. For example, in 1909, an *encargado* cut the wages of four platform drivers who refused to work on Sunday on Hacienda Eknakán, owned by Luisa Hubbe de Molina, sister-in-law of Olegario Molina, Yucatán's richest planter and leader of the dominant oligarchical faction. When violence erupted, the national guard was sent in to restore order. The captain of the *Guardia Nacional* was wounded by the peons and forced to withdraw. One *sirviente* was killed and eight others injured in the outburst. Reinforcements from Mérida had to be rushed in to quell the disturbance.[100]

Episodes were also reported in which *jornaleros* protested abusive treatment and punishment on the estates. Late in 1907, 110 *peones* marched from Hacienda Oxcúm, owned by Don Olegario's son-in-law, Avelino Montes, to the *palacio municipal* of nearby Umán to protest the imprisonment of three co-workers. National guard troops were dispatched from Mérida, 6 kilometers away. Undaunted by the show of force, six Oxcúm peons later assaulted the estate's hated overseer. The national guard was called in again, this time to arrest the six offenders.[101] Several years later, in March 1911, a notoriously brutal labor regime on the Hacienda Catmis, a more isolated southern estate near Peto, provided the spark for perhaps the most violent episode of *acasillado* protest under the ancien régime. Exploding in cathartic rage, peons destroyed machinery and carved up the *hacendado* and members of his family and staff.[102]

In each case, violence came in response to specific local causes, and was isolated and quickly repressed by state authorities. As we have noted, such riots and jacqueries had occurred from time to time throughout the Porfirian *auge*. Between 1907 and 1911, however, their incidence appears to have steadily increased. Indeed, by 1911, both within the henequen zone and on its less controllable fringes, such *motines* and a sharp increase in related acts of rustling and theft were becoming principal forms of resistance among the Maya *campesinado*.[103]

Mobilization and Demobilization in the Henequen Zone

Jornaleros were not alone in their disaffection. Despite the fabulous wealth generated by the henequen boom, the first decade of
this century was a veritable "summer of discontent" for the vast
majority of regional producers, merchants, urban workers, and free
villagers who found themselves subordinated in one way or another
to the dominant oligarchical *camarilla* based on the *parentesco* of
Olegario Molina and Avelino Montes. This ruling faction—or "di-
vine caste," as it was called and came to call itself early in the century—had homogeneous interests, a relatively closed membership,
and owing in part to its collaboration with the principal buyer of
raw fiber, the International Harvester Company, such control over
the economic and political levers of power in the region that it was
able to thwart the opportunities of rival planter groups in late Porfirian society.[104]

By late 1909, when Molinista repression foreclosed the electoral
road to them, these frustrated planters, now organized into two
rival *camarillas*, perceived rebellion as the only means to force a
more equitable reappointment of the spoils of henequen monoculture. Just as Madero's national coalition would topple Díaz's oligarchy, the *científicos*, so these dissident planter factions—popularly
known as "Morenistas" and "Pinistas," after their respective stan-
dard-bearers, Delio Moreno Cantón and José María Pino Suárez—
now hoped to break the stranglehold of Molina's *casta divina*. Each
elite *camarilla* rapidly attempted to construct a loose coalition
reaching into the middle-class intelligentsia, the small urban working and artisan class, and—perhaps most important, but until now
not fully explained—into the Maya *campesinado*. In short order,
Yucatán's bickering summer of discontent festered into several violent seasons of upheaval that shook the oligarchical order between
1910 and 1913.[105] Although they never played a leading or autonomous role in these cycles of unrest, the *peones acasillados* of the
henequen zone often made up a significant component of the rebel
bands.

Of course, the judicial records permit us to speak with greater
confidence about the character of the mobilizations than about the
motivations of the peons who joined or refused to join them.
Indeed, many students of social movements wonder whether we
can ever really accurately—let alone retrospectively, with incomplete data—determine individual motivations. Particularly within
the context of riots and rebellions, the insurgents may not even be
conscious at the moment they join a band of what motivates them.

One Yucatecan peon, Marcos Chan, tersely remarked at his trial: "They asked me if I wanted to join them and I said yes."[106] How can we begin to know what went through his mind? How can we know if he would have acted differently a day or a week later if presented with the same choice? So subjective do some "structuralists" find the exercise of assessing motivation that they completely discourage asking the question "why" people acted and seek only to understand "how" they acted.[107]

Certainly these critics raise a valid point. A careful reading of the court records suggests that individual peons may have joined or refused to join insurgent bands for a plethora of conscious (often interlinked) motivations: economic gain, familial attachments, and an urge for revenge among them. But beyond these surface motivations there were, no doubt, other unconscious, psychologically based factors that likely entered into individual behavior choices. For example, psychologists have documented the collective lowering of thresholds of inhibition in mobs and other crowd phenomena: in fact, some episodes of Yucatecan insurgency resembled public fiestas, in which entire groups of people accompanied by the community band defected en masse.[108] And what role did gender issues play in motivation? In certain cases we found mothers and wives egging on their male relations, in effect challenging the *machismo* of their men: "C'mon, why don't you kill that *cabrón* now that you have the chance; you can bet he wouldn't go soft with you!"[109]

Certainly a variety of conscious and unconscious motivations and variables, as well as numerous other contingencies, come into play when we are called to ponder the question of why individuals participate in riots and rebellions (indeed, in any social action). But ultimately, in grappling with these acts of resistance and rebellion, we feel compelled to attempt a general explanation of why they took place and why peons decided to join them—to offer at least a proximate cause, to pass through the eye of the needle, as it were.

In order to do this, we are obliged to look beyond the individual insurgent's own belief about his actions; rather we must read this belief against the structural considerations that affected the individual as a member of a group (or groups) and as part of a larger social formation. Effectively, this means that the full range of "external" power relations must be considered, in addition to people's own "internal" perceptions of their conditions and behavior.[110] The dynamic relations of domination within the henequen zone during the final years of the Porfiriato have been analyzed in some detail. We have suggested that increasingly miserable condi-

tions of life and a feeling among *sirviente* that the estate might no
longer be able to guarantee the security for which they had traded
their autonomy likely persuaded many peons to participate in the
insurgency of the period. Similarly (although it was not our central
concern here), we have touched upon the severe threat that the
expansion of the fiber estate posed to the existence of the poor but
free peasant villages within the henequen zone and on its less controllable fringes. This threat, coupled with deteriorating economic
conditions and widening political space after 1910, suggests why so
many of the *cabecillas* (chiefs) and recruits of these revolts came
from these embattled, but more tactically mobile villages and small
towns, particularly those on the zone's periphery.[111]

The British social historian E. P. Thompson provides what might
well be the best guidepost for the challenging task of penetrating
the *mentalidad* of peons (or villagers) during episodes of insurgency:

> The consciousness of a worker is not a curve that rises and falls
> with wages and prices; it is an accumulation of a lifetime of expe-
> rience and socialization, inherited traditions, struggles successful
> and defeated. . . . It is this weighty baggage that goes into the
> making of a worker's consciousness and provides the basis for his
> behavior when conditions ripen . . . and the moment comes.[112]

Thompson's insight may be profitably read against the evidence
of the period. Agitators like Tomás Pérez Ponce and the *cabecillas*
who followed generally received an ambiguous reception when
they arrived on henequen estates seeking adherents or recruits
between 1909 and 1913. Bad as conditions had become, many
peons still eschewed a strategy of confrontation. Likely, they
believed that, as in the past, such actions were doomed to failure,
and that the spoils that might temporarily be won were not worth
the loss of the modicum of security that the estate still provided—
not to mention the potential loss of life and limb. Moreover, not all
henequeneros had abandoned paternal incentives; certainly, conditions varied from estate to estate. No doubt, many peons favored a
strategy of extracting what security they could and resisting the
demands of monoculture in more "routine," less risky ways.[113]

Still, many *sirvientes* took the risk and joined popular insurgencies. With much of the social science literature on rural revolt, we
would concur that however aggrieved or "motivated," *campesinos*
generally wait for evidence that power-holders are weak and/or
divided before they will take the risks attending insurrection. News
of such opportunities for revolt is often brought to peasants by out-

side agents—typically dissident elites—or by "hingemen," local rural brokers with some cultural experience in the dominant society that complements, indeed often enhances, their standing in the subordinate rural society.[114] In the case of Yucatán, independent plot-holders, petty traders, bandits, artisans, teachers, and former militia officers frequently became *cabecillas* and hingemen, linking dissident planter-led parties and local concentrations of peons and villagers. Typically Morenista and Pinista planters and middle-class intellectuals, based in Mérida, would plan a revolt—frequently timing their regional *complot* to coincide with a national-level conspiracy—and then, through an extended network of middlemen, including local contacts and spies known popularly as *"orejas,"* would mobilize sympathetic elements (and often coerce less than sympathetic ones) in rural towns, villages, and haciendas.[115]

During the round of elite plotting and intrigue that took place on the eve of the Madero rebellion, one Morenista leader declared: "I have more faith in the people in the countryside than I do in the Meridanos."[116] Throughout 1910 and early 1911, the tenuous alliance between dissident elite factions in the cities and influential rural brokers in the interior continued to grow, as the elites secured arms and cash, and the local *cabecillas* recruited in their *pueblos* and on neighboring estates. Yet the Morenistas and the Pinistas soon came to reconsider the wisdom of their mobilization of villagers and peons. By the spring of 1911, the latest round of local riots and revolts began to spin out of control.

What the elites did not fully consider as they constructed these rudimentary insurgent networks was that the incipient rural rebels also had their own agendas, which were not congruent with the elite's own, rather limited, political projects. Gradually, from the aborted Candelaria *conjura* in October 1909, through the failed rebellion in Valladolid during the late spring of 1910, to the more freewheeling revolts that periodically rocked the state during 1911 and 1912, locally based popular mobilization and protest had begun to evolve a life of its own, one that took little heed of political posturings or fraudulent election returns. Yucatán's competing elites had opened a Pandora's box, and try as they might they could never successfully harness the rage that exploded in peripheral areas like the Puuc[117] and the eastern *partido* of Temax.

Here on the fringes of monoculture throughout 1911 and 1912, haciendas were overrun by marauding bands who "liberated" peons and property alike—occasionally from the very Morenista and Pinista elites who had initially fomented the mobilization. On some estates, jacqueries erupted from within. In a variety of *cabe-*

ceras, rebels dynamited the houses and stores of local notables, attacked the *cuarteles* of national guard detachments, and summarily "brought to justice" abusive *jefes politicos,* municipal authorities, and hacienda personnel.[118] They held Halachó, a good-sized *cabecera,* in the Puuc, for two days and began naming new municipal authorities.[119] Occasionally popular *cabecilla*-led bands, joined by local peons, raided *casas principales,* then smashed *desfibradoras* and tore up stretches of Decauville tram tracks in the best Luddite fashion. After years of exploitation and racial degradation, Maya *sirvientes* suddenly found themselves enthusiastically discussing their actions in the *tienditas* and at Saturday night *bailes populares:* "I lit the dynamite that blew up the *caldera,*" offered Fulano; "I knocked down the *albarradas* (stone markers) around the new *plantel,*" commented Mengano; "Imagine," interjected Zutano, "All these fine clothes were paid for with the loot he extracted from the *costillas* (ribs) of our *pueblo.*"[120] At various junctures in 1911 and 1912, such popular insurgency on the henequen zone's periphery threatened to spill over and paralyze production in the heart of the zone.[121]

Thus, much in the manner of the great nineteenth-century peasant rebellion known as the *Guerra de Castas,* the mass participation of subaltern groups in these Maderista-era rebellions began to infuse the struggle with a local resistance to elite domination that became cause for concern among the elites who had initially precipitated these mobilizations. Once again on the nation's far periphery, Maya villagers and peons had begun to turn elite feuds to their own advantage: fleeing estates, targeting the property and persons of abusive individuals and, in certain noteworthy cases, building clienteles that would later be consolidated into *cacicazqos* when the Revolution gained a foothold in Yucatán after 1915. It was this structural dichotomy between popular and elite politics and mentalités, manifest in these Maderista-era seasons of upheaval (as it had been during the Caste War), that constituted the most dramatic challenge to the hegemony of monoculture prior to the arrival of General Alvarado's revolutionary army.[122]

Ultimately, several structural factors, as well as short-term emergency strategies by *henequeneros* and the state, explain why in Yucatán political conflict and popular insurgency stopped short of the generalized rebellion that occurred in many other parts of the republic.[123] First of all, in Yucatán the old order had certain built-in advantages that permitted it to contain festering popular discontent and pull itself back from the brink. The peninsula's remote location—there were no roads connecting with central Mexico until well after World War II—impeded communications with revolutionary chiefs in Mexico's core and in the north, and made coordination of joint campaigns virtually impossible. Second, we have seen that henequen monoculture's coercive and highly regulated system of domination, maintained both by landowners and the state, worked to impede collaboration between villagers and peons and to keep local outbreaks isolated.

Moreover, the "social memory" or *mentalidad* of the *henequeneros* class might be viewed as another structural factor. The *henequeneros'* obsession with the haunting specter of another Caste War gave them second thoughts about a full-scale mobilization of railway and dock workers, let alone Maya villagers and peons. Although Morenista and Pinista planters itched to defeat the Molinista *camarilla*, the majority feared that arming the rural masses would undermine the elaborate mechanisms of social control that had so successfully underwritten the *auge*. Now, if *hacendado*-led mobilizations ignited a social revolution, Yucatán's contending elites might well lose their properties, their social world, and even their lives in another *Guerra de Castas*. That certain elites would take such a chance and arm *campesinos* throughout the state, demonstrates the divisiveness of the dominant class in late Porfirian Yucatán, as well as the sense of desperation of some *henequeneros*.

Nevertheless, even with the structural obstacles arrayed against it, popular insurgency was reaching dangerous new levels late in 1912. This obliged the *henequeneros*, at least in the short run, to make some concessions to peons, or, as Knight puts it for Mexico as a whole, "to wheedle and promise, as well as to repress."[124] Like earlier slave rebellions in the Caribbean or the U.S. South, Yucatán's popular seasons of upheaval elicited the drafting of a reform agenda by progressive planters (the *liga de Acción Social*) and actual material concessions on some estates, even as they provoked more severe measures of control on others.[125] In general, we have found that after 1913, the local courts, which were still planter controlled, were more responsive to addressing (and occasionally even redressing) the most egregious abuses against peons.[126]

The state played a complementary role in demobilizing bands of rampaging *campesinos*. General Victoriano Huerta's imposition of military rule early in 1913 institutionalized a political stalemate among Yucatán's three contending elite factions, in effect forcing upon them an *entente* that preserved the social peace. With the issue of state power resolved, at least temporarily, justice was meted out alternately with Porfirian shrewdness and verve. The executive and judiciary amnestied and coopted certain influential *cabecillas* and led others to the *paredón*. By mid-1913, the countryside had been demobilized.

Nevertheless, the passage of a state decree abolishing debt peonage a year later attests to how tenuous the social peace really was in Yucatán. The decree was never implemented and seems to have been issued only expediently, to buy time in the wake of a new burst of riots and violence preceding the fall of Huerta. In fact, the plantocracy was at pains to maintain calm in the countryside right up until its hegemony was shattered when General Alvarado's 8,000-man Constitutionalist army invaded the peninsula in March of 1915.

The reform-minded Alvarado, in turn, quickly came to appreciate that the *campesinado* had been changed by its participation in the recent seasons of upheaval. The general's newly installed military tribunal received waves of petitions from hacienda peons demanding that their *patrones* raise their salaries and improve their working conditions. In one colorful instance, the rendering of a positive decision by Alvarado's court was not enough to satisfy the leader of a delegation of peons. He continued to rail against the haughtiness and cruelty of his *mayordomo* until he was found in contempt and forcibly removed from the tribunal. Indeed, Alvarado's presiding *comandante* wrote the general that the man was being detained until he cooled off, in order that "such incendiary behavior not promote class antagonisms harmful to production and the spirit of work"—the guiding principles of Alvarado's progressive, modernizing regime.[127]

While such assertive behavior may have caught the authoritarian *caudillo* off guard, the planters had watched it brew for some years now. The judicial records and regional press between 1910 and 1915 reveal a variety of complaints by overseers and masters that their *sirvientes* no longer doffed their hats or kissed their hands.[128] Tomás Pérez Ponce's warning to the planters had been a prophetic one: in spite of all the elaborate precautions that they had taken over the course of the Porfirian *auge* to subordinate and control their *sirvientes, henequeneros* discovered painfully during the Madero period just how "concientes" their peons were. While free villagers provided the leadership and first ranks of the popular movements in Yucatán, many resident peons took advantage of the *apertura* accompanying the fall of Díaz to violently contest the monocultural regime. Their participation in these short-lived revolts represented an appropriate political response at a dynamic and strategic conjuncture. Whereas northwestern *peones acasilla-*

dos had usually chosen "quieter," "everyday" acts of resistance to contest what had previously appeared to be an unshakable structure of domination, many now took greater risks to deal a more telling blow against the social regime that bound and exploited them. By 1913, in remote, oligarchical Yucatán, as in other parts of Mexico, traditional modes of behavior and deference were giving away to a new assertiveness and empowerment—to what Knight has called "a new plebeian insolence."[129] Yucatán's popular seasons of upheaval failed, but in important respects they laid the groundwork for more extensive mobilizations under subsequent, radical revolutionary regimes.[130]

Abbreviations

AGEY   Archivo General del Estado de Yucatán, Mérida.

CGPD   Colección General Porfirio Díaz, Mexico City.

Notes

We gratefully acknowledge support by the National Endowment for the Humanities, the Center for U.S.-Mexican Studies (University of California, San Diego), the American Philosophical Society, the University Research Committee of Appalachian State University, and the Institute of Latin American Studies of the University of North Carolina, Chapel Hill, in the research and writing of this chapter. An earlier draft of this chapter was published as "El monocultivo henequenero y sus contradicciones: Estructura de dominación y formas de resistencia en haciendas yucatecas a fines del Porfiriato," *Siglo XIX* 3:6 (July–December 1988), 215–277. We are grateful to the editor of that journal for permission to publish the present version.

1. *El Ciudadano*, 10 June 1911, no. 34.

2. We should clarify the several terms we use throughout this essay to characterize rural workers in Yucatán during the Porfiriato. *Jornalero de campo* is a generic term for an agricultural worker, and we use it freely to embrace henequen workers, tenants, and part-time workers (specifically called *luneros*, or Monday men in Yucatán). *Peones acasillados* or *sirvientes* are permanent resident peons tied to the henequen estate through the mechanism of debt. *Campesinos* or *comuneros*, on the other hand, are "free" peasants who may or may not own land, but live in village communities and are not tied by debt to neighboring haciendas.

3. Roughly within a radius of 70 to 80 kilometers of Mérida.

4. *El Ciudadano*, 10 June 1911, no. 34.

5. The Morenistas, who enjoyed political power during the gubernatorial *cuatrenio* of General Francisco Cantón (1897–1901), had waited impatiently for over a decade to retake the state house. Porfirio Díaz had selected Olegario Molina and his *camarilla* (the Molinistas) to govern the

state in uninterrupted fashion from 1902 to 1911; later, when the Morenistas were again excluded by the imposition of a new Maderista governor, José María Pino Suárez in 1911, they rebelled. For the origins of the Morenista discontent during the last years of the Porfiriato, see Gilbert M. Joseph and Allen Wells, "Yucatán: Elite politics and rural insurgency," in Thomas Benjamin and Mark Wasserman (eds.), *Provinces of the Revolution: Essays on regional Mexican history, 1910–1929* (Albuquerque, 1990), 93–131.

6. This chapter represents but a portion of an ongoing book-length study of politics and society in Yucatán, entitled *Summer of Discontent, Seasons of Upheaval: Elite politics and rural rebellion in Yucatán, 1890–1915.*

7. Eric Wolf, *Peasant Wars of the Twentieth Century* (New York, 1969), Introduction. John Tutino writes more explicitly: "In places where the most radical economic changes of the Porfiriato occurred, where established peasant communities were suddenly incorporated into the export economy as export producers, there was little revolutionary insurrection after 1910." John Tutino, *From Insurrection to Revolution in Mexico: Social bases of agrarian violence, 1750–1940* (Princeton, 1986), 296.

8. A recent portrayal of the Yucatecan *peón acasillado* as quiescent is found in Alan Knight, *The Mexican Revolution* (Cambridge, England, 1986), I:89. For a general discussion of the literature during the Porfiriato, see Allen Wells, *Yucatán's Gilded Age: Haciendas, henequen and International Harvester, 1860–1915* (Albuquerque, 1985), ch. 6.

9. This point is developed more fully in Allen Wells, "From Hacienda to Plantation: The transformation of Santo Domingo Xcuyum," in Jeffrey Brannon and Gilbert M. Joseph (eds.), *Land, Labor and Capital in Modern Yucatán: Essays in regional history and political economy* (Tuscaloosa/London, 1991), 112–142, 269–275. On the transformation of the Caribbean sugar industry, see Manuel Moreno Fraginals, "Plantations in the Caribbean: Cuba, Puerto Rico, and the Dominican Republic in the late nineteenth century," in Manuel Moreno Fraginals, Frank Moya Pons, and Stanley L. Engermann (eds.), *Between Slavery and Free Labor: The Spanish-speaking Caribbean in the nineteenth century* (Baltimore, 1985), 3–21; Christian Schnakenbourg, "From the Sugar Estate to Central Factory: The Industrial Revolution in the Caribbean (1840–1905), in Bill Albert and Adrian Graves (eds.), *Crisis and Change in the International Sugar Economy, 1860–1914* (Norwich/Edinburgh, 1984), 83–93.

10. In a seminal article written over three decades ago, Eric Wolf and Sidney Mintz attempted rigorously to define haciendas and plantations. A "hacienda" is "an agricultural estate operated by a dominant landowner . . . and a dependent labor force, organized to supply a small-scale market by means of scarce capital in which the factors of production are employed not only for capital accumulation but also to support the status aspirations of the owner." By contrast, a "plantation" is "an agricultural estate operated by dominant owners (usually organized into a corporation) and a dependent labor force organized to supply a large-scale market by means of abundant capital, in which the factors of production are employed primarily to further capital accumulation without reference to the status needs of the owners." Although this typology of ideal types simplifies complex institutions, it points up key differences in property ownership, capi-

talization, markets, and the desire for conspicuous consumption by land-owners. See Wolf and Mintz, "Haciendas and Plantations in Middle America and the Antilles," *Social and Economic Studies*, 6 (1957), 380–412.

11. Wolf and Mintz, "Haciendas and Plantations," 390.

12. Wolf and Mintz, "Haciendas and Plantations," 401.

13. Georg Hegel's penetrating insight in *The Phenomenology of the Mind* is particularly appropriate here: personal power taken to an extreme ultimately contradicts itself, "for total domination can become a form of extreme dependence on the object of one's power, and total powerlessness can become the secret path to control of the subject that attempts to exercise such power." Hegel cited by Orlando Patterson in *Slavery and Social Death: A comparative study* (Cambridge, Mass., 1982), 2.

14. Philip Mason, *Patterns of Dominance* (London, 1971), 11.

15. Mason, *Patterns of Dominance*, 11.

16. The following discussion of idioms of power draws heavily on Patterson, *Slavery and Social Death*, ch. 1.

17. Karl Marx, *Capital*, I:77, cited in Patterson, *Slavery and Social Death*, 19.

18. Designed by *henequeneros* to limit the workers' mobility and autonomy, the three mechanisms were often so mutually reinforcing that it is sometimes difficult to delineate where one begins and the other leaves off. Institutions like the hacienda store (*tienda de raya*), for example, served many functions. On one level, the *tienda* gave *henequeneros* a sure-fire mechanism for raising workers' debts (coercion). Yet, on another level, by providing basic foodstuffs and household needs, it diminished the need for resident peons to leave the property to purchase goods, thereby minimizing the chances of potentially disruptive contact between *peones acasillados* and neighboring villagers and agitators (isolation). Finally, through the sale of corn, beans, and other staples, it ensured subsistence for resident peons (security). In sum, the *tienda de raya* was a perfect vehicle for appropriating labor in a scarce market, since it facilitated dependency and immobility, while conveying a measure of convenience and security to landless peons. For heuristic purposes, however, we will break down these interrelated mechanisms, to clarify how the idiom of power was conveyed to the *jornaleros* and villagers who lived in the henequen zone. Their strategic convergence, however, must always be kept in mind.

19. Wells, *Yucatán's Gilded Age*, 161–162.

20. Moisés González Navarro provides several examples of blatant land-grabbing by government officials and *hacendados* who used vacant land (*terrenos baldíos*) laws to usurp lands from neighboring *pequeños propietarios* and *ejidos*. Even Secretary for *Fomento* Olegario Molina was accused of fraudulently acquiring 2,179 hectares of land in Tizimín *partido*. González Navarro, *La vida social*, IV:193, in Daniel Cosío Villegas (ed.), *Historia Moderna de Mexico*, 10 volumes (México, 1956–1972).

21. Antonio Betancourt Pérez, *Revoluciones y crisis en la economía de Yucatán* (Mérida, 1953), 52.

22. Víctor Suárez Molina, *La evolución económica* (Mérida, 1977), I:160.

23. For a comprehensive examination of Yucatán's ecology, see Eugene M. Wilson, "Physical Geography of the Yucatán Peninsula," in Edward H.

Moseley and Edward D. Terry (eds.), *Yucatán: A world apart* (Tuscaloosa, 1980), 5–40.

24. For a discussion of henequen's poor soil quality and low corn yields, see Malcolm K. Shuman, "The Town Where Luck Fell: The economics of life in a henequen zone pueblo," Ph.D. dissertation, Tulane University (New Orleans, 1974), ch. 5.

25. If Maya *jornaleros* were relatively easy to find, think how simple it was to apprehend runaway contract workers. González Navarro cites the case of a Huastecan *enganchado* who had been enticed to leave his native San Luis Potosí and come to Yucatán with his family by promises of high wages by unscrupulous agents. When he discovered that his daily wage was 48 centavos a day instead of the two pesos originally promised by the agents, he tried to return home and was captured on two separate occasions by the authorities and returned to his estate. On his third try, the persistent *indio* successfully escaped but had to leave his family behind in Yucatán. González Navarro, *La vida social*, IV:225.

26. Nancy Farriss makes a strong case for the migratory propensity of the colonial Yucatec Maya in *Maya Society under Colonial Rule: The collective enterprise of survival* (Princeton, 1984), esp. ch. 12.

27. For a discussion of the new scholarly work on the Caste War, *cruzob*, and *pacíficos* since the publication of Nelson Reed's *The Caste War of Yucatán* (Stanford, 1964), see Gilbert M. Joseph, *Rediscovering the Past at Mexico's Periphery: Essays on the history of modern Yucatán* (Tuscaloosa, 1986), chs. 2 and 3.

28. Correspondence between Díaz and Ignacio Bravo, the federal general he entrusted with the *cruzob* campaign, reveals that both were bent on fully eliminating the *cruzob*, who they perceived to be a pernicious race. Ignacio Bravo to Porfirio Díaz, 23 June 1905 (letter), Legajo 30, Caja 25 009756, CGPD. See also Marie Lapointe, *Los mayas rebeldes de Yucatán* (Zamora, Michoacán, 1983).

29. Wells, *Yucatán's Gilded Age*, 105, note 30.

30. Wells, *Yucatán's Gilded Age*, 159.

31. The mastermind of the partition of Quintana Roo was Manuel Sierra Méndez, a trusted *científico* and the brother of education minister Justo Sierra. Interestingly, Sierra Méndez, a native Yucatecan, argued strenuously (but ultimately in vain) that the partition should be temporary and that following the pacification of the *cruzob*, the territory should be reincorporated into Yucatán. See Sierra Méndez to Díaz, "Apuntes sobre la campaña de indios presentados en cróquis explicatorio de ellos," 12 July 1897 (memorandum), Legajo 22, Caja 17, CGPD. For a detailed discussion of the partition, see Wells, *Yucatán's Gilded Age*, ch. 4.

32. CGPD, passim.

33. Occasionally, Díaz would ask his hand-picked governors to look into specific complaints from individuals, but it is difficult to discern whether the problems themselves were ever addressed. CGPD, passim.

34. Wells, *Yucatán's Gilded Age*, 176–177.

35. Henry Baerlein, *Mexico: Land of unrest* (Philadelphia, 1914), 19–20, 182; Alan Knight, "Mexican Peonage: What was it and why was it?" *Journal of Latin American Studies*, 18:1 (1986), 61.

36. Méndez, quoted by Wells, *Yucatán's Gilded Age*, 178.

37. Sometimes whipping was not applied "in moderation." Pedro Chin, a *jornalero* on San Antonio, died as a result of *flagelación* by the *mayordomo* of the hacienda. González Navarro, *La vida social*, IV:225–226.

38. Quoted in Knight, "Mexican Peonage," 61.

39. Ramo de Justicia, "Flagelación a Julio Pérez," 1912, AGEY.

40. Wells, *Yucatán's Gilded Age*, 157.

41. Knight, "Mexican Peonage," 51.

42. Ramo de Justicia, "Incidente promovido por el apoderado de D. Dolores Guerra de Mendoza pidiendo la suspensión del remate de la hacienda Xcuyum," 1895, AGEY; quotation from *El Crédito del Estado*, 15 July 1895, no. 12; and Wells, *Yucatán's Gilded Age*, 157.

43. A recent study of peonage in Puebla and Tlaxcala during the Porfiriato has noted that some *acasillados* actually sought to increase their indebtedness, because they believed that the size of the debt was indicative of social status on the hacienda. Herbert J. Nickel, *Peonaje e inmovilidad de los trabajadores agrícolas en México: La situación de los peones acasillados en las haciendas de Puebla y Tlaxcala*. Translated by Catalina Valdivieso de Acuña (Bayreuth, 1980), 54–55. While data for this for Yucatán are lacking, it is clear that many *acasillados* built up extremely high debts during the Porfiriato. Whether they considered the debt irrelevant and as a result borrowed as much as the estate management allowed, or if they, indeed, saw debts as a symbol of status within the hacienda community, is unclear.

44. Ramo de Justicia, "Acusación presentada por el C. Juan Bautista Chan contra el Juez Segundo de Paz de Tahmek," 1912, AGEY.

45. One point that was not addressed in the grievance brought by Chan was how a *jornalero* could pay such an enormous sum, since the daily wage in the *campo* at that time was approximately 1 peso a day. Without financial assistance from another quarter, it would have been difficult to raise that much money to liquidate a debt.

46. On payment of token or scrip, see Edwin C. Leslie and A. F. Pradeau, *Henequen Plantation Tokens of the Yucatán Peninsula* (Washington, D.C., 1972).

47. For an interesting discussion of debts and wages on henequen and maize haciendas in Yucatán during the *auge*, see Christopher J. Gill, "Regional Variations and Patterns of Resistance: The henequen boom in Yucatán, 1880–1915," Senior Honors Thesis, University of Michigan (Ann Arbor, 1987), ch. 3. Also see Wells, "From Hacienda to Plantation"; and Joseph, *Rediscovering the Past*, 59–81.

48. Ramo de Justicia, "Tercer cuento de administración de la finca San Pedro sujeta al juicio hipotecario promovido por la representación de Don Perfecto Doña Mercedes Irigoyen de Herrera," 1911, AGEY; Ramo de Justicia, "Segunda cuenta de administración que rinde el depositario de la finca sujeta al juicio que siguen los Sres. Avelino Montes S. en C. contra Marcos Díaz Cervera," 1911, AGEY.

49. Ramo de Justicia, "Cuenta de administración de la Compañía Agrícola del Cuyo y anexas, S.A.," 1910, AGEY.

50. The concept is discussed at length in Tutino, *From Insurrection to Revolution*.

51. Knight, "Mexican Peonage," 64.

52. For a discussion of the multiple roles played by *casas exportadoras* during the boom, see Joseph and Wells, "Corporate Control of a Monocrop Economy: International Harvester and Yucatán's henequen industry during the Porfiriato," *Latin American Research Review*, 17:1 (1982), 69–99.

53. Fernando Arjona, *Breves apuntes sobre la pelagra* (Mérida, 1989).

54. Knight, "Mexican Peonage," 64.

55. Although the predominantly Maya settlements of *acasillados* lacked even the attenuated organizational and cultural resources of the embattled village communities within the zone, as we will see, they were not entirely bereft of cultural resources and affines—most notably their families and religion.

56. This is also a classical illustration of the isolation mechanism at work. Alejandra García Quintanilla, personal communication, 20 November 1987.

57. Ramo de Justicia, "Diligencias en la averiguación del delito atentado contra la libertad de la menor Eleuteria Ek, denunciado por Micaela López," Caja 658, 1908, AGEY.

58. Farriss, *Maya Society under Colonial Rule*, 268–269.

59. E.g., see Alberto García Cantón, *Memorias de un ex-hacendado henequenero* (Mérida, 1965).

60. Rebecca J. Scott, *Slave Emancipation in Cuba: The transition to free labor, 1860–1899* (Princeton, 1985); Manuel Moreno Fraginals, *El ingenio: Complejo económico social cubano del azúcar*, 3 vols. (Havana, 1978); For reviews of the historiographical debate over Cuban working conditions, see Franklin W. Knight, "The Caribbean Sugar Industry and Slavery," *Latin American Research Review*, 18:2 (1983), 219–229; and Wells, "The Terrible Green Monster: Recent literature on sugar, coffee, and coerced labor in the Caribbean," *Latin American Research Review*, 23:2 (1988), 189–205.

61. Scott, *Slave Emancipation in Cuba*, 19.

62. González Navarro, *La vida social*, IV:529–530.

63. The social history of the Maya family has yet to be written. Our general observations for the late nineteenth and early twentieth century henequen zone are derived from a variety of sources, including the AGEY's Ramo de Justicia and the regional press, and are informed by the ethnohistorical (e.g., Farriss, *Maya Society under Colonial Rule*) and ethnographic literature.

64. Sir Moses Finley prefers to characterize slaves as "outsiders" rather than as "deracinated." *Ancient Slavery and Modern Ideology* (New York, 1980), ch. 2. Space limitations and incomplete data prevent us here from reconstructing patterns of resistance for the henequen zone's imported workers (Yaquis, Asians, central Mexican deportees, and *enganchados*). Non-Maya labor and resistance (and the difficulties attending its study) are discussed in Joseph, *Rediscovering the Past*, ch. 4 and esp. pp. 80–81.

65. From James Harrington's *A System of Politics*, quoted in Eugene D. Genovese, *From Rebellion to Revolution: Afro-American slaves in the making of the modern world* (Baton Rouge, 1979), vi.

66. James C. Scott, *Weapons of the Weak: Everyday forms of peasant resistance* (New Haven, 1985), and "Resistance without Protest and with-

out Organization: Peasant opposition to the Islamic *zakat* and the Christian tithe," *Comparative Studies in History and Society,* 29:3 (1987), 417–452. Scott acknowledges that his emphasis on "quiet" forms of struggle builds on previous studies by scholars working in other areas on similar agrarian themes: e.g., Genovese, *Roll, Jordan, Roll: The world the slaves made* (New York, 1972); and Michael Adas, "From Avoidance to Confrontation: Peasant protest in precolonial and colonial Southeast Asia," *Comparative Studies in History and Society,* 23:2 (1981), 217–247. For a provocative state-of-the-art survey of peasant resistance in its various forms, see Steve J. Stern, "New Approaches to the Study of Peasant Rebellion and Consciousness: Implications of the Andean experience," in Stern (ed.), *Resistance, Rebellion, and Consciousness in the Andean World: Eighteenth to twentieth centuries* (Madison, 1988), 3–25.

67. Scott, "Resistance without Protest," 450.

68. Eric J. Hobsbawm, "Peasants and Politics," *Journal of Peasant Studies,* 1:1 (1973), 12; also cited in Scott, "Resistance without Protest," 424.

69. Scott, "Resistance without Protest," 422; cf. *Weapons of the Weak,* ch. 7. Scott points out, for example, that the accumulation of thousands of individual acts of tax evasion, theft, or desertion can seriously disrupt elite establishments and even destabilize regimes.

70. While hacienda records and mortgage suits (*juicios hipotecarios*) inform us about such categories as wages, debts, production, machinery, property values, and goods on hand in the *tienda de raya,* they say little about the day-to-day interaction of management and labor on the estates. The Ramo de Justicia contains isolated references to shirking, foot dragging, etc. (see note 73, below); but note the absence of such detail in *hacendado* memoirs: e.g., García Cantón, *Memorias de un ex-hacendado henequenero,* and Gustavo Molina Font, *La tragedia de Yucatán* (México, 1941).

71. A perfect example of this vague impressionism is found in the 1971 documentary "Mexico: The frozen revolution," in which an old-timer is interviewed about the nature of labor relations during the henequen boom. Citations of the sparse oral historical literature are found in Joseph, *Rediscovering the Past,* ch. 4 and Epilogue.

72. Joseph and Wells, "Verano de descontento, estaciones de sublevación: Hacia un análisis de la política de élites y la rebelión rural en Yucatán, 1890–1915," in Othón Baños Ramírez (ed.), *Sociedad, estructura agraria y estado en Yucatán* (Mérida, 1990), 233–256.

73. E.g., for shirking: Ramo de Justicia, "Denuncia que hace Lorenzo Díaz de varios hechos delictuosos cometidos por Federico Trejo," 1910, AGEY; Ramo de Justicia, "Toca a las diligencias practicadas en averiguación de las lesiones que presenta Francisco Lugo contra Temístocles Correa Gutiérrez," 1914, AGEY; Ramo de Justicia, "Toca a la causa seguida a Antonio Puc y socios por los delitos de motín y destrucción de propiedad," 1911, AGEY.

74. E.g., for rustling and theft: Ramo de Justicia, "Causa seguida a Severiano Baas y socios, por el delito de abigeato," 1911, AGEY; Ramo de Justicia, "Toca a la causa seguida a Hermenegildo Nah y socio por los delitos de

robo y destrucción de propiedad ajena por incendio," 1912, AGEY. For arson and sabotage: Ramo de Justicia, "Toca a la causa a Santiago May por el delito de destrucción en propeidad ajena por incendio," Caja 674, 1908, AGEY; Ramo de Justicia, "Denuncia que hace Manuel Ríos de destrucción de propeidad ajena," 1914, AGEY.

75. E.g., Ramo de Justicia, "Cuenta de administración de la fincas San José Kuché y su anexa San Francisco correspondiente a un més corrido de 25 de julio a hoy agosto 25 de 1897," 1897, AGEY. The comparative slavery literature contains an abundant discussion of feigned illness as resistance. The contemporary travelers' accounts of Baerlein (*Mexico: Land of unrest*) and John Kenneth Turner (*Barbarous Mexico* [Chicago, 1910]), among others, contain a variety of planter complaints regarding the laziness and drunkenness of their peons.

76. *Diario Oficial*, 9 January 1908, no. 3096.

77. For examples of deaths attributed to *congestión alcohólica*, see Ramo de Justicia, "Toca a las diligencias practicadas en averiguación del motivo de la muerte de Manuel J. Chávez," 1913, AGEY; Ramo de Justicia, "Toca a las diligencias practicadas en averiguación del motivo de la muerte de Victoriano Chan," 1913, AGEY.

78. Thus far, we have not attempted to quantify the criminal data in the Ramo de Justicia. The unorganized and incomplete nature of the data will render such an exercise extremely difficult and problematic.

79. Scott, *Weapons of the Weak*, 302.

80. A perfect case in point is found in Ramo de Justicia, "Toca a la causa seguida a Manuel Baas por el delito de destrucción en propiedad ajena por incendio," 1912, AGEY. Baas claimed he was in a "total alcoholic stupor" when he burned the hut down. He was given a very lenient sentence.

81. E.g., see the characterization of Maya *campesinos* in Ramo de Justicia, "Toca a la causa seguida a Hermenegildo Nah y socio," 1912, AGEY; and "Toca a la causa seguida a Visitación González y Magdelena Alcocer de González por injurias al funcionario público y resistencia a la autoridad," 1914, AGEY.

82. For a fascinating discussion of the colonial judicial system's attitude toward alcohol abuse, which has yielded important insights for the present work, see William B. Taylor, *Drinking, Homicide and Rebellion in Colonial Mexican Villages* (Stanford, 1979).

83. One *mecate* equals approximately one acre. Ramo de Justicia, "Toca la causa seguida a Santiago May," Caja 674, 1908, AGEY.

84. A sampling of some of the other unusually severe verdicts for destruction of property include Ramo de Justicia, "Toca a la causa seguida a Felipe Medina por el delito de destrucción ajena por incendio," 1908, AGEY (7 years, 9 months for burning a house); Ramo de Justicia, "Toca a la causa seguida a Pedro May y socios por el delito de robo," 1912, AGEY (2 years, 6 months for stealing a modest quantity of corn); Ramo de Justicia, "Toca a la causa seguida a Francisco Várguez por destrucción de propiedad ajena," 1914, AGEY (1 year for illegally making *milpa* on eight hectares of an *hacendado*'s *monte*). Cf. the lenient sentence for murder in Ramo de Justicia, "Toca a la causa seguida a Manuel Fernández y Antonio Tos por el delito de homocidio," Caja 680, 1908, AGEY (5 years, 6 months for killing another *jornalero*).

85. The eighteenth-century English courts similarly upheld the sanctity of private property. See Douglas Hay, "Property, Authority, and the Criminal Law," in Douglas Hay, Peter Linebaugh, John G. Rule, E. P. Thompson, and Carl Winslow (eds.), *Albion's Fatal Tree: Crime and society in eighteenth-century England* (London, 1975), 17–64.

86. Carlos R. Menéndez, *Noventa años de historia de Yucatán (1821–1910)* (Mérida, 1937), 137.

87. For a frustrated investigation, see Ramo de Justicia, "Toca a la causa seguida a Pedro Chí por el delito de destrucción de propeidad ajena por incendio," 1912, AGEY.

88. Scott, "Resistance without Protest," 452.

89. For a discussion and critique of Scott's argument, see Gilbert M. Joseph, "On the Trail of Latin American Bandits: A reexamination of peasant resistance," *Latin American Research Review*, 25:3 (1990), esp. 25–33.

90. A number of contemporary observers, including General Salvador Alvarado and Yucatecan poet Antonio Mediz Bolio, commented on the endemic incidence of suicide among peons during the *auge*. See James C. Carey, *The Mexican Revolution in Yucatán, 1915–1924* (Boulder, 1985), 78, note 29. It appears that the high incidence of suicides was not limited to Yucatán during the late Porfiriato. Contemporary Mexican observers blamed the rash of suicides on gambling, cynicism, superstition, bad business, unrequited love, and impugned honor. González Navarro, *La vida social*, IV:428–29. Interestingly, the researches on the colonial Yucatec Maya by Nancy Farriss and Grant Jones have not uncovered a high incidence of Maya suicides (personal correspondence, winter and spring 1988). This may reflect differences in social conditions and/or differences in record keeping. The Porfirian state was likely more disposed and certainly far better equipped to investigate violent deaths in the *campo*, filing a detailed *diligencia* on each episode.

91. Ramo de Justicia, "Diligencias en averiguación de la muerte de José María Eb, vecino que fué de la hacienda San José, 1912, AGEY. Pellagra sometimes drove its victims to socially aggressive acts, including murder. Typically the victims were close relations or fellow peons. In 1912, Hermenegildo Puc beat Desideria Canul senseless in her hammock with his sandals. Puc believed that Cunal was a witch *(hechicera)* who had cast a spell on him, "darkening his sight and refusing to make him well." The court ruled that pellagra had so diminished the defendant's senses that he was not responsible for his actions. Instead, Puc was taken to the Ayala Asylum in Mérida for treatment. Ramo de Justicia, "Toca a la causa seguida a Hermenegildo Puc por el delito de homicidio," 1912, AGEY.

92. We are indebted to Alejandra García Quintanilla, who first alerted us to this ritualistic Maya method of suicide. E.g., Ramo de Justicia, "Diligencias con motivo del suicidio de Candelario Cauich, sirviente de la finca Chunkanán," 1913, AGEY; Ramo de Justicia, "Diligencias con motivo de la muerte de Vcente Cen," Caja 714, 1908, AGEY; Ramo de Justicia, "Diligencias en el suicidio del que se cree llamarse Valerio Godoy, verificada en la Hacienda Vista Alegre," Caja 296, 1897, AGEY; Ramo de Justicia, "Toca a las diligencias practicadas en averiguación de la muerte de Enrique Canché Piña," 1912, AGEY. These are poignantly descriptive episodes, but

multiple references from the Ramo de Justicia could be provided for any year in this study.

93. Quote from J. Eric S. Thompson, *Maya History and Religion* (Norman, Okla., 1970), 301. Sylvanus Morley relying on Bishop Diego de Landa does not specify a special heaven for suicide victims, but insists that suicides by hanging, warriors killed in battle, sacrificial victims, women who died in childbirth, and priests all automatically went directly to the Maya paradise. Morley, *The Ancient Maya*, 3rd edition, revised by George W. Brainerd (Stanford, 1956), 194.

94. Thompson, *Maya History and Religion*, 301; and Morley quoted, *The Ancient Maya*, 203.

95. Morley, *The Ancient Maya*, 194.

96. Of course, the southeastern rebel Maya's creation of a Speaking Cross cult following the Caste War provides the most celebrated example of the reformulation of established forms and symbols within an ongoing, dynamic process of resistance.

97. For the debate over the panic, see Knight, *The Mexican Revolution*, I:64–65, and Friedrich Katz, "Mexico: Restored republic and Porfiriato," in *Cambridge History of Latin America* (Cambridge, England, 1986), V: 62–68. More detailed discussions of the panic's economic and political consequences are in Joseph and Wells, "Yucatán: Elite politics and rural insurgency," and "Summer of Discontent: Economic rivalry among elite factions during the late Porfiriato in Yucatán," *Journal of Latin American Studies*, 18:2 (1986), 255–282.

98. Gill, "Regional Variation," chs. 3 and 4, provides an acute microhistorical analysis of the involution of labor conditions, the tightening of credit, and the threat to the peon's security of subsistence during the final years of the Porfiriato. Particularly promising is his preliminary discussion of planter appropriation of household labor (pp. 39–40, 60–61).

99. For an insightful critique of such models, see Rod Aya, "Theories of Revolution Reconsidered: Contrasting models of collective violence," *Theory and Society*, 8 (1979), 39–99; and Aya, "Popular Intervention in Revolutionary Situations," in Charles Bright and Susan Harding (eds.), *Statemaking and Social Movements: Essays in history and theory* (Ann Arbor, 1984), 318–343.

100. *La Revista de Mérida*, 20 September 1909.

101. *La Revista de Mérida*, 5 September 1907; *El Imparcial*, 6 September 1907.

102. On the Catmís incident and its aftermath, see *Diario Oficial*, *La Revista de Mérida*, and *Diario Yucateco* for 8–12 March 1911.

103. There is, however, a danger of *overemphasizing* both the novelty and threat presented by these violent, late Porfirian episodes. Isolated outbursts flared up *throughout* the Old Regime. It was only by about 1910, however, that they began to be reported consistently—and frequently in exaggerated fashion—by the opposition press, which was tied to the elite factions contesting the hegemony of the Molinista oligarchy. Long a repressed and cowed institution, the opposition press' coverage of unrest in the henequen zone dramatically increased when political space opened following the outbreak of the national Madero rebellion. See José Luis Sierra

Villarreal, "Prensa y lucha política en Yucatán, 1895–1925" mimeographed ms., 1984.

104. The political economy of henequen during the late Porfiriato and particularly International Harvester's predominant influence on it—a subject that has generated substantial debate both inside and outside of Yucatán—are treated in depth in Gilbert M. Joseph and Allen Wells, "Corporate Control of a Monocrop Economy: International Harvester and Yucatán's henequen industry during the Porfiriato," *Latin American Research Review*, 17:1 (1982), 69–99; and *Yucatán y el International Harvester Company* (*Mérida*, 1986).

105. For a more detailed reconstruction of the 1909–1913 period, see Joseph and Wells, "Yucatán: Elite politics and rural insurgency."

106. Ramo de Justicia, "Toca a la causa seguida a Juan Jiménez y socios por el delito de provocación al delito de rebelión," 1913, AGEY.

107. E.g., see Joseph Foweracker, *Personal Networks and Political Strategies: The making of democracy in Spain* (Cambridge/New York, 1989); and Theda Skocpol, *States and Social Revolutions* (Cambridge, England, 1979), esp. 16–18.

108. E.g., see *La Revista de Mérida*, 16 May 1911.

109. Ramo de Justicia, "Toca a la causa seguida a Luis Uc y socios por los delitos de amenaza de injurias," 1913, AGEY.

110. Taylor, *Drinking, Homicide, and Rebellion*, 128–142, and Stern, "New Approaches," provide insightful discussions of the relationship between peasant consciousness and structural relations of power in the genesis of rural insurgencies.

111. The mobilization and *mentalidad* of Yucatán's "free" peasant villages are treated in greater detail in Joseph and Wells, "Seasons of Upheaval: The crisis of oligarchical rule in Yucatán, 1909–1915," in Jaime E. Rodríguez (ed.), *The Revolutionary Process in Mexico: Essays on political and social change, 1880–1940* (Los Angeles, 1990), 161–185.

112. Thompson cited in Peter Winn, *Weavers of Revolution: The Yarur workers and Chile's road to socialism* (New York, 1986), v.

113. Episodes that reveal much of this ambiguity and complexity are Ramo de Justicia, "Toca a la causa seguida a Pedro Chí," 1912, AGEY; Ramo de Justicia, "Toca a la causa seguida a Juan Jiménez," 1913, AGEY.

114. E.g., see Tutino, *From Insurrection to Revolution*, 13–32. The term "hingeman" originally appeared in Peter Brown, *Society and the Holy in Late Antiquity* (Berkeley, 1982).

115. For a case that dramatically documents these networks of recruitment, replete with hingemen and *orejas*, see Ramo de Justicia, "Causa seguida a José Policarpo Mendoza y socios por el delito de rebelión," 1912, AGEY; Ramo de Justicia, "Toca a la causa seguida a Luis Uc y socios por los delitos de amenaza de injurias," 1913, AGEY.

116. Ramo de Justicia, "Causa seguida contra Alfonso Cámara y Cámara y socios por el delito de rebelión," 1909, AGEY.

117. A range of hills that runs from Maxcanú to Muna in Ticul *partido*.

118. For *ajusticiamientos populares*, e.g., see Ramo de Justicia, "Causa seguida contra Pedro Crespo y socios por el delito de homocidio, rebelión y robo," 1911, AGEY; and Ramo de Justicia, "Causa seguida a Bernabé Esca-

lante por suponérsele presunto cómplice del delito de homocidio, para la continuación respecto de Juan Osorio, Juan Campos y socios," 1912, AGEY.

119. Ramo de Justicia, "Toca a la causa a José Dolores Cauich y socios por los delitos de rebelión, robo y destrucción de propiedad ajena," 1912, AGEY.

120. E.g., see Ramo de Justicia, "Causa seguida a Guillermo Canul y socios por los delitos de daño y destrucción de propiedad ajena," 1912, AGEY.

121. *Henequeneros* felt so threatened by the spread of violence in mid-1911 that they petitioned Governor Pino Suárez and the federal government for additional military protection. Pino ultimately requested and received from Mexico City two additional federal battalions and a shipment of machine guns. *Diario Oficial,* 8 July 1991, no. 4180, and 16 August 1911, no. 4213.

122. For a comparison of the two cycles of popular rebellion, see Gilbert M. Joseph, "The United States, Feuding Elites, and Rural Revolt in Yucatán, 1836–1915," in Daniel Nugent (ed.), *Rural Revolt in Mexico and U.S. Intervention* (La Jolla, Calif., 1988), 167–197.

123. The following discussion of the demobilization of popular insurgency draws heavily on Joseph and Wells, "Seasons of Upheaval."

124. Knight, *The Mexican Revolution,* I:221.

125. E.g., see Genovese, *From Rebellion to Revolution,* 110–113; for the *Liga de Acción Social,* see Ramón Chacón, "Yucatán and the Mexican Revolution: The pre-constitutional years, 1910–1918," Ph.D. dissertation, Stanford University, (Stanford, 1981), 118–131.

126. This suggests parallels with the plantation regime of the antebellum U.S. South, whereas Genovese and others have shown, the law fulfilled something of a hegemonic function, providing at least the appearance of a disinterested standard of justice in the minds of the subordinate class. See Genovese, *Roll, Jordan, Roll,* 25–49. For example, see the court system's judicious handling of the notorious "San Nicolás" case of *henequenero* abuse of peons (consistent use of leg-irons, floggings with wire, etc.): Ramo de Justicia, "Toca a la causa seguida a Pedro Pinto y socios por los delitos de lesiones y atentados contra la libertad individual," 1914, AGEY.

127. Ramo de Justicia, "Diligencias contra Juan Córdova," 1915, AGEY.

128. E.g., see Ramo de Justicia, "Incendio en la finca Texán," 1914, AGEY; and *La Revista de Yucatán,* 31 March 1914.

129. Knight, *The Mexican Revolution,* I:169.

130. For the more extensive mobilizations under Alvarado, socialist governor Felipe Carrillo Puerto, and, later, under Cardenismo, see Gilbert M. Joseph, *Revolution from Without: Yucatán, Mexico and the United States, 1880–1924,* rev. ed. (Durham, N.C., 1988), xi–xxv, 93–98.

9

Planter Control and Worker Resistance in Northern Peru, 1880–1921

Michael J. Gonzales

Over the past decade or so, Latin Americanists have developed a more sophisticated understanding of forms of social protest and social control in plantation societies in the postslavery era. Gone is the universal stereotype of planters using debt peonage, corporal punishment, and paternalism to dominate submissive, ignorant peons. This view, so conveniently adapted from Indianist novels, liberal journalists, and Hispanophobe Anglo-Saxon travelers, has been shown to represent only the extreme case. Without denying the inherent inequalities in the relationship of production, it is now possible to identify a wide range of labor conditions on the plantations of Latin America ranging from semislavery to nearly free wage labor. Such diversity is not remarkable when one considers the differences that existed in levels of technical development, political economy, demography, seasonal labor requirements, ecology, and culture. One could also mention the diversities that existed in intelligence, temperament, and managerial philosophies of individual planters.[1]

It is also important to recognize that peons were historical actors with considerable capacity to shape their own histories. The pioneering work of Eugene Genovese on slave society in the antebellum United States showed how slaves improved their lot through various methods of passive and active resistance short of violence and rebellion; James Scott and others have studied how peasants in southeast Asia employed "everyday forms of resistance" to gain both individual and group advantage.[2] Latin American peons could also be resourceful, clever, and persistent in resisting planter domination and in earning modest improvements for themselves.

On the sugarcane plantations of northern Peru, the rationalization of production after 1890 decisively shaped labor relations. Several decades of land consolidation and technical development created "factories in the field," with large numbers of semipermanent wage laborers who gradually became proletarians. Although such practices as debt peonage persisted, they became ineffective, controversial, and ultimately counterproductive over time. In some cases, planters sought to maintain their dominance over laborers through paternalistic gestures, but this tactic only masked the adversarial relationship between labor and capital and was eventually doomed to fail. An important period in Peruvian working-class history occurred during and after World War I, when a number of serious strikes occurred on sugarcane plantations in response to falling real wages.

This more violent and open type of confrontation forced planters to rely more on the state for social control. This meant calling in the army to crush strikes, arrest labor leaders, and in general create a more repressive society.[3]

Structure of the Sugar Industry

Unlike other major sectors of the national economy, sugarcane remained largely in the hands of domestic capitalists during the late nineteenth and twentieth centuries. Foreign investors shied away from a commodity whose value was falling on the world market in favor of investing in high-growth industries, such as copper and petroleum. Despite the risks involved, many Peruvians who invested in sugar survived the crisis and later profited handsomely during the boom period from 1915 to 1920.[4]

By the turn of the twentieth century, the most important area of sugar production was the northern coast, specifically the Moche, Chicama, Saña, and Lambayeque valleys. Although these valleys differed in size, each had fertile soils, adequate sources of irrigation water, and access to nearby ports.

Sugar had been cultivated in the region since the sixteenth century,[5] but production dramatically increased after 1870. Leading the way were a handful of large plantations primarily owned and operated by a new generation of growers, many of whom were of recent immigrant origin.[6] Table 1 shows the location of major plantations and the nationality of their owners.

Most of these planters purchased their estates with money that they had earned in mercantile or mining activities within Peru. For example, Juan Gildemeister earned his initial fortune in nitrates, the Larco brothers ran a prosperous retail business in Lima before buying Roma, the Aspíllagas hauled freight between Lima and Callao in the mid-nineteenth century, and the Pardos were important guano merchants. Nevertheless, when the crisis struck the Peru-

Table 1. North Coast Sugar Estates

Estate	Valley	Owner	Nationality	Family Origin
Casa Grande	Chicama	Gildemeister	Peruvian*	Germany
Cartavio	Chicama	W. R. Grace	U.S.A.	Ireland
Roma	Chicama	Larco	Peruvian	Italy
Laredo	Moche	Chopitea	Peruvian	Spain
Cayaltí	Saña	Aspíllaga	Peruvian	Chile
Pomalca	Lambayeque	de la Piedra	Peruvian	Ecuador
Tumán	Lambayeque	Pardo	Peruvian	Peru
Pucal	Lambayeque	Izaga	Peruvian	Peru

*This family clearly bridges simple categories of national and foreign, exhibiting some characteristics of a multinational firm but with most of its investments and influence in Peru. See Michael Gonzales, *Plantation Agriculture and Social Control in Northern Peru, 1875–1933* (Austin, Texas, 1985), 27.

vian economy in the 1870s and 1880s, planters were forced to borrow heavily from foreign merchants, especially the British import-export houses that marketed their sugar abroad. Firms such as Graham, Rowe & Company and Henry Kendall & Sons loaned planters hundreds of thousands of pounds sterling. Once the crisis had passed, planters once again relied heavily on domestic sources of capital as well as profits.[7]

Planters used this capital to rationalize production through land consolidation and technological innovation. Planters found many of their competitors willing to sell out following the disastrous War of the Pacific, during which some estates had suffered serious damage. On other occasions, planters used their positions as administrators of local water districts to pressure small farmers and Indian communities into divesting themselves of their land. These tactics were particularly successful during the early twentieth century, when sugar planters enjoyed considerable political influence on both the local and national level.

All of the estates listed in Table 1 grew considerably in size during these years. For example, in the Lambayeque Valley, Pomalca expanded from 2,400 hectares in 1870 to 7,267 hectares in 1924, and Tumán grew from 3,734 hectares in 1907 to 6,632 hectares in 1924. In the Chicama Valley, Casa Grande experienced the most spectacular growth, expanding from a modest 724 hectares in 1850 to 40,848 hectares in 1927. In the latter year the Gildemeisters acquired Roma, thereby doubling the size of their estate.[8]

Coinciding with land consolidation was massive investment in technical development. Arthur Lewis has written that sugar was the only plantation crop to undergo a scientific revolution before World War I.[9] There were two fundamental reasons for this: cane had to be processed where it was grown, and the world price of sugar fell in the 1880s, which necessitated lowering the cost of production. Planters largely succeeded in doing this, as witnessed by the dramatic increase in global sugar production in the late nineteenth and early twentieth centuries.[10] Peruvian planters kept pace, for the most part, by importing the latest technology from Europe. During the period from about 1870 to 1920, the principal improvements were railroads for shipment to port, portable railroads for cane hauling, steam-powered plows for cultivation, and steam-powered mills for processing.

Railroads were desirable, of course, because they increased planters' shipping capacity and gave them more control over shipping schedules. By the 1890s, planters in the Lambayeque Valley had already commissioned the construction of a series of railroads linking their estates with nearby ports, and around the turn of the century, the Aspíllagas financed the construction of a line connecting Cayaltí with Eten. By 1905, railroads had also been built joining the estates Casa Grande, Cartavio, and Laredo with the port of Salaverry, and the plantations Roma and Chiquitoy with the harbor of Huanchaco. All of these railroads, with the exception of the Casa Grande–Cartavio line, were privately owned and operated.[11]

It was also during this period that planters constructed larger steam-powered mills with improved crushing capacity. This was probably the most important step in the modernization of the sugar industry. The first steam-powered mills went up at Facalá (1860s), Pomalca (1873), and Cayaltí (1877); but the big surge in mill modernization occurred in the late nineteenth and early twentieth centuries with the assistance of loans from British merchant houses. The lead in mill size and design was eventually taken by Casa Grande and Caravio, whose facilities were reputedly equal to any in Java and Hawaii.[12]

The construction of larger mills necessitated the cultivation of more cane and new methods for hauling it to the mill more rapidly. These challenges were largely overcome through the introduction of a series of technical innovations in the field, especially portable railroads and steam-powered plows. Portable railroads were based on the principle of moving track from one harvested field to another, which reduced hauling time and allowed the use of larger carts.[13]

The cultivation of sugarcane was labor intensive and time consuming. Planters frequently complained about the inefficiency and expense of using oxen-driven plows. The introduction of steam-powered Fowler plows meant that plowing could be done more rapidly and efficiently with fewer workers.[14]

Taken together, these technical innovations transformed the sugar industry by increasing production and reducing the cost of production. At Casa Grande, for example, the Gildemeisters reduced the cost of producing one quintal (100 pounds) of sugar (FOB) from £14.11.5 sterling in 1924 to £4.1.80 in 1932.[15] Casa Grande also led the way in sugar production, as shown in Table 2.

The dramatic increase in production made sugar one of Peru's leading exports and allowed planters to earn fortunes during the boom period of World War I. During these years of peak sugar prices, planters netted between $300,000 and $1,000,000 per year, largely depending on the amount of sugar that they had to sell. The

Table 2. Sugar Production on North Coast Estates, Selected Years (quintales)

Estate	Year	Production
Casa Grande	1916	1,000,000
	1923	1,185,802
Cartavio	1898	175,000
	1916	800,000
Tumán	1916	200,000
	1919	350,000
	1920	396,737
	1923	443,000
Pomalca	1914	168,962
	1915	166,917
	1921	231,000
Cayaltí	1895	100,000
	1912	202,539
	1915	256,595
	1922	370,000

SOURCES: AH to AH, 26 February 1898, AFA; AAB to RAB and BAB, 6 March 1917, AFA Balance of July 1941, Gildemeister folder, AFA; BAB to ABB, RAB, and his mother, 10 April 1916, AFA; AAB to RAB and BAB, 23 December 1916, AFA; Administrador General to Gerente, 24 December 1926, Correspondencia Tumán, AFA; Libro de Contabilidad de la Hacienda Pomalca, 1914–1917, AFA; Douglas Horton, *Haciendas and Cooperatives: A preliminary study of Latifundist agriculture and agrarian reform in Northern Peru*, research paper, no. 53 (Madison, Wis.: Land Tenure Center, 1973), 21.

dramatic fall in the price of sugar in 1920s, on the other hand, resulted in falling profits and after 1924 in severe losses.[16]

Labor Recruitment

On the plantations, the dramatic increase in the scale of production significantly affected the demand for labor and patterns of social control. Despite the technological improvements outlined above, the cultivation of sugar remained largely labor intensive. As north coast plantations expanded in size, there was a corresponding increase in the need to hire laborers for interrelated production tasks involving planting, weeding, irrigation, harvesting, hauling, and processing. The dramatic increase in the number of workers, in turn, forced planters to invest more in services and to devise more sophisticated methods of social control.

The increased demand for workers created special problems for planters, because their traditional source of labor, the coolie trade, had been severed by the British and Chinese in 1874.[17] Planters hoped that they could find an alternative source of servile labor but were forced to make do with those Chinese already within the country until the chaos of the 1870s and 1880s subsided. Planters subsequently experimented with Japanese contract laborers but were unable to import them in large numbers. As a result, growers became committed to the large-scale recruitment of local peasants, a process referred to as *enganche* (hooking).[18]

Throughout Latin America, the recruitment of peasants to work as wage laborers in commercial agriculture and mining is linked to the expansion of the export economies in the late nineteenth century. In most cases, traditional sources of servile labor, especially African slavery and Chinese indentured labor, were unavailable or insufficient to accommodate increased scales of production created by new technologies. The transition to wage labor in Peru and elsewhere has been a controversial topic among historians. Many scholars have characterized *enganche* as little better than slavery, emphasizing the use of force, deception, and debt peonage to create large, stable work forces.[19] Recent research on northern Peru, however, shows that many peasants migrated to the coast on their own volition, chose to remain on plantations for long periods of time, and gradually became full-time wage laborers. Planters still resorted to extraeconomic coercion and debt peonage but complained that indebted laborers sometimes absconded with their money and that physical violence frequently created more problems than it solved. As a result, some planters, at least, found manipulation and accommodation more effective methods of social control.[20]

Although some laborers were recruited from coastal villages and Indian communities, the vast majority of wage earners came from the highland regions of La Libertad and Cajamarca. As Lewis Taylor has shown, peasants migrated to the coast to escape overpopulation, banditry, and food shortages, as well as to earn higher wages. Migration was a logical economic decision.[21]

Labor recruitment was in the hands of labor contractors *(enganchadores)* who were generally highland merchants or *hacendados*. In most cases, peons voluntarily signed short-term labor contracts at Sunday markets, where contractors had set up temporary stalls, or at the contractor's office in town. Peons always received an advance on their wages, which they agreed to repay with their labor. On other occasions, however, contractors applied special leverage to force *serranos* to migrate. For example, some merchants had peasants repay outstanding debts by working on the coast, and *hacendados* sometimes demanded that tenants and subtenants fulfill labor obligations by cutting cane.[22]

With the advent of *enganche*, the problem of labor supply was largely solved. Plantations such as Cayaltí, which had been struggling with less than 500 tired and increasingly unproductive Chinese laborers, managed to recruit nearly 1,000 *serranos* by the turn of the century.[23] As estates grew in size and production capacity, the number of workers kept pace. For example, during World War I, Cayaltí, Laredo, and Cartavio all had around 1,500 laborers, and Casa Grande, the largest producer, had a work force of 5,000.[24] Although this was far from a perfect labor market, competition for labor among estates and contractors caused wages to rise and made selling one's labor an attractive alternative to peasant agriculture (see Table 3).

The most controversial aspect of labor contracting concerns the importance of debt peonage, traditionally seen as bondage or semislavery.[25] The infrastructure of debt peonage existed on all north coast plantations. All contracted workers received advances on their wages and purchased goods on credit, and some of them accumulated substantial debts; it was expected that workers would not leave plantations until their debts had been repaid.[26] Planters had some incentive for creating permanent work forces because, owing to the use of irrigation, production could continue year-round. Nevertheless, forcing *serranos* to remain on plantations for decades proved impractical and counterproductive.

The principal impediments to bondage were worker illness,

Table 3. Daily Cash Wage at Cayaltí

Year	Daily Cash Wage (soles)	Value in Dollars
1897	.40	.1940
1899	.50	.2425
1902	.40	.1940
1905	.50	.2425
1907	.50	.2425
1914	.60	.2832
1919	1.12	.5510
1923	1.30	.5265
1932	1.30	.4589

SOURCES: For 1897–1914: AH to AH, 7 July 1897, AFA; AH to AH, 29 April 1899, AFA; AH to Los Contratistas, 1 May 1899, AFA; N. Salcedo to AH, 4 March 1902, AFA; BAB to AH, 4 February 1903, AFA; VAT to AH, 3 September 1907, AFA; VAT to AH, 28 August 1907, AFA; VAT to AH, 8 May 1912, AFA; letter dated 19 July 1919, AFA. For 1919–1930: Unsigned letter dated 19 July 1919, AFA; IAA to RAB, 16 November 1923, AFA; unsigned letter dated 13 June 1932, AFA.

NOTE: Wages were determined by piecework, and included a daily ration of beef, rice, and salt.

exhaustion, and runaways. The northern coast was beset with a variety of serious illnesses, including malaria, smallpox, and influenza, as well as with dysentery and other intestinal disorders associated with unsanitary conditions, the hot climate, and lack of potable water. Moreover, plantation labor was taxing, and there were always some debilitating accidents, especially in the mill and in cane hauling. Over time, the combination of exertion and illness normally caused reductions in worker productivity and necessitated replenishment of the work force with new recruits. Under these circumstances, the least productive workers, regardless of their indebtedness, were dismissed from the estate.[27]

The effectiveness of debt peonage was also undermined by the growing size and social complexity of the plantations. Work forces that numbered in the thousands were difficult to police, and planters could not prevent hundreds of indebted peons from running away. Runaways generally fled to nearby estates, where they could receive additional advances from a different contractor, or to the highlands. In either case, runaways could be difficult to capture. Contractors were reluctant to return peons once they had advanced them money, and highlanders were sometimes protected by friends and relatives in the sierra. Moreover, efforts to recover debts through the courts were not always successful.[28]

Social Control

The transition to wage labor changed the character, composition, and size of plantation communities and presented planters with new problems in controlling the work force. Not only had the size of the work force increased, but highlanders tended to migrate with their wives or to acquire companions on the coast. Plantations, in the process, became less like work camps and more like communities.

Faced with a more complex environment, some planters developed a philosophy of control that depended less on force and more on accommodation and manipulation. It did not signal a change of attitude on their part toward laborers but the need to devise more sophisticated methods of controlling larger, more mobile, and potentially militant work forces. Among other things, these planters found it necessary to invest more money in housing, health care, and recreation. As a case study, the following discussion will focus mostly on the plantation Cayaltí.

The construction of new housing was clearly predicated on the desire to attract laborers to settle down on plantations with their families. The Aspíllagas even referred to new housing as the "bond that tied workers to the estate."[29] These new dwellings were small adobe structures with a separate living room, kitchen, and patio; and they were a major improvement over the large, dilapidated dormitories in which workers were traditionally housed.[30]

The Aspíllagas also believed that the daily provision of a pound of beef and a pound of rice to each worker was a powerful inducement to work for the estate. They also sold food and clothing at competitive prices at the estate store, and they permitted vendors onto the plantation to sell a variety of goods. Contrary to the traditional image of the company store, which portrays owners as selling goods on credit at inflated prices to entrap workers in a web of indebtedness, the store at Cayaltí did not sell on credit until the 1930s, and only then as a relief measure.[31]

After the turn of the century, planters also made a greater effort to provide medical care for workers. At Cayaltí, this involved hiring a full-time doctor, maintaining a modest hospital, and vaccinating against key diseases. Sugar workers were exposed to a wide variety of communicable diseases that, if they became epidemic, could paralyze production and result in significant losses in profits. Diseases found in the region read like a list of the greatest killers of all time: bubonic plague, smallpox, measles, influenza, cholera, malaria, pneumonia, yellow fever, typhoid, typhus, and tuberculosis. Some of these maladies periodically carried away hundreds of workers. Nevertheless, the incidence of disease decreased as time progressed, and smallpox and the plague were almost completely eradicated from the plantations.[32]

In addition to these programs, the Aspíllagas offered workers financial rewards for hard work and permanency. These took the form of bonuses and prizes awarded at the annual New Year's party. For example, in 1906 the Aspíllagas gave $2.40 to "about ten workers who had most distinguished themselves in their conduct and work," including migrants who had been on the estate the longest. This bonus represented about ten days' wages. In later years, the Aspíllagas continued to give bonuses for longest periods of employment, as well as for the best attendance records. Through these modest expenditures, planters were encouraging workers to accept a life of wage labor and to sever ties with peasant agriculture.[33] This process of proletarianization, which these programs facilitated, was well advanced by the 1920s. For example, in 1924, 56.2 percent of Cayaltí's workers had severed all ties with contractors and had settled permanently on the coast.[34]

The provision of material incentives and improved services made coastal plantations relatively easier places in which to live and did not leave workers completely at the mercy of the cash nexes of the labor market. Nevertheless, the transition from highland peasant to coastal sugar worker was a difficult one that involved changes in life style, work patterns, climate, and, ultimately, world view. Problems in adjustment gave rise to widespread alcoholism and absenteeism, both of which proved impossible to overcome completely. Nevertheless, planters attempted to placate and divert workers by providing them with a variety of recreational facilities, a yearly party, and free transportation to local festivals. These tactics undoubtedly helped to relieve tension and to prevent the outbreak of violence. Interestingly, similar strategies of control were employed by some early industrialists in Great Britain and in the United States to cure adjustment problems experienced by first-generation industrial workers.[35]

There is no question that these programs were paternalistic in nature, but it is less certain that they constituted paternal regimes. Paternal orders are characterized by close personal relationships between employers and employees, and considerations of kinship, loyalty, and duty influence hiring practices. In purely capitalist regimes, by contrast, the marketplace, on-the-job performance, and cash in exchange for labor determine employment and retention.

Cayaltí represents a mixture of these extremes. The Aspíllagas wanted their workers to accept routinized wage labor, similar to industrial capitalism, but they also wanted to prevent the formation of a working class. Their complex array of services were designed to attract migrant labor and to mask the inherent inequities in the relationship of production and forestall class formation. There is no evidence that the Aspíllagas provided services out of considerations of kinship, Christian sensibility, humanitarianism, or other signposts of paternal orders.

Moreover, there were practical impediments to the formation of such orders. For example, contracted workers may have already established paternalistic relationships with contractors, who were frequently elites from peons' home villages. As intermediaries between employers and employees, contractors frequently relieved planters of the necessity of dealing directly with workers. In addition, there was a high turnover in the work force owing to illness, exhaustion, and re-migration. It would have been impossible for planters to develop close personal relationships with a majority of *serranos*. In all likelihood, such relationships only existed between the Aspíllagas and their veteran administrative staff and a handful of workers who had been on the estate for several decades.

This interpretation, which has been developed at greater length in my book,[36] has been disputed by Eric Langer. He believes that a "profoundly paternalistic regime" based on "traditional" forms of reciprocity and exchange existed at Cayaltí. As he puts it: "It is probable that the highland migrants, presumably imbued with Andean peasant culture with its heavy emphasis on reciprocity and redistribution, forced coastal landowners to maintain their paternalistic regimes."[37] However, Langer does not present any documentary evidence to support his position, and there is no evidence that he has read any plantation records.

Although it is logical to assume that first-time migrants arrived on coastal plantations with their peasant culture intact, it is clear that they underwent acculturation on the estates. Highlanders experienced changes in work routine, diet, dress, housing, and recreation, and they also came into contact with people from the coast, who were frequently mulattoes.[38] Moreover, Cayaltí's migrants were more susceptible to acculturation because they were Spanish-speaking *mestizos* (half-Spanish, half-Indian), as opposed to Quechua-speaking Indians.

Nor were peons forced by the Aspíllagas into redistributing wealth based on notions of cultural reciprocity, as Langer has suggested.[39] When peons sought higher wages and improved services, they basically had two options: they could migrate or run away to another estate that paid more, or they could threaten to strike. There were a number of strikes on north coast estates after 1912, and this increasingly aggressive behavior suggests the emergence of a rural proletariat as opposed to the existence of paternal orders based on cultural reciprocity.

Confrontation

Strike activity on the northern coast was most intense on the largest, most technically advanced plantations, especially those in the Chicama Valley, where working conditions and managerial philosophies most closely resembled a factory system. Two of the largest strikes in the history of the nation occurred at Casa Grande in 1912 and at Roma in 1921. Violence spread to neighboring estates, and the army was eventually called in to crush the strikes and force peons back to work.

By contrast, the Aspíllagas avoided major confrontations with their workers during this period. There were several reasons for this success. First, Cayaltí modernized at a slower pace than Chicama Valley estates and therefore experienced fewer disruptions in labor relations resulting from the rationalization of production. Second, the Aspíllagas were willing to grant modest wage increases and other concessions when threatened with a major strike, and this usually appeased workers. Third, the Aspíllagas tightened plantation security during periods of labor unrest in the region to prevent outside agitators from entering the estate. Fourth, the Aspíllagas had a better rapport with their workers than their counterparts in the Chicama Valley. Nevertheless, when labor unrest did occur at Cayaltí, the Aspíllagas were never motivated by considerations of cultural reciprocity in seeking settlements.

The history of labor conflict in the Chicama Valley has already been discussed in some detail and there is no need to present a lengthy analysis here.[40] The principal causes of the 1912 strike at Casa Grande were uncompensated increases in work loads and the hiring of several German employees, who did not speak Spanish fluently and were called "imperious" by the Aspíllagas.[41] The early stages of the strike were characterized by a great deal of worker violence, which quickly spread to neighboring plantations. Considerable damage was done before the army, local police, and militia brought an enforced peace to the region.[42]

Despite this repression, strikers gained some concessions by remaining organized and refusing to return to work.[43] In the end, the Gildemeisters agreed to a general wage increase (a major concession), the construction of a larger hospital and new housing, and the firing of a German employee who had trampled on the Peruvian flag.[44]

Labor conflict became more generalized in the region during World War I. The basic cause was the inherent inequity of rising profits for planters and falling real wages for workers. There were a number of strikes throughout these years, with the largest ones occurring in 1917 at Casa Grande and Pomalca, the latter a major producer in the Lambayeque Valley. In each case, strikers demanded higher pay and lower food prices at the plantation store. At Casa Grande a strike committee was formed, but workers remained divided and planters quickly called in the army to force them back to work. At Pomalca, planters refused to negotiate with strikers over wages, and only offered them temporary relief in the form of cash bonuses and some free food. When strikers refused to accept these minor concessions, troops were called in to crush the strike.[45]

Confrontation in the Chicama Valley culminated in a massive strike in 1921 at Roma, the second largest plantation in the region and the property of Victor Larco Herrera. Larco's problems began when he rescinded a 33 percent wage hike he had granted the previous year. This was a serious blunder on his part, as workers not only demanded full reinstatement of their wages but an additional increase. They also added a long list of demands, including formal recognition of a secretly formed union, the *Sociedad Obrera de Auxilios Mutuos y Caja de Ahorros de Roma*. When Larco rejected every demand, the entire work force walked out in solidarity.[46]

The strike quickly spread to neighboring estates and gained the support of railway workers and longshoremen in Salaverry. Local authorities attempted to crush the movement using strong-arm tactics, including arresting union leaders, but this only strengthened strikers' resolve.[47] The gravity of the crises compelled President Augusto B. Leguía to send a labor negotiator to the scene who, perhaps surprisingly, reached a settlement that incorporated most of the strikers' demands.[48]

The strike broke out again, however, when Larco closed down his mill and fired mill workers, among them several union leaders. Roma's strikers were supported by sugarcane workers from other estates and workers from nearby towns, who formed the *Sindicato Regional de Trabajo* (Regional Labor Union). Planters refused to negotiate with the union, however, and Leguía once again dispatched an arbitrator to reach a settlement. The government favored an eight-hour day but rejected wage increases and recognition of the union.[49] This position was unacceptable to both parties, and the strike continued. Leguía now had to choose between supporting sugar planters, who included some of his strongest political enemies, or sugar workers and their union, who constituted a threat to the political and economic status quo. In the end, he chose to force strikers back to work at gunpoint. Ironically, however, Victor Larco Herrera was one of the biggest losers. He lost enormous sums of money because of the strike, and following a major flood in 1925, he sold out to Gildemeisters.[50]

The willingness of Chicama Valley planters to use the army to crush strikes can be contrasted with the Aspíllagas' handling of labor disputes at Cayaltí. They sought to avoid outside intervention and adopted a more conciliatory attitude toward workers' grievances, reasoning that it was better to grant concessions than to risk a major confrontation that could poison labor relations for decades.

Between 1912 and 1921, the Aspíllagas were periodically faced with labor disputes on their estate. During the 1912 strike at Casa Grande, Cayaltí's cane cutters demanded higher wages and smaller work loads. The Aspíllagas noted that neighboring plantations were paying 15 to 20 *centavos* an hour more than Cayaltí, and they chose to increase wages by 10 centavos per hour and to decrease work loads. Antero Aspíllaga Barrera wrote to his brothers that "it was not possible to continue as before without at the least losing workers and having conflicts, which prudence and our own understood interests counsel us to avoid."[51] This was clearly a decision based on economic self-interest, as opposed to cultural reciprocity.

In the years that followed, the Aspíllagas became increasingly concerned with plantation security. A riot on the estate in 1915, in which drunken mobs sacked stores, caused them to increase the size of the plantation police force and screen new recruits. They were especially concerned with preventing outside agitators from entering the estate, and they also urged local authorities to crush socialist clubs and other threatening organizations in nearby towns.[52]

The Aspíllagas also recognized that a major cause of labor unrest in the region was the spiraling cost of living. Therefore, in 1917 they ordered the plantation store to sell staples at cost, noting that it would be "stupid and dangerous" to do otherwise.[53] Despite this effort, however, two years later the cost of living rose by 73 percent on the estate. On four different occasions between March and November, workers demanded wage increases to offset rising prices, and each time the Aspíllagas granted them modest raises

and sold them foodstuffs at a loss. The Aspíllagas were still sufficiently concerned, however, to arrange for troops to be stationed
on the estate as a show of force. This extreme step suggests a deterioration in labor relations and directly denies the existence of a
paternal order at Cayaltí.[54]

In 1921, the same year that the Regional Labor Union was
created in La Libertad, an attempt was made to form a mutual aid
society at Cayaltí. The proposed charter seemed harmless enough,
but Antero Aspíllaga Barrera rejected it for the following reasons.

> The terms of the organization and of the proposal we find correct
> and mannerly for the administration and the estate. The end that
> they pursue appears very saintly and good. But as behind the
> cross one can find the Devil, and we have already agreed, for no
> manner or reason should we consent or permit institutional
> societies, clubs on the estate that carry the danger that they can
> serve as foci of pretentious intriguers and . . . pernicious agita-
> tors.[55]

From the Aspíllagas' perspective, it was best to nip working-class
organizations in the bud. Although they preferred to control workers through manipulation and accommodation, they always maintained the threat of repression and violence. As Ismael Aspíllaga
Anderson once said: "I listen to everyone, but in one hand [I have]
bread and assistance, and in the other authority."[56]

Conclusion

The course of technical change in the Peruvian sugar industry, by
dramatically increasing the scale of production, unleashed a series
of significant social, economic, and political changes. The rationalization of production concentrated export activity into the hands of
a few large estates and generated fortunes for a few sugar barons
and corporations during the early decades of the twentieth century. It depended on securing adequate sources of capital, expanding the size of estates, buying new technology, and increasing the
size of the work force. The necessity of finding more workers
resulted in the advent of *enganche* and the migration of tens of
thousands of peasants to sugarcane plantations. The traditional
interpretation of labor contracting emphasizes the importance of
coercion and debt peonage. The evidence from northern Peru
shows that coercion was relatively unimportant and that debt
peonage was inefficient and oftentimes counterproductive. Work-

ers helped to undermine debt peonage by running away in large numbers after accumulating large debts, sometimes from more than one source.

As the size of the plantation work force doubled and tripled, planters were faced with new problems of social control. With the Chinese, they had relied upon corporal punishment, drug addiction, and indebtedness. With Peruvians, the Aspíllagas, at least, found accommodation and manipulation more effective methods of control, although they never completely abandoned violence and repression.

These tactics, however, could only delay confrontation between labor and capital. During World War I, planters' profits soared, while workers' real wages declined. A series of strikes erupted on sugarcane plantations, with the largest ones occurring on Chicama Valley estates, especially Casa Grande and Roma. These estates had achieved the highest levels of technical development, and working conditions and managerial philosophies most closely approximated a factory system. The largest strikes occurred in 1912 and 1921 and could only be suppressed through the intervention of the army.

Strike activity was less intense in the Lambayeque region, especially on the plantation Cayaltí. The Aspíllagas fashioned a paternalistic veneer in labor relations that helped to mask the inherent inequities in the system of production. This was easier to achieve on a plantation that lagged behind in the modernization process. The Aspíllagas were also willing to grant workers modest wage increases when threatened by a strike. Nevertheless, Cayaltí was never immune from labor unrest, and in later years it suffered several major strikes. The inherent inequities in the system of production eventually dissolved the paternalist constraints on social conflict.[57]

Abbreviations

Much of the information for this chapter comes from the Aspíllaga family's private correspondence, which is now housed in the Archivo del Fuero Agrario (AFA) in Lima. The names of the principal correspondents are referred to in the notes by their initials, except in those cases where they simply signed the title of the family firm, Aspíllaga Hermanos. Names of correspondents and titles are abbreviated as follows:

RAF	Ramón Aspíllaga Ferrebú
AAB	Antero Aspíllaga Barrera
RAB	Ramón Aspíllaga Barrera

BAB	Baldomero Aspíllaga Barrera
IAB	Ismael Aspíllaga Barrera
VAT	Víctor Aspíllaga Taboada
IAA	Ismael Aspíllaga Anderson
LAA	Luis Aspíllaga Anderson
GAA	Gustavo Aspíllaga Anderson
AAA	Antero Aspíllaga Anderson
AH	Aspíllaga Hermanos

Notes

1. Kenneth Duncan and Ian Rutledge (eds.), *Land and Labour in Latin America: Essays on the development of agrarian capitalism in the nineteenth and twentieth centuries* (Cambridge, England, 1977); Arnold J. Bauer, "Rural Workers in Spanish America: Problems of peonage and oppression," *Hispanic American Historical Review*, 59:1 (1979), 34–64; Michael J. Gonzales, "Capitalist Agriculture and Labour Contracting in Northern Peru, 1880–1905," *Journal of Latin American Studies*, 12 (1980), 291–315; Michael J. Gonzales, *Plantation Agriculture and Social Control in Northern Peru, 1875–1933*, Latin American Monograph No. 62 (Austin, Texas, 1985).

2. Eugene D. Genovese, *Roll, Jordan, Roll: The world the slaves made* (New York, 1972); James C. Scott, "Everyday Forms of Peasant Resistance," *Journal of Peasant Studies*, 13:2 (1986), 5–35.

3. See Michael J. Gonzales, "Planters and Politics in Peru, 1885–1919," *Journal of Latin American Studies*, 23:3 (1991), 515–541.

4. Gonzales, *Plantation Agriculture*, ch. 2.

5. Susan E. Ramírez, *Provincial Patriarchs: Land tenure and the economics of power in colonial Peru* (Albuquerque, 1986).

6. Gonzales, *Plantation Agriculture*, ch. 2.

7. Gonzales, *Plantation Agriculture*, ch. 2.

8. Gonzales, *Plantation Agriculture*, Table 7 (p. 46), Table 10 (p. 49).

9. W. A. Lewis (ed.), *Tropical Development 1880–1914* (London, 1970), 19.

10. Food and Agricultural Organization, *The World Sugar Economy in Figures, 1880–1959* (Rome, 1961), Tables 1 and 6.

11. Federico Costa y Laurent, *Reseña historia de los ferrocarriles del Perú* (Lima, 1908), 28; Alejandro Garland, *Las vias de communicación y la futura red ferroviaria del Perú* (Lima, 1906), 24; BAB to AAB, October 1900, AFA, Lima (no day given).

12. Antonio Raimondi, *El Perú* (Lima, 1965), vol. 1, 323; George R. Fitz-Roy Cole, *The Peruvians at Home* (London, 1884), 129; RAB to AAB, 21 September 1875, AFA; AH to AH, 7 October 1887, AFA. Cayaltí's mill contained the latest technological improvements. See Noel Deerr, *The History of Sugar* (New York, 1949), vol. 1, 559–560; VAT to AH, 23 February 1915,

AFA; AAB to RAB and BAB, 3 January 1917, AFA; AAB to RAB and BAB, 6 March 1917, AFA. An excellent summary of Peruvian sugar technology is Bill Albert, "The Course of Technical Change in the Peruvian Sugar Industry, 1860–1940," unpublished paper, 1985.

13. AAB to AH, 2 January 1882, AFA; AH to AH, 4 November 1887, AFA; AAB to RAB and IAB, 10 October 1884, AFA; AH to AH, 5 August 1887, AFA; Thomas Colston to RAB, August 1886, AFA (no day given); BAB to AAB, October 1900, AFA (no day given).

14. Gonzales, *Plantation Agriculture*, 58.

15. Gildemeister Folder, AFA.

16. Gonzales, *Plantation Agriculture*, 37–41.

17. Robert L. Irick, *Ch'ing Policy Toward the Coolie Trade, 1847–1878* (San Francisco, 1982); Arnold J. Meagher, "The Introduction of Chinese Laborers to Latin America: The 'coolie trade,' 1847–1874," Ph.D. dissertation, University of California (Davis, 1975); Humberto Rodríguez, *Hijos del Celeste Imperio en el Peru (1850–1900). Migración, agricultura, mentalidad y explotación* (Lima, 1989); Michael J. Gonzales, "Chinese Plantation Workers and Social Conflict in Peru in the Late Nineteenth Century," *Journal of Latin American Studies*, 21 (1989), 385–424.

18. For longer discussions of *enganche* in northern Peru see, Gonzales, "Capitalist Agriculture," passim, and Gonzales, *Plantation Agriculture*, ch. 7. More generally see Tom Brass, "The Latin American *Enganche* System: Some revisionist reinterpretations revisited," *Slavery & Abolition*, 11:1 (1990), 74–103.

19. For example, Peter F. Klarén, *Modernization, Dislocation and Aprissmo: Origins of the Peruvian Aprista Party, 1870–1932* (Austin, 1973), 26–28; and Ernesto Yepes del Castillo, *Perú, 1820–1920* (Lima, 1972), 209–213.

20. Gonzales, *Plantation Agriculture*, chs. 7 and 8.

21. Lewis Taylor, "Earning a Living in Hualgayoc, 1870–1900," in Rory Miller, (ed.), *Region and Class in Modern Peruvian History* (Liverpool, 1987), 103–125.

22. C. D. Scott, "Peasants, Proletarianisation and the Articulation of Modes of Production: The case of sugar-cane cutters in northern Peru, 1940–1969," *Journal of Peasant Studies*, 3 (1976), 328, for recruiting at markets; Miguel Coronad to Catalino Coronado, 22 August 1918, AFA; Cesar Coronado to Catalino Coronado, 16 October 1918, AFA; AH to AH, 27 June 1889, AFA; AH to AH, 31 July 1889, AFA; AH to AH, 11 October 1892, AFA; interview with Galindo Bravo, CAP Pucalá, 27 June 1975. Bravo was originally contracted to work at Pucalá by Catalino Coronado. In 1975 he was a member of the Consejo de Administración, CAP Pucalá.

23. AH to AH, 19 January 1900, AFA.

24. AH to AH, 11 September 1914, AFA; VAT to AH, 23 February 1915, AFA; AAB to RAB and BAB, 19 January 1917, AFA.

25. For example, Klarén, *Modernization*, 26–28; Yepes del Castillo, *Perú, 1820–1920*, 209–213.

26. AH to AH, 16 October 1885, AFA; AH to AH, 23 September 1888, AFA; AH to AH, 3 October 1889 (a), AFA; AH to AH, 15 November 1889, AFA; AH to AH, 16 November 1891, AFA; AH to AH, 16 February 1892,

AFA; Gonzales, *Plantation Agriculture*, 129–130; Bravo interview; Coronado correspondence, AFA; Francisco Pérez Cespedes to AH, 24 November 1902, AFA; AH to AH, 13 December 1897, AFA.

27. Gonzales, *Plantation Agriculture*, 140–141.

28. AH to AH, 18 November 1887, AFA; BAB to AH, 6 December 1902, AFA; Negrete Hnos. to AH, 12 June 1903, AFA; VAT to AH, 11 August 1908, AFA; VAT to AH, 14 October 1908, AFA; BAB to AH, 26 October 1908, AFA; Manuel Coronado to Catalino Coronado, 16 November 1917, AFA; Gregorio Castaños to Catalino Coronado, 10 March 1909, AFA; *enganche* contract between Eduardo Tiravante and Hacienda Pomalca, 20 June 1917, AFA (it was written into this contract that if a peon owed money to more than one contractor, the contractor "in possession" of the peon would have to pay off the outstanding debts to other contractors). Manuel E. Espinoza to José Ignacio Chopitea, 24 August 1903, AFA; Administrador de Tumán to Señor Sub-Prefecto e Intendiente de Policia de la Provincia de Chiclayo, 20 April 1907, AFA; Gonzales, *Plantation Agriculture*, 141; Manuel Torres to Carlos Gutierrez, 1 July 1907, AFA; M. Coronado to Catalino Coronado, 25 June 1915, AFA; Manuel Coronado to Catalino Coronado, 15 November 1918, AFA; V. Mires to Catalino Coronado, 25 March 1910, AFA; J. Orrego to Catalino Coronado, 22 September 1911, AFA; J. Orrego to Catalino Coronado, 10 March 1909, AFA; J. Orrego to Catalino Coronado, 15 September 1916, AFA; V. Mires to Catalino Coronado, 22 April 1910, AFA; Scott, "Peasants," 328.

29. AH to AH, 25 July 1887, AFA.

30. Gonzales, *Plantation Agriculture*, 148–152.

31. Gonzales, *Plantation Agriculture*, 152–155.

32. Gonzales, *Plantation Agriculture*, 155–159.

33. Gonzales, *Plantation Agriculture*, 166.

34. LAA to RAA, 16 October 1924 (b), AFA.

35. Herbert G. Gutman, *Work, Culture, and Society in Industrializing America* (New York, 1976), 20–21; Sidney Pollard, *The Genesis of Modern Management* (London, 1965), ch. 5; Gonzales, *Plantation Agriculture*, 160–167.

36. See Gonzales, *Plantation Agriculture*, 167–170.

37. Erik D. Langer, "Debt Peonage and Paternalism in Latin America," *Peasant Studies*, 13:2 (1986), 121–127.

38. Gonzales, *Plantation Agriculture*, ch. 8.

39. Langer, "Debt Peonage and Paternalism in Latin America," 121–27.

40. See Klarén, *Modernization*, ch. 2; Bill Albert, *An Essay on the Peruvian Sugar Industry, 1880–1920 and the Letters of Ronald Gordon, Administrator of the British Sugar Company in Cañete, 1914–1920* (Norwich, 1976), 106a–109a, 203a–218a.

41. Letter dated 18 April 1912, in Pablo Macera, (ed.), *Cayaltí, 1875–1920: Organización del trabajo en una plantación azucarera del Perú* (Lima, 1973), 224.

42. Albert, *Peruvian Sugar Industry*, 106a; *La Reforma*, 8–11 April, 15 April, 18 April, 19 April 1912.

43. This interpretation is based largely on the reporting of the Trujillo daily *La Reforma*. This paper may be considered a controversial source,

because it was owned by Víctor Larco Herrera, the proprietor of the plantation Roma and an adversary of the Gildemeisters. It could be argued that Larco exaggerated strikers' gains to embarrass his German-Peruvian rivals, especially since his plantation had escaped unharmed. However, this would have also involved the risk of publicizing nonexistent advances that could have encouraged his workers to seek equity. False reporting would also have opened the door to lawsuits by the powerful Gildemeisters, as well as by the Peruvian and German officials reported as parties to the agreement. Thus, there seems to be more reason to believe the authenticity of the reporting than to dismiss it as inherently biased.

44. *La Reforma,* 30 April, 1 May, 6 May 1912.

45. *La Reforma,* 1–2 June, 4 June, 6 June 1917; *La Industria,* 4 June 1917; Albert, *Peruvian Sugar Industry,* 187a–188a, 171a; RAB to AAB, 12 July 1917, AFA; RAB to AAB, 17 July 1917, AFA.

46. Joaquín Díaz Ahumada, "Las luchas sindicales en el valle de Chicama," in *Imperialismo y el agroperuano* (Lima, n.d.), 28–29. Díaz Ahumada presents a full list of demands.

47. Díaz Ahumada, "Las luchas sindicales," 29–34.

48. Díaz Ahumada, "Las luchas sindicales," 31–32; Klarén, *Modernization,* 46.

49. Lauro A. Curletti, *El problema industrial en el valle de Chicama. Informe del Ministro de Fomento* (Lima: Biblioteca Peruana de Historia Economica, 1972), 31–33.

50. Gonzales, *Plantation Agriculture,* ch. 3.

51. AAB to AH, 13 November 1912, AFA.

52. Gonzales, *Plantation Agriculture,* 182–184.

53. AAB to RAB and BAB, 29 April 1917, AFA.

54. Macera, (ed.), *Cayaltí, 1875–1920,* 231; Albert, *Peruvian Sugar Industry,* 197a; letter dated 7 August 1919, AFA; letter dated 28 August 1919, AFA.

55. AAB to VAT, 26 December 1921, AFA.

56. IAA to RAB, 18 September 1923, AFA.

57. Gonzales, *Plantation Agriculture,* 185–188.

10

Reflections

Edward D. Beechert

History does nothing. It possesses no immense wealth, it wages no battles. It is man, real living man, that does all that, that possesses and
fights; history is not a person apart, using man as a means for its own
particular aims; history is nothing but the activity of man pursuing his
aims.

—*The Holy Family: A Critique of Critique*

Writing the history of the worker is difficult at best and at times
seems beset with formidable obstacles. Workers leave few diaries
and convenient records to guide the historian. Some reflection on
the methodology of examination is essential to illuminate an otherwise obscure path. The contributors to this volume have avoided
the temptation to seek out the conflicts and the moments of high
drama and substitute these for an understanding of the labor process under the guise of class history. This trap of negativism
produces a false history and often results in a sense of disappoint-
ment, since the conflicts seem to lead nowhere. The failure of the
worker to move through struggle to a higher level of class consciousness has disappointed many historians.[1] Analysts of the
United States scene have repeatedly expressed regret and bewil-
derment over the failure of the American worker to develop a
sense of class consciousness despite an amazing record of militancy
and struggle. Since Werner Sombart explored the question of why
there was no socialism in America, countless other scholars have
examined the question in microscopic detail. Some have found the
worker to be an "uprooted" immigrant, despondent and lonely.
Others have found the answer in cultural differences—the "pre-
industrial" culture of the immigrant worker in an industrial setting.[2]

One way to possibly escape this trap of negativism is to consider
the problem of the worker from the viewpoint of the changing relations of production. By the fact of the worker supplying the labor
power in the productive process, a degree of control is exerted over
production. In greater or lesser degree, this control over the pro-

ductive process is the site of an ongoing tension between the capitalist and the worker—a condition that is unceasing and unchanging. The struggle for control is perhaps the most dialectical of all the processes of class. To put it more sharply, by examining the framework of the production process—"the sum of the productive forces"—we have the possibility of gaining an understanding of the circumstances that people create, and, in turn, are shaped by.[3]

The issue of control over working conditions—whether industrial, agricultural, or domestic—takes forms other than those of confrontation. We are perhaps too aware of instances of conflict because of their record-generating tendency. The same applies to peasant rebellions. As James C. Scott has pointed out:

> The events that claim attention are the events to which the state and the ruling classes accord most attention in their archives. Thus, for example, a small and futile rebellion claims an attention out of all proportion to its impact on class relations while unheralded acts of flight, sabotage, theft which may have far greater impact are rarely noticed. The small rebellion may have a symbolic importance for its violence and for its revolutionary aims but for most subordinate classes historically such rare episodes were of less moment than the quiet, unremitting guerilla warfare that took place day-in and day-out [and which finds little echo in the records].[4]

More subtle forms of resistance to job control are indeed more common or typical of worker responses. These range from passive resistance to organized efforts to determine the outcome of work rules. Struggle may be nothing more than an expression of frustration or anger. It may also represent a stage of awareness or an expression of class consciousness. The point is that the events or circumstances that make up the social relations of production need to be examined, especially the circumstances following the climax of a struggle.[5] For plantation agriculture, the systems of labor deployment and capital structure of plantation organization are the decisive factors that channel the workers' responses to the social relation of production.

In relating these observations to the theme of this volume, one must make a distinction between *labor* (in the sense of work done) and *labor power* (the capacity to do work). It is in the latter sense that the worker has the potential to exert some measure of control, and this is particularly so in the case of large groups of immigrant plantation workers. While they can and do translate their immediate situation into terms of work control, they have difficulty in transcending racial and ethnic barriers. Until such time as the common work experience transforms those ethnic differences, one cannot expect to find the degree of communication and understanding represented by class consciousness.[6] Class comes into existence at the point where a grouping begins to acquire an awareness in and for itself. Before that point, the working class can be said to exist only *in* itself and not *for* itself.[7]

The contest for control of the workplace, then, is an unremitting one. It is clear from the varied examples presented in this volume that the contest is shaped by the range of opportunities characteristic of each of the political economies in which the workers find themselves. The social relations of production are conditioned by the changing circumstances of experience, technology, political structures, and, importantly, by world market conditions.

The forms of resistance available to plantation workers were conditioned first of all by the decision to seek work in distant places. Grinding poverty, denial of access to land, political and ethnic exploitation, a desire to acquire worldly goods to enhance one's status, and, perhaps most important, the opportunity for regular, long-term employment, were the sources of the decision to leave.

The comparative view of the wide range of plantation systems covered here suggests that generalizations must be conditioned by the particular circumstances of each system. Abstract ideas of rebellion or resistance find little support. Nor does the evidence indicate that the immigrant workers were naïve or romantic in thinking that immigration would lead to wealth. The degree of communication of such workers with the homeland rules out that type of analysis. A combination of resistance and accommodation appears in each of the areas surveyed. The ability of the worker to modify the process of production through resistance and accommodation is often lost in the outbreaks of violence that occurred in all the examples and which have tended to dominate the thinking of observers. Such a phrase as "new system of slavery" to describe indentured labor obscures the behavior. Examples such as those of Samoa and the Solomon Islands, marginal economies at best when market prices for copra were depressed, were extreme cases. The more viable plantation systems operated in a dynamic context with their workers and the market.

The phenomenon of workers remaining after the expiration of their contracts occurred in each of the examples. The need of the planters for experienced workers, the frequent disruption of the supply of workers in the late nineteenth and early twentieth centuries, and the unwillingness of many workers to go back to the poor conditions they had left are all evident in the numbers of those remaining. Many Pacific Islanders in Queensland resisted repatriation between 1901 and 1906, supported by planters who found their experience valuable. Fiji and Trinidad furnish examples of many Indian workers returning to India to retrieve families and returning to the sugar estates, clearly finding this a better situation than that of the homeland. In Hawaii, the Japanese and Filipino workers remained in large numbers. Many of the Japanese who left Hawaii went to the mainland United States rather than return to Japan. The degree of cultural adaptation exhibited suggests that the workers found sufficient opportunity to establish themselves.

Organized resistance among the plantation workers was rare and seldom successful. Hawaiian workers, after the arrival of United States legal institutions, were able to mount large-scale industrial strikes in 1909, 1920, and 1924. Although not immediately successful, the severity of the strikes and the degree of solidarity displayed by the workers forced dramatic changes in the operational patterns of the plantations. Workers in Peru faced the power of the national government in similar situations, although here, too, the complexity of the plantation economy forced the planters to rationalize their system. Profit potential rather than class power took precedence in both cases.

In most cases, a facade of law was established to provide protection for the immigrant workers. Trinidad and Fiji are good examples of the class bias in the administration of labor law. In Fiji, the colonial government "did not take seriously its role as the trustee of the indentured laborer's rights." In Trinidad, the planters effectively maintained legislative pressure to control wages of time-expired workers and retain control of punitive measures. Magistrates generally shared the class interests of the planters. In Hawaii, even though the planters and their magistrate cohorts operated under a more liberal and restrictive constitutional system, the resulting degree of identity between planters and the legal system was very similar to that obtaining elsewhere. Many Protectors of Immigrants took the view that the complaints they investigated, whether from Melanesian, Japanese, Chinese, Norwegian, Indian, or Filipino, were rooted in the defects of the workers, who were largely "unreliable" and "untrustworthy," and above all, "ungrateful." The class interests of those appointed to such positions together with the remoteness from the seat of power made such officers prone to support the local ruling class more often than not.

Despite the many theories of ethnic conflicts contributing to the lack of solidarity among workers, the evidence suggests that the commonality of the work experience often reduced these differences. Among Indian workers, the experience of the long journey from the homeland created a bond of brotherhood, which was an important source of mutual support in the difficult work situation. This support extended to financial arrangements of "box money" among Indians and the "tanomoshi" mutual financing associations among Japanese workers. At the same time, cultural resilience was one of the more persistent sources of resistance. "Culture defined an area to which Indians, after defeats at the work place, could retreat to heal and bind the wounds before sallying forth again." The number of workers was generally sufficiently large to encourage and make possible the maintenance of cultural practices. In Hawaii, where the vagaries of labor supply made for a multiethnic labor force, the workers were generally housed in camps more or less segregated. This practice strengthened the retention of language and cultural practices. The Pacific Islanders' resort to expressions of ethnicity, the use of talismans, sorcerers, songs, and dances reinforced their spirit.[8]

World market conditions and technology were important factors conditioning the labor situation. This is particularly true of Hawaii and Peru, where the rationalization of agriculture and production processes brought far-reaching changes in labor deployment and handling. At the other end of the economic scale, periods of severe profitability restraints were experienced in both Samoa and the Solomon Islands. The degraded working conditions of those two places during such times are closely related to economic marginality.

The studies presented here may lead to more examinations and the opportunity for comparative study. While labor is certainly one of the important factors in the organization of production, it too often has been treated as an abstraction. We have seen here the resilience, the adaptability, and the persistence of workers. These qualities are as important in shaping the outcomes as those of capital and land.

Notes

1. This brings to mind the wry observation that "[t]here is a germ of truth in the charge that the New Zealand left is constantly winning in books the battles it lost long ago in reality." W. H. Oliver, "A Destiny at Home," *New Zealand Journal of History*, 21:1 (1987), 13.

2. Herbert G. Gutman, *Work, Culture and Society in Industrializing*

322   Edward D. Beechert

America (New York, 1976); Eric Foner, "Why There Is No Socialism in America," History Workshop Journal, 17 (1984), 57–80.

3. Karl Marx, The Eighteenth Brumaire of Louis Bonaparte (New York, 1936), 13; David Montgomery, Worker Control in America (New York, 1971), 4, 14.

4. James C. Scott, "Everyday Forms of Peasant Resistance," Journal of Peasant Studies, 13:2 (1986), 5.

5. Stanley Aronowitz, "Trade Unionism in America," in B. Silverman, B. and M. Yanowitch (eds.), The Worker in Post-Industrial Capitalism (Chicago, 1974), 420.

6. Richard Edwards, Contested Terrain: The transformation of the workplace in the twentieth century (New York, 1979), 16–22.

7. Istvan Mezaros, Aspects of History and Class Consciousness (London, 1971), 9; Joseph Gabel, False Consciousness: An essay on reification (London, 1975), ch. 1.

8. Adrian Graves, Cane and Labour: The political economy of the Queensland sugar industry (Edinburgh, 1993), 192–194.

Index

Names of academic writers cited have been set in **bold**.

Indians from the Indian subcontinent have been distinguished from indigenous Indians in Latin America thus: East Indians, Guatemala Indians, Maya Indians, Peruvian Indians, Yacqui Indians.

Absconding. *See* Desertion
Abusive language, 76, 78–79 (Table 1), 83, 145–146, 202 (Table 1), 210
Alcohol abuse: in Hawaii, 7–8; in Peru, 306; in Queensland, 30, 70, 78–79 (Table 1), 83, 92; in Solomon Islands, 133; in Yucatán, 259, 262, 271, 274
Ali, Ahmed, 195
Allardyce, K. J., investigates labor matters in Solomons, 150
Althusser, Louis, 218
Alvarado, General (Mexican soldier), 282, 284
American Civil War, 2, 47, 50, 103, 105
Andrews, C. F. (Indian nationalist sympathizer), 187, 213
Angel Asturias, Miguel (novelist), author of *El Senior Presidente,* 222
Anglo-German Agreement (1886), 115
Anson, Henry (Agent-General of Immigration, Fiji), 195, 198

Anthony, Joseph (Queensland planter), 92
Armed forces, use of, against workers: in Guatemala, 17, 221–224; in Hawaii, 58; in Peru, 298, 308, 309–310; in Yucatán, 21, 277
Arson. *See* Sabotage
Ashley, F. N. (Resident Commissioner of British Solomon Islands Protectorate), 159
Asians. *See* Hawaii; Workers, by nationality
Aspíllagas (planter family in Peru), 299 (Table 1); involvement in freighting, 298; and 1912 and 1921 strikes, 308–311; paternalism of, 19–20, 21, 306–307; and railway development, 300; social control of workers, 305–311. *See also names of individual family members*
Aspíllagas Anderson, Ismael, on worker control, 311
Aspíllagas Barrera, Antero: opposes

323

Contributors

Edward D. Beechert, Professor Emeritus, University of Hawaii, has been writing on the political economy of Hawaii, with special emphasis on labor, since 1963. His most recent work is *Honolulu: Crossroads of the Pacific* (1991), a history of the port of Honolulu. He is currently editing the autobiographies of two of the Hawaii Smith Act trial victims of 1953, to be published by the University of Hawaii Center for Biography. He is the author of *Working in Hawaii: A labor history* (1985).

Judith A. Bennett has taught in Papua New Guinea, Australia, and New Zealand. She lectured on Pacific Islands History at Massey University and more recently at the University of Otago in New Zealand, after a five-year stint as warden of a residential college, St. Margaret's, in Dunedin. Her main research area has been the Solomon Islands, culminating in *Wealth of the Solomons: A history of a Pacific archipelago, 1800–1978* (1987). Her current interest is forestry history, especially its social and economic aspects.

Stewart Firth, Associate Professor of Politics at Macquarie University, Sydney, is the author of *New Guinea under the Germans* (1982) and *Nuclear Playground* (1987), and coauthor of *Papua New Guinea: A political history* (1979). He previously taught at the University of California at Santa Cruz, the University of Papua New Guinea, and the University of Hawaii. His current research focuses on contemporary developments in the Pacific, and he is active in various peace movements.

341

Michael J. Gonzales is Professor of History and Director of the Center for Latino and Latin American Studies at Northern Illinois University. He is the author of *Plantation Agriculture and Social Control in Northern Peru, 1875–1933* (1985), as well as numerous articles and essays on Peruvian history. While retaining a strong interest in the history of plantation agriculture, he is currently working on the history of the copper industry in Mexico from approximately 1890 to 1940, with particular attention to labor relations and political economy.

Gilbert M. Joseph is Professor of Latin American History at Yale University. He is the author of *Revolution from Without: Yucatán, Mexico and the United States, 1880–1924* (1982), *Rediscovering the Past at Mexico's Periphery* (1986), *Yucatán y la International Harvester* (with Allen Wells, 1986), and numerous articles on the Mexican revolution, regional history, and the history of crime and protest. Currently he is completing a book with Allen Wells entitled *Summer of Discontent, Seasons of Upheaval: Elite politics and rural insurgency in Yucatán, 1890–1915*, and editing a volume on state formation and popular culture in modern Mexico.

Brij V. Lal is a Senior Research Fellow at the Australian National University. He previously worked at the University of the South Pacific and, from 1983 to 1992, at the University of Hawaii at Manoa, where he taught Pacific Islands and world history. His publications include *Girmitiyas: The origins of the Fiji Indians* (1983), *Power and Prejudice: The making of the Fiji crisis* (1988), and *Broken Waves: A history of the Fiji Islands in the twentieth century* (1992). His research interests lie in the social history of plantations and in contemporary Pacific Islands history.

David McCreery is Professor of History at Georgia State University and Director of the Latin American Studies Consortium for the University System of Georgia. He has also taught at the University of New Orleans, Tulane University, and the University of Wisconsin, and in Brazil at the Federal University of Minas Gerais. In addition to articles on Guatemalan and Central American history, he is the author of *Development and the State in Reforma Guatemala* (1983) and a book on rural Guatemala, 1760–1940, soon to be published by Stanford University Press. His current research is on Brazil.

Clive Moore has taught at the James Cook University of North Queensland and the University of Papua New Guinea, and he is now Senior Lecturer in History at the University of Queensland. In addition to numerous journal articles on the Queensland labor trade and Australian Aborigines, he is the author of *Kanaka: A history of Melanesian Mackay* (1985) and coauthor of *Colonial Intrusion: Papua New Guinea, 1884* (1984). He has also edited

The Forgotten People: A history of the Australian South Sea Island community (1979) and is coeditor of *Labour in the South Pacific* (1990). He is presently completing a book on British New Guinea.

Doug Munro is Reader in History/Politics at the University of the South Pacific. His earlier research in Pacific history focused on Tuvalu (the former Ellice Islands), resulting in several journal articles and the coauthorship of *Te Aso Fiafia: Te tala o te Kamupane Vaitupu, 1877–1887* (1987) and *Te Tala of Niuoku: The German plantation at Nukulaelae atoll, 1865–1890* (1990, 2nd ed., 1991). He has a secondary interest in Tasmanian convict history that stems from having worked for the Port Arthur Conservation Project. Increasingly, his research has centered on labor migration and plantation studies. He is coeditor of *Labour in the South Pacific* (1990).

Allen Wells is Professor of History at Bowdoin College, Maine. He previously worked at Appalachian State University. In addition to *Yucatán's Gilded Age: Haciendas, henequen and International Harvester, 1860–1915* (1985), he has written widely on Mexican history. Currently he is completing a book with Gilbert M. Joseph entitled *Summer of Discontent, Seasons of Upheaval: Elite politics and rural insurgency in Yucatán, 1890–1915.*

Production Notes

Composition and paging were done on the
Quadex Composing System and typesetting
on the Compugraphic 8400 by the design
and production staff of University of
Hawaii Press.

The text typeface is Century Book and the
display typeface is Goudy Old Style.

Offset presswork and binding were done by
The Maple-Vail Book Manufacturing Group.
Text paper is Writers RR Offset, basis 50.